DYNAMICS OF RAILWAY BRIDGES

DYNAMICS OF RAILWAY BRIDGES

L. FRÝBA

Institute of Theoretical and Applied Mechanics,
Academy of Sciences of the Czech Republic, Prague, Czech Republic

English co-edition published by Thomas Telford Services Ltd, Thomas Telford House, 1 Heron Quay, London E14 4JD, and Academia, publishing house of the Academy of Sciences of the Czech Republic, Vodičkova 40, Prague, Czech Republic.

First published 1996

Distributors for Thomas Telford books are
USA: American Society of Civil Engineers, Publications Sales Department, 345 East 47th Street, New York, NY 10017-2398
Japan: Maruzen Co Ltd, Book Department, 3-10 Nihonbashi 2-chome, Chuo-ku, Tokyo 103
Australia: DA Books and Journals, 648 Whitehorse Road, Mitcham 3132, Victoria

Distributors for in Eastern Europe, China, Northern Korea and Mongolia by Academia

A catalogue record for this book is available from the British Library

ISBN 0 7277 2044 9

© L. Frýba, 1996
Translation © S. Kadečka, 1996

All rights, including translation reserved. Except for fair copying, no part of this publication may be reproduced, stored in a retrieval system or transmitted in any form or by any means, electronic, mechanical, photocopying or otherwise, without the prior written permission of the Books Publisher, Publications Division, Academia, Vodičkova 40, 112 29 Prague, Czech Republic and Thomas Telford Services Ltd, Thomas Telford House, 1 Heron Quay, London E14 4JD, UK.

Typeset, printed and bound in the Czech Republic.

To the memory of my parents

Contents

Preface	11
Notation	13
1. Introduction	19
1.1 Object and history of the dynamics of railway bridges	19
1.2 Deterministic vibration	22
1.3 Dynamic coefficient	24
1.4 Stochastic vibration	26
2. Theoretical bridge models	29
2.1 Beams	29
2.1.1 Mass beams	29
2.1.2 Massless beams and other special cases	30
2.1.3 Continuous beams	31
2.1.4 Boundary and initial conditions	32
2.2 Plates	33
2.3 Complex systems	34
2.3.1 Trusses	34
2.3.2 Frames	35
2.3.3 Curved bars	35
2.4 Lumped masses and models with continuously distributed mass	37
2.5 Bridge deck modelling	38
2.5.1 Cross-beam effect	38
2.5.2 Ballast	41
2.6 Modelling other factors	42
2.6.1 Variable cross section	42
2.6.2 Prestressed concrete bridges	43
2.6.3 Influence of elastic foundation, shear and rotatory inertia	43
2.7 Modelling of railway bridges	45
3. Modelling of railway vehicles	47
3.1 Moving vertical forces	48
3.1.1 Constant forces	48
3.1.2 Harmonic variable force	49
3.1.3 Continuous load	49
3.1.4 Random load	50
3.2 Mass elements	51
3.2.1 Lumped mass	52
3.2.2 Rigid plates and bodies	53
3.3 Springs and damping elements	54
3.3.1 Linear spring	55
3.3.2 Non-linear spring	55
3.3.3 Stop	56
3.3.4 Viscous damping	57

3.3.5 Friction	57
3.3.6 Elastic-plastic element	58
3.4 Modelling of a bridge and a running train	59
3.4.1 Initial assumptions	59
3.4.2 Equations of motion	61
3.4.3 Movement of the vehicle along the bridge	63
4. Natural frequencies of railway bridges	66
4.1 Calculation of natural frequencies	66
4.1.1 Beams	67
4.1.2 Continuous beams	73
4.1.3 Plates	77
4.1.4 Natural frequencies of loaded bridges	77
4.2 Experimental results	80
4.2.1 Dynamic bridge stiffness	80
4.2.2 Statistical evaluation of natural frequencies	85
4.2.3 Empirical formulae	90
5. Damping of railway bridges	94
5.1 Damped vibrations of a beam during the passage of a force	94
5.1.1 Viscous damping proportional to the velocity of vibration	95
5.1.2 Dry friction	99
5.1.3 Complex theory of internal damping	104
5.2 Experimental results	107
5.2.1 Statistical evaluation of logarithmic decrement of damping	107
5.2.2 Empirical formulae	110
6. Influence of vehicle speed on dynamic stresses of bridges	112
6.1 Constant velocity of motion	112
6.2 Variable velocity of motion	116
7. Influence of track irregularities and other parameters	120
7.1 Periodic irregularities	121
7.1.1 Impact of wheel flats	123
7.1.2 Cross-beam and sleeper effects	124
7.1.3 Isolated irregularities	126
7.2 Random irregularities	126
7.3 Further parameters	129
8. Horizontal longitudinal effects on bridges	130
8.1 Motion of a disc rolling along a beam taking into account adhesion	130
8.1.1 Solution	132
8.1.2 Influence of some parameters	136
8.2 Quasistatic model	139
8.2.1 Solution	141
8.2.2 Influence of some parameters	146
8.2.3 Experiments on bridges	156
8.3 Starting and braking forces on bridges	158
9. Horizontal transverse effects on bridges	162
9.1 Beam	162
9.1.1 Vertical vibration	163
9.1.2 Horizontal vibration	166

9.1.3 Torsional vibration	168
9.2 Thin-walled bar with vertical axis of symmetry	170
9.2.1 Approximate solution	171
9.2.2 Dynamic solution	173
9.2.3 Thin-walled bar with two axes of symmetry	175
9.3 Experiments on bridges	175
9.4 Coefficients of variation for horizontal and torsional vibrations	177
9.5 Centrifugal forces	181
10. Traffic loads on railway bridges	183
10.1 Axle forces	185
10.1.1 Measurements of axle forces	185
10.1.2 Vertical axle forces	189
10.1.3 Horizontal transverse wheel forces	195
10.2 Axle spacing	195
10.3 Velocities	197
11. Statistical counting methods for the classification of random stress-time history	199
11.1 Statistical counting methods	200
11.1.1 Sampling method	200
11.1.2 Threshold method	201
11.1.3 Peak counting methods	201
11.1.4 Level crossing methods	202
11.1.5 Stress range counting methods	203
11.1.6 Multiparametric methods	203
11.2 Rain-flow counting method	204
11.2.1 Justification of the counting method from the $\sigma(\varepsilon)$ diagram	205
11.2.2 Counting rules	207
11.2.3 Algorithm for a computer	208
11.3 Appreciation of counting methods	212
11.4 Statistical evaluation of stresses	213
11.4.1 Stress extremes during train passage	213
11.4.2 Statistical evaluation of results of the sampling method	215
12. Stress ranges in steel railway bridges	217
12.1 Theoretical calculation of stress spectra	217
12.1.1 Characteristic trains	217
12.1.2 Traffic loads	221
12.1.3 Bending moment spectra	222
12.2 Experimental stress spectra	229
12.2.1 Empirical formula for the number of stress ranges	231
12.2.2 Probability density of stress ranges	231
12.2.3 Number of stress cycles per year	235
12.2.4 Number of stress cycles	239
12.3 Growing traffic loads	240
12.4 Influence of overloading	241
12.5 Other factors	242
12.6 Appreciation of stress spectra	243
13. The assessment of steel railway bridges for fatigue	244
13.1 Theory of fatigue damage accumulation	244
13.2 Method of equivalent damage	247

13.3 Limit state for fatigue	250
13.4 Fatigue assessment of bridges according to limit states theory	251
13.5 Propagation of fatigue cracks	253
13.6 Fatigue life and interval of railway bridge inspections	258
13.6.1 Residual service life for initial crack a_0	258
13.6.2 Service life assuming fatigue damage accumulation	258
13.6.3 Interval of railway bridge inspections	260
Appendix	262
14. Thermal interaction of long-welded rail with railway bridges	**262**
14.1 Theoretical model of long-welded rail and bridge	262
14.1.1 Basic assumptions	262
14.1.2 Basic equations and their solution	263
14.1.3 Examples	273
14.2 Comparison of theory with experiments	275
14.3 Expansion length of bridges with long-welded rail	278
14.3.1 Maximum and minimum temperatures	278
14.3.2 Strength condition	280
14.3.3 Gap condition	282
14.3.4 Mutual displacement condition	285
14.3.5 Stability condition	288
14.3.6 Admissible expansion length of bridges	288
14.4 Horizontal forces in bridges due to temperature changes	291
14.5 Effect of some parameters	305
14.5.1 Rail displacement	305
14.5.2 Rail force	305
14.5.3 Mutual displacement	306
14.5.4 Force in fixed bearings	307
14.5.5 Uniform load subjecting the bridge with curved rail	307
14.5.6 Strength condition	307
14.5.7 Gap condition	307
14.5.8 Mutual displacement condition	309
14.6 Conclusions for the application of long-welded rail on bridges	309
Bibliography	311
Author index	322
Subject index	326

Preface

The author has been concerned for years with the effects of moving loads on structures and has dealt with this subject in the book *Vibration of Solids and Structures under Moving Loads*, Academia Publishers, Prague, and Noordhoff International Publishing, Groningen, 1972, [68].

The book *Dynamics of Railway Bridges* represents a continuation of the author's interest in this field with the applications to the vibration of railway bridges. It is the result of the theoretical and experimental activity of the author in the course of the solution of various research problems, of scientific as well as of applied character, and in the course of his activity for international research programs ORE D 23, D 101 and D 128 of the Office for Research and Experiments (ORE), now European Rail Research Institute (ERRI), of the International Union of Railways (UIC).

The book summarizes the author's theoretical contributions to the dynamics of railway bridges as well as the results of experiments on many bridges in Czech and Slovak Republics and abroad. It is divided into 14 chapters dealing systematically both with the bridge and with railway vehicles travelling along it. It sums up the basic dynamic characteristics of railway bridges (natural frequency and damping) and describes the influence of the most important parameters, such as the speed of vehicles, track irregularities, etc. Apart from the vertical effects of vehicles, attention is also given to horizontal longitudinal and horizontal transverse effects on bridges.

The book pays special attention to traffic loads and their railway bridge response. These factors influence principally the fatigue of bridges and their life expectancy as well as the safety, maintenance and economy of railway traffic.

The problem of thermal interaction of long-welded rails with the bridge is considered in the appendix. This is a static rather than a dynamic problem, but it is very important for the passenger's environment and ecology.

It is expected that railway traffic based on the principle of wheels rolling along rails has a maximum speed of about 500 km/h. However, further development includes magnetically levitated vehicles which will attain even higher speeds. It is interesting that the structures, along which the air cushion vehicles move, may be investigated using the methods explained in [68] and in the present book, see [12], [124], [126], [170].

In spite of the large scope of the book it was impossible to include some other problems either forming part of dynamics of bridges, such as the effect of wind load [114] and earthquakes [150] (which, however, are not of great importance in Central European conditions) or which are at the boundaries of this discipline, such as noise [161] (noise generated by vehicles passing over bridges). With one exception the book does not include any computer programs as these are created individually according to the relevant hardware available.

The purpose of the book is to present a well founded survey of the dynamic behaviour of railway bridges, to present abundant experimental data obtained on numerous bridges, and to describe the methods which have been successfully applied in this field. The book is intended for civil and railway engineers, and for scientists and students concerned with the problem of the behaviour of bridges during the passage of vehicles.

In conclusion, I should like to thank all my former and present fellow-workers as well as Academia and Thomas Telford Publications who enabled this book to be written.

Ladislav Frýba

Notation

a	constant
a	axle distance
a	crack length
a	acceleration or deceleration of motion
b	constant
b	constant of damping or dry friction
b	annual increment of traffic load
b	relative rail and bridge displacement
b_i	coeffiecient in adhesion computation
c	velocity of motion
c	constant
c	crack width
c_1, c_2	speed of propagation of longitudinal or transverse waves
d	constant
d	sleeper distance
d_n	axle distance
f	frequency
f_g	deflection due to uniformly distributed load
f_j	natural frequency
$\bar{f}_j$	natural frequency of loaded bridge
$f(x)$	function
$f(x, t)$	load per unit length
$f(x, t)$	probablity density
$\mathring{f}(x, t)$	centred value of random function $f(x, t)$
$g = 9.81 \text{ m s}^{-2}$	acceleration due do gravity
$g_1(x)$	initial beam deflection
$g_2(x)$	initial beam velocity
h	difference in height
h	plate thickness
$i = 1, 2, 3, \ldots$	
$j = 1, 2, 3, \ldots$	
k	constant
k	spring stiffness
k	gradient of Wöhler fatigue curve
l	span of the bridge
l	length

m	mass
m	coefficient
m_i	lengths ratio
$n = 1, 2, 3, \ldots$	
n	number of stress cycles per year
n	number of spans of a continuous beam
n_i	number of stress cycles in the ith class
n_T	number of trains per year
p	horizontal transverse uniformly distributed load
q	generalized displacement
q	longitudinal load
q	number of tracks on the bridge
r	radius of wheel, curvature, inertia
r	span ratio in a continuous beam
$r(x)$	track irregularity
s	amplitude
s	auxiliary variable
s	standard deviation
t	time
t	temperature
$u(x, t)$	displacement in the direction of axis x
$u(t)$	function expressing the motion of a force
v	velocity (km h^{-1})
$v(x, t)$	displacement in the direction of axis y
v_{st}	static midspan deflection due to self-weight
v_0	midspan deflection due to force F applied at midspan
var (τ)	function of time
$w(x, t)$	displacement in the direction of axis z
$w(x, y, t)$	deflection of the plate
x	coordinate
y	coordinate
y	vertical unevenness
z	coordinate
z	horizontal unevenness
A	constant
A	cross section area
B	constant
B	parameter of accelerated or decelerated motion
B	dynamic bridge stiffness
C	constant
C_g	centre of gravity (centroid)

Notation

C_s	centre of flexure
$C_f(x, t)$	variation coefficient of random function $f(x, t)$
$C_{xy}(x_1, t_2)$	covariance of random functions $x(t)$ and $y(t)$
D	constant
D	fatigue damage
D	bending stiffness of a plate
D	standard deviation
D^2	variance
D_n	dimensionless axle distance
E	modulus of elasticity
E	experiment
E_{jn}	constant
$E[f(x, t)]$	mean value of random function $f(x, t)$
F	force
F	vehicle weight
F	track frequency
$F(t)$	time-variable force
F_{cfg}	centrifugal force
G	weight of the bridge
G	modulus of elasticity in shear
$G(x, s)$	influence or Green function
$G_{rr}(\Omega)$	power spectral density of unevennesses (one-sided)
$H(t)$	horizontal force
H_y	horizontal transverse force
I	moment of inertia
K	spring force
K	stress intensity factor
L	length
L	life of the bridge
L	length of unevenness
L	expansion length of the bridge
L_i	interval of bridge inspections
M	bending moment
M	mean value
N	horizontal longitudinal force
N	normal force
N	number of axles
N	number of stress cycles
N_i	ultimate number of stress cycles
N_T	number of trains per bridge life
P	probability
S_i	horizontal force in rails

$S_{ff}(q_1, q_2, \omega_1, \omega_2)$	power spectral density of random function $f(x, t)$, (two-sided)
T	traffic load per year (in millions of tonnes)
T	dimensionless coefficient
T	period of natural vibration
T	theory
$T(t)$	tensile force
$U(x)$	Heaviside unit function
$V(x, t)$	variation coefficient
V_{nm}	matrix of dependence of the nth and the mth forces
W	cross section modulus
$W(t)$	resistance to motion
X_i	horizontal longitudinal force below bridge bearings
Y	horizontal transverse force
Z_i	vertical reaction under bridge bearings
$Z^2_{nm}(z)$	function
α	velocity parameter
α	angle
α	coefficient of thermal extension
β	damping parameter
γ	coefficient of reliability
γ	coefficient of internal damping
γ	load factor
δ	dynamic coefficient
$\delta(x)$	Dirac delta function
ε	strain
$\varepsilon = 0$ or 1	
$\varepsilon \ll 1$	very small quantity
$\zeta(t)$	yawing
$\eta(t)$	pitching
ϑ	logarithmic decrement of damping
$\varkappa$	weight parameter
λ_i	dimensionless stress range
λ_j	value dependent on natural frequency
λ_T	fatigue load factor
μ	mass per unit length of a beam or per unit area of a plate
μ	friction coefficient
υ	Poisson's number ($\upsilon < 1$)
ξ	dimensionless length coordinate
$\xi(t)$	rolling
ρ	arm of a force
$\sigma(x, t)$	stress

Notation

$\sigma_v^2(x, t)$	deflection variance
τ	dimensionless time
τ	auxiliary time variable
φ	dynamic increment
φ	dimensionless parameter
φ	sectorial coordinate
φ	angle
φ_j	phase
ω	circular frequency
ω	circular frequency of passage of force
ω_b	circular frequency with damping
ω_d	circular frequency of damped vibrations
ω_j	natural circular frequency
Γ	Gamma function
$\Delta\sigma$	stress range
Δt	rail temperature difference from fixing temperature
ΔT	bridge temperature difference from fixing temperature
Σ	summation
Φ	dimensionless friction force
$\Phi(x)$	distribution function of the normal Gauss distribution
Ω	circular frequency of harmonically variable force
Ω	circular frequency of nonuniform motion
Ω	track frequency

Subscripts

adm	admissible value
b	damping
cfg	centrifugal
com	comparative
cr	critical value
d	damping
ev	equivalent
h	horizontal
$i = 1, 2, 3 \ldots$	
$j = 1, 2, 3 \ldots$	
kt	yield limit
n	standard load
w	horizontal transverse
y	horizontal
F	fixed bearing
M	movable bearing

T	traffic
ξ	torsional
φ	torsional

Superscripts

′, ″, IV	derivatives with respect to length coordinate
′	damped vibration
˙, ¨	derivatives with respect to time
–	loaded
*	complex quantity

Units of measurements
International System (SI)

Lengths	:	metre (m), millimètre (mm), kilometre (km)
Force	:	newton (N), kilonewton (kN), meganewton (MN)
Mass	:	kilogram (kg), tonne (t)
Time	:	second (s), hour (h), year
Frequency	:	hertz (Hz)
Circular frequency	:	(s^{-1})
Velocity	:	$(m\ s^{-1})$, $(km\ h^{-1})$
Stress	:	megapascal (MPa) = $N\ mm^{-2}$
Temperature	:	degrees centigrade, Celsius, (°C)

1. Introduction

1.1 Object and history of the dynamics of railway bridges

The dynamics of railway bridges is a scientific discipline forming part of applied mechanics and its subdivision dynamics of structures. It is concerned with the study of deflections and stresses in railway bridges. The loads are represented by the moving wheel and axle forces, by means of which railway vehicles transmit their load and inertia actions to railway bridges. A survey of the dynamic effects of vehicles on railway bridges is given in Fig. 1.1.

Thus the dynamics of railway bridges involves the response of bridges to the movement of vehicles and to the influence of a number of parameters which

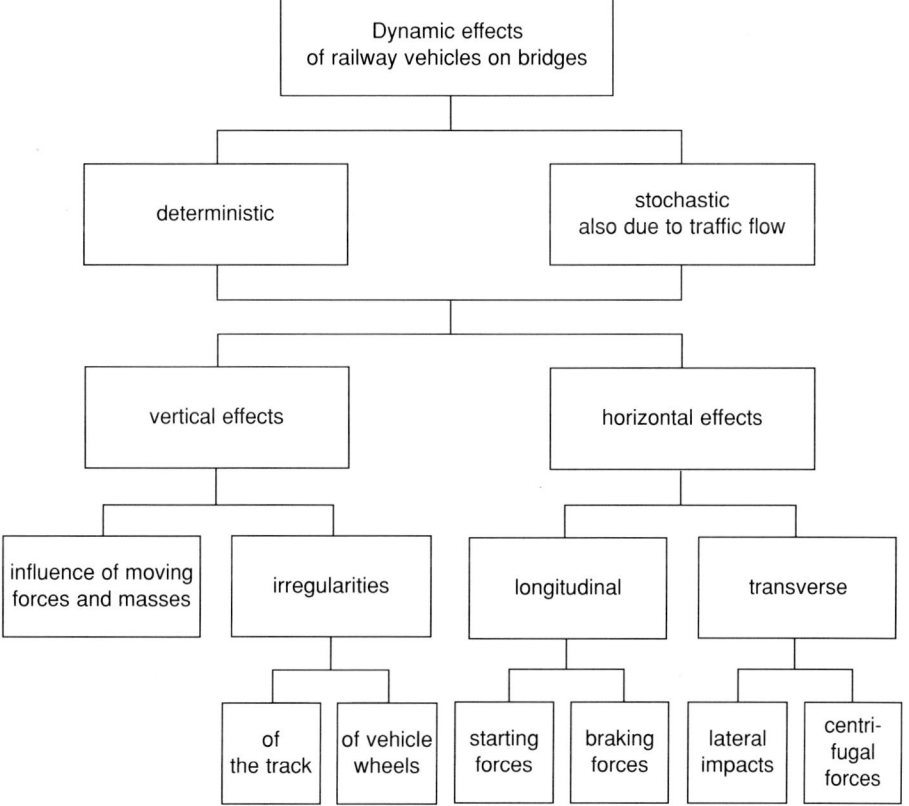

Fig. 1.1. Dynamic effects of railway vehicles on bridges.

increase dynamic strains or stresses. The most important parameters influencing the dynamic stresses in railway bridges are: the frequency characteristics of bridge structures (i.e. the length, mass, and rigidity of individual members), the frequency characteristic of vehicles (i.e. the sprung and unsprung masses, the stiffness of springs), the damping in bridges and in vehicles, the velocity of vehicle movement, the track irregularities, and so on.

The vehicles affect the bridges not only by vertical forces, but also by movements which generate longitudinal and transverse horizontal forces.

This results in an increase or decrease of bridge deformations when compared to that due to static forces. In design practice, these effects are described by the dynamic coefficient (or dynamic impact factor) which, however, only states how many times the static effects must be multiplied in order to cover the additional dynamic loads. Because of its simplicity the dynamic coefficient cannot characterize the effect of all the above-mentioned parameters, but it does generally ensure the safety and reliability of bridges.

The fatigue assessment of bridges has resulted in the derivation of a new approach. This assumes the magnitude and number of stress cycles generated in the bridge by the passage of all trains during its service life. This approach, which is closer to reality, has yielded valuable data for the fatigue assessment of bridges, for the estimation of their fatigue life and for the determination of inspection intervals.

At the boundary between the statics and the dynamics of bridges, the problem of thermal interaction of bridges with rails occurs. Because the temperature changes also depend on time, the solution of these problems has been also included in this book, although the thermal effects do not generate any vibration in bridges.

In addition to a wide range of problems of the railway bridge dynamics, some commonly used experiments are also described; these are to check the reliability of bridges in practice and also to verify the feasibility of new theories in research. Over the years, a certain methodology of these experiments has developed, the keeping of which greatly contributes to the comparability of the individual experiments.

Scientific and research studies in the field of railway bridge dynamics have yielded a number of measures and instructions which have been incorporated into national or even international standards for the design and analysis of railway bridges [211] to [213].

This brief survey of the problems of railway bridge dynamics confirms that this discipline has a rich history and great attention has been paid to the study of bridge dynamics all over the world. Indeed, the problem of vehicle movement along railway bridges was the second problem (after the study of impact of two colliding solids) with which structural dynamics was concerned. The problem arose during the construction of the first railways in England in the first half of the 19th century, when the engineers split into two groups. One believed that the passage of a railway locomotive along the bridge would generate an impact, while the other was of the opinion that the structure would not have enough time to become deformed during the engine passage.

For this reason, this very early period gave rise to the first experiments by R. Willis [222] and to the first theoretical studies by G. G. Stokes [201] which suggested that the actual effect of a moving railway locomotive on the bridge would lie somewhere between those two extreme opinions mentioned above. Since that time the dynamics of railway bridges has received consistent attention in most technically developed countries all over the world.

From the number of theoretical pioneers of this discipline mention should be made of H. Zimmermann [230], A. N. Krylov [130], and particularly S. P. Timoshenko [207], who solved the two fundamental problems of motion of a constant and of an harmonically variable force along a beam.

Between the two World Wars the dynamics of railway bridges was given greatest attention in the former USSR [55] and in Great Britain [175]. In this connection particular mention should be made of the classical work by Prof. C. E. Inglis [106] who explained both theoretically and experimentally the effect of steam railway locomotives on the vibration of railway bridges. His research exercised a decisive influence on the subsequent development of the whole discipline.

In late Czechoslovakia, Prof. V. Koloušek [120] solved the effects of steam railway locomotives on the statically indeterminate continuous, frame and arch railway bridges. The author has also contributed, together with other Czech and Slovak specialists [8], [63] to [80], [142], [191].

In the former USSR, there originated at least three schools studying this discipline both theoretically and experimentally. Their number included Prof. N. G. Bondar [22], [23], [208] and his followers in Dnepropetrovsk, Prof. Ju. G. Kozmin [22], [23] in St. Petersburg, and I. I. Kazej [115] and colleagues in Moscow.

Also in the USA, the dynamics of railway bridges has been studied at several universities, particularly the Northwestern University, the University of Illinois, Massachusetts Institute of Technology, Stanford University, Michigan State University and elsewhere [29], [52], [180], [202], [206].

From other countries mention should be made of Poland [46], [90], [128], [173], the Federal Republic of Germany [25], [27], [124], [170], [185], [188], [192], [196], Switzerland [9], [30], [97], [100] to [102], [138], France [31], [32], Great Britain [175], Sweden [43], [44], [98], [156], [157], the Netherlands [137], [200], Japan [99], [108], [140], [141], [147], [226], [227], and India [26], [99], [227].

Several international organizations have been concerned with research on bridge dynamics. For example, the OSZhD (Organizacija sotrudničestva železnych dorog – Organization of Railways Cooperation) dealt with this problem between 1960 and 1970 and with the methodology of loading tests from 1980 to 1985. The Office for Research and Experiments (ORE) of the International

Union of Railways (UIC) has had in its research program a number of problems which form part of railway bridge dynamics. These are particularly the questions ORE D 23, 101, 128, 154, 160 and others dealing with the problem. The research has been continuing by the successor of the last organization, i.e. by the European Rail Research Institute (ERRI).

The author of the present book had the opportunity of being able to cooperate or to lead (ORE D 128 and ERRI D 191) the research in both international organizations mentioned above. The results of this activity appear in the present book.

This brief survey of state-of-the-art railway bridge dynamics and of its individual problems shows the importance of this discipline for structural engineering in general, and for the design and analysis of bridge structures in particular. Correct understanding of the problems of bridge dynamics contributes to economic design of new structures and to the rational exploitation of bridges in service.

1.2 Deterministic vibration

The response of a railway bridge to the passage of any vehicle manifests itself as vibration. It will be advisable, therefore, to recapitulate first the fundamental concepts of the vibration of mechanical systems [8], [42], [120], [176].

Deterministic vibration is motion which can be predicted at any moment. The basic model from the field of vibrations is a system with one degree of freedom (Fig. 1.2) whose motion is described, according to Newton's second law and D'Alembert's principle, by the differential equation

$$m \frac{d^2 v(t)}{dt^2} + b \frac{dv(t)}{dt} + k\, v(t) = F(t) \qquad (1.1)$$

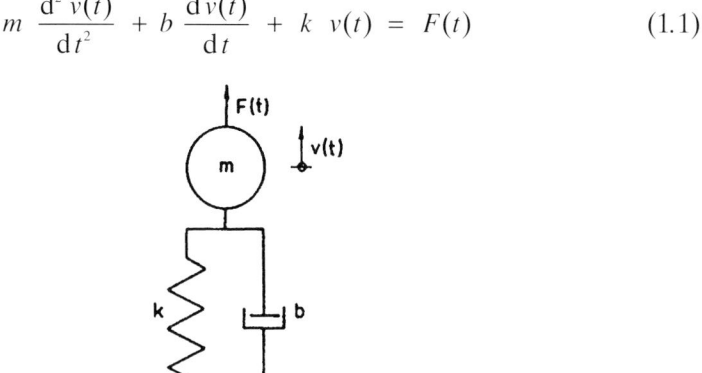

Fig. 1.2. System with one degree of freedom with Kelvin–Voigt absorber of vibrations.

where $v(t)$ — displacement of a body of mass m at time t,
       m — lumped mass of the system,

1.2 Deterministic vibration

$b = 2m\omega_b$ – damping force at unit velocity (according to Voigt's hypothesis of damping proportional to the velocity of vibration),

k – stiffness or rigidity of the spring (force per unit length of spring deformation), assumed constant,

$F(t)$ – external force dependent on time t.

The fundamental concepts in the field of vibrations are the natural circular frequency of the system (1.1)

$$\omega_0 = \left(\frac{k}{m}\right)^{1/2}, \qquad (1.2)$$

the circular frequency of damped vibrations with subcritical damping ($\omega_b < \omega_0$)

$$\omega_d^2 = \omega_0^2 - \omega_b^2, \qquad (1.3)$$

the natural frequencies of undamped or damped vibrations derived from it

$$f_0 = \frac{\omega_0}{2\pi} \qquad (1.4)$$

and

$$f_d = \frac{\omega_d}{2\pi}, \qquad (1.5)$$

respectively, and the period of natural vibrations

$$T_0 = \frac{1}{f_0} \qquad (1.6)$$

and

$$T_d = \frac{1}{f_d}, \qquad (1.7)$$

respectively, this being the shortest time after which the vibration repeats.

The damping of the system (1.1) is characterized most frequently by the logarithmic decrememt of damping, ϑ, which is defined as the natural logarithm of the ratio of any two successive amplitudes of like sign after time T_d. In the case of damping proportional to vibration velocity, this ratio is constant. In practice, it is often determined on the basis of n successive vibrations

$$\vartheta = \frac{1}{n} \ln \frac{s_0}{s_n} \qquad (1.8)$$

where s_n is the amplitude after the nth cycle (see Fig. 1.3).

The relation of the logarithmic decrement of damping to the constant b or ω_b in equation (1.1) is

$$\vartheta = \frac{\omega_b}{f_d} = \frac{b}{2m \, f_d}. \qquad (1.9)$$

The quantity ϑ is sometimes replaced by the dimensionless damping ratio

$$\beta = \frac{\omega_b}{\omega_0} = \frac{b}{2m\,\omega_0} = \frac{f_d}{\omega_0}\,\vartheta \approx \frac{\vartheta}{2\pi}\,. \qquad (1.10)$$

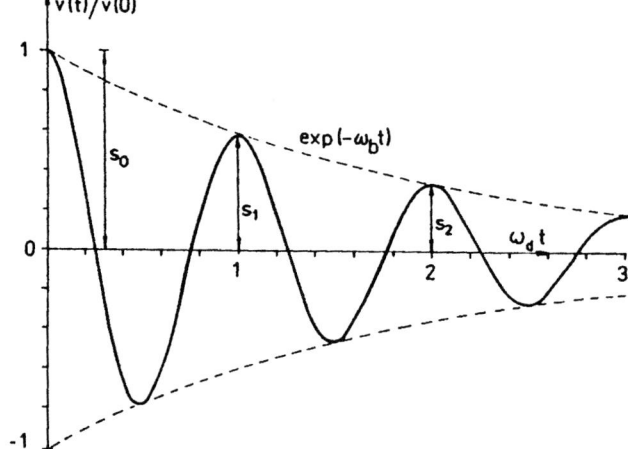

Fig. 1.3. Time-history of natural damped vibration of a system with one degree of freedom at subcritical damping.

1.3 Dynamic coefficient

The dynamic coefficent (also known as the dynamic impact factor or dynamic magnification or dynamic amplification) is usually defined as a dimensionless ratio of the maximum dynamic displacement to the static displacement. In the simplest case, shown in Fig. 1.2 and described by equation (1.1), it can be obtained as follows:

Analysis of the forced vibration of a system with one degree of freedom requires the solution of equation (1.1) under the action of a force

$$F(t) = F\,\sin\omega t\,. \qquad (1.11)$$

At $t \to \infty$, the sustained vibration takes the form of

$$v(t) = A\,\sin\omega t + B\,\cos\omega t\,. \qquad (1.12)$$

After the substitution of equations (1.11) and (1.12) in equation (1.1) and a comparison of coefficients of the individual terms we obtain the following expressions for the constants A and B:

$$A = \frac{F(\omega_0^2 - \omega^2)}{m\left[(\omega_0^2 - \omega^2)^2 + 4\omega^2\omega_b^2\right]},$$

$$B = \frac{-2F\omega\omega_b}{m\left[(\omega_0^2 - \omega^2)^2 + 4\omega^2\omega_b^2\right]}. \qquad (1.13)$$

1.3 Dynamic coefficient

The maximum amplitude of forced stationary vibrations is then the vector sum of A and B, so

$$s_0 = \left(A^2 + B^2\right)^{1/2} = \frac{F}{m}\left[\left(\omega_0^2 - \omega^2\right)^2 + 4\omega^2\omega_b^2\right]^{-1/2}. \qquad (1.14)$$

The static displacement of a mass on the spring of stiffness k subjected to the force F, according to the definition of the spring stiffness (Fig. 1.2 and equation (1.2)), is

$$v_{st} = \frac{F}{k} = \frac{F}{m\omega_0^2}. \qquad (1.15)$$

The dynamic coefficient for the simplest case of forced vibration of a system with one degree of freedom is then defined as the ratio of the maximum dynamic displacement (1.14) and the static displacement (1.15)

$$\delta = \frac{s_0}{v_{st}} = \left[\left(1 - \omega^2/\omega_0^2\right)^2 + 4\omega^2\omega_b^2/\omega_0^4\right]^{-1/2} \qquad (1.16)$$

and depends, consequently, on the ratio ω/ω_0 and on $\beta = \omega_b/\omega_0$. The function (1.16) is represented graphically by the resonance curve in Fig. 1.4.

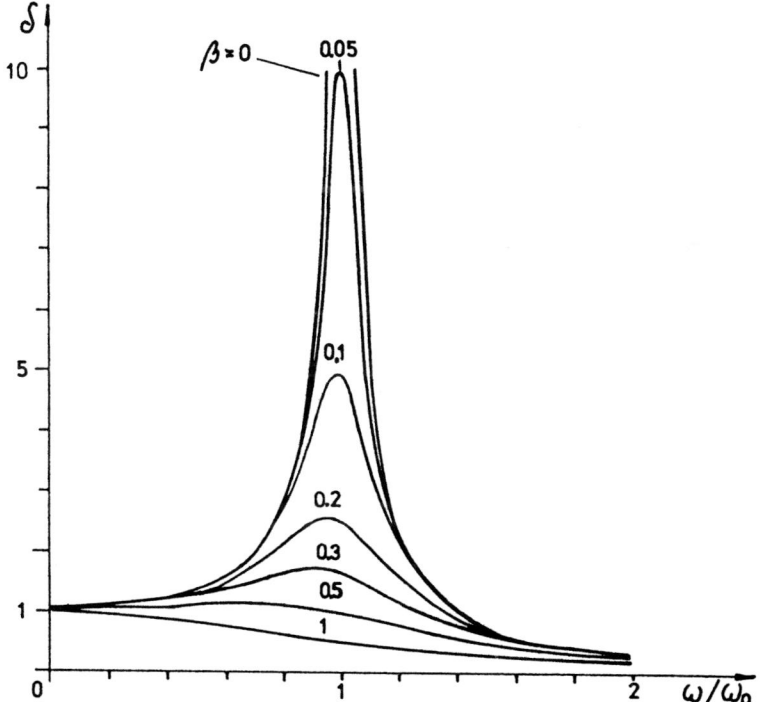

Fig. 1.4. Resonance curve of amplitudes of damped stationary vibrations (dynamic coefficient δ plotted against the excitation frequency ω/ω_0 for various damping ratios β of a system with one degree of freedom).

The maximum dynamic coefficient may be calculated from the condition that the expression $(1 - \omega^2/\omega_0^2) + 4\omega^2\omega_b^2/\omega_0^4$ (the denominator in equation (1.16)) attains its minimum value. When the derivative of this expression is made equal to zero the maximum dynamic coefficient is obtained when

$$\omega^2 = \omega_0^2 - 2\omega_b^2. \tag{1.17}$$

This means that for small damping the highest effects arise when the frequency ω is approximately the natural frequency ω_0; that is

$$\omega \approx \omega_0. \tag{1.18}$$

This frequency produces strong (resonance) vibrations of the system.

If we substitute the excitation frequency (1.17) into equation (1.16), we obtain the maximum values of the dynamic coefficient (see also equations (1.9) and (1.10)):

$$\delta_{max} = \frac{\omega_0^2}{2\omega_b \omega_d} \approx \frac{\omega_0}{2\omega_b} = \frac{1}{2\beta} = \frac{\pi}{\vartheta}. \tag{1.19}$$

Equation (1.19) shows that maximum dynamic effect in resonance conditions depends chiefly on the damping characteristics of the system (it is indirectly proportional to the logarithmic decrement of damping).

1.4 Stochastic vibration

Random or stochastic vibration represents a motion the characteristics of which can be determined only with a certain probability. The number of principal concepts from the field of stochastic vibration includes (for details see [20], [41], [42], [176]):

The *mean value* (or mathematical expectation) of a stochastic process $x(t)$ is defined as a statistic moment of the first order

$$E[x(t)] = \int_{-\infty}^{\infty} x \, f(x, t) \, dx \tag{1.20}$$

where $f(x, t)$ is the first-order probability density function.

The *variance* (dispersion) is a central moment of the second order

$$D^2[x(t)] = E\left[\{x(t) - E[x(t)]\}^2\right] =$$

$$= \int_{-\infty}^{\infty} \{x - E[x(t)]\}^2 f(x, t) \, dx. \tag{1.21}$$

The *standard deviation* is the square root of the variance (often denoted by RMS)

1.4 Stochastic vibration

$$D[x(t)] \quad (1.22)$$

and the *variation coefficient* expresses the dimensionless ratio of the standard deviation to the mean value

$$C(t) = \frac{D[x(t)]}{E[x(t)]}. \quad (1.23)$$

The characteristics of a random process (1.20) to (1.23) are called statistical characteristics of the first order. The statistical characteristics of the second order express the relation of the processes $x(t)$ and $y(t)$ in two moments, t_1 and t_2 according to Fig. 1.5.

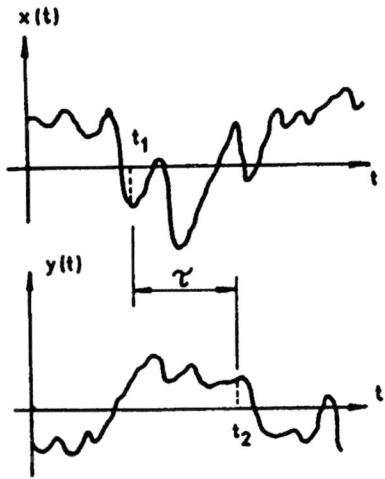

Fig. 1.5. Correlation of random processes $x(t)$ and $y(t)$ at times t_1 and t_2.

The *correlation function* is the mean value at times t_1 and t_2 (generally statistical moment of two-dimensional distribution):

$$R(t_1, t_2) = E[x(t_1) x(t_2)]. \quad (1.24)$$

In the case of two processes $x(t)$ and $y(t)$, the correlation function is defined as (Fig. 1.5):

$$R_{xy}(t_1, t_2) = E[x(t_1) y(t_2)]. \quad (1.25)$$

If $x(t) = y(t)$, we speak about the *autocorrelation function* $R_{xx}(t_1,t_2)$, i.e. (1.24).

The random function $x(t)$ is often resolved into its mean value $E[x(t)]$ and the *centred random value* $\mathring{x}(t)$

$$x(t) = E[x(t)] + \mathring{x}(t). \quad (1.26)$$

The correlation function of a centred random process is called the *covariance*

$$C_{xy}(t_1, t_2) = E[\mathring{x}(t_1) \mathring{y}(t_2)]. \quad (1.27)$$

In the special case of $x = y$, and $t_1 = t_2 = t$, according to equation (1.21)

$$C_{xx}(t, t) = E\left[\{x(t) - E[x(t)]\}^2\right] = D^2[x(t)]. \tag{1.28}$$

The covariance (1.27) and the correlation function (1.25) are bound by the following relation

$$C_{xy}(t_1, t_2) = R_{xy}(t_1, t_2) - E[x(t_1)]\, E[y(t_2)]. \tag{1.29}$$

In *stationary processes* (in the wider meaning of the term), the mean value and the variance are constants, while the correlation function depends only on the difference $\tau = t_2 - t_1$.

Another important concept is the *power-spectral-density* of a random process which characterizes its frequency composition. It is defined mathematically as a Fourier integral transformation of the correlation function.

For stationary processes, therefore, the following Wiener–Khinchin mutual relations hold true:

$$S_{xy}(\omega) = \int_{-\infty}^{\infty} R_{xy}(\tau)\, e^{-i\omega\tau}\, d\tau,$$

$$R_{xy}(\tau) = \frac{1}{2\pi} \int_{-\infty}^{\infty} S_{xy}(\omega)\, e^{i\tau\omega}\, d\omega \tag{1.30}$$

where S_{xy} is the cross power-spectral-density of random functions $x(t)$ and $y(t)$. Also the one-side power-spectral density will be used in Sect. 7.2.

A special case of stationary random processes is represented by *ergodic processes* which, apart from being stationary, must comply also with the condition that all statistical characteristics calculated from a single realization of a random process must equal the characteristics calculated for the whole set of sample functions of the respective random process. For instance, the loading of a railway bridge by a running train with unequal wagons (with the exception of the first and the last vehicles), track irregularities, wind loads, etc., are approximately considered as ergodic processes.

Random processes, which do not comply with the conditions of stationarity, are called *non-stationary processes.* In the case of bridges, their number includes such transient processes as the loading by one vehicle with random effects or running over random irregularities, earthquake, and so on.

2. Theoretical bridge models

Railway bridges are generally long structures which is reflected also in the theoretical models used in their analysis. In principle, theoretical models of railway bridges are of two types: those with continuously distributed mass and those with mass concentrated in material points (lumped masses), or their combinations. The choice of an adequate model depends on the particular case and on the purpose of the analysis.

2.1 Beams

The railway bridge model most frequently used is a beam which models well and simply the linear character of the structure which has small transverse dimensions when compared with its length.

2.1.1 Mass beams

If the mass of the bridge structure is comparable with or considerably higher than the mass of the vehicles, it cannot be neglected. This is the case for medium and large span bridges. In this way the mass beam is necessary (see Fig. 2.1) which is used most frequently for theoretical idealization. The equation of motion of the beam expresses the equilibrium of forces per unit length:

$$EI \frac{\partial^4 v(x,t)}{\partial x^4} + \mu \frac{\partial^2 v(x,t)}{\partial t^2} + 2\mu\omega_b \frac{\partial v(x,t)}{\partial t} = f(x,t) \qquad (2.1)$$

where $v(x, t)$ – vertical deflection of the beam at the point x and at time t,
E – modulus of elasticity of the beam,
I – moment of inertia of beam cross section,
μ – mass per unit length of the beam,
ω_b – circular frequency of viscous damping, see equation (1.9),
$f(x, t)$ – load at point x and time t per unit length of the beam.

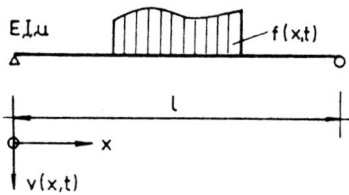

Fig. 2.1. Mass beam of span l.

Differential equation (2.1) was derived by Bernoulli and Euler assuming the theory of small deformations, the validity of Hooke's law, Navier's hypothesis and the Saint-Venant principle. Equation (2.1) assumes constant cross section and mass per unit length of the beam and the damping according to the Kelvin–Voigt model is considered proportional to the velocity of vibration.

Apart from the differential equation (2.1) the behaviour of the beam can also be described by the following integral-differential equation

$$v(x, t) = \int_0^l G(x, s) \left[f(s, t) - \mu \frac{\partial^2 v(s, t)}{\partial t^2} - 2\mu\omega_b \frac{\partial v(s, t)}{\partial t} \right] ds, \quad (2.2)$$

which follows from the theory of influence lines (see [68]). In equation (2.2):

$G(x, s)$ – influence function of the beam also called Green's function. It is the deflection of the beam at point x due to a unit force applied at point s,

l — span of the beam.

Both methods – that using equation (2.1) and that using equation (2.2) – are equivalent.

A current method of analysis is that using equation (2.1) which is applied in all analytical and numerical methods of applied mathematics. Equation (2.2) provides some advantages in those cases where the influence function $G(x, s)$ is known, e.g. from the structural analysis. The advantage of the second method is that the theory of integral equations of Fredholm type makes it sometimes possible to estimate the error by considering a finite number of successive approximations (see [174]).

2.1.2 Massless beams and other special cases

If the mass of the bridge structure or of its element is substantially lower than the mass of the vehicle, it is possible to neglect it entirely; thus we obtain a massless beam as shown in Fig. 2.2. This idealization is used for small span bridges, and longitudinal or transverse girders, which fulfil the above conditions. The equation of motion of such a beam can be obtained from equation (2.1) or equation (2.2) for $\mu \to 0$:

$$EI \frac{\partial^4 v(x, t)}{\partial x^4} = f(x, t) \quad (2.3)$$

or

$$v(x, t) = \int_0^l G(x, s) f(s, t) \, ds, \quad (2.4)$$

where the individual symbols have the same meaning as in equations (2.1) and (2.2).

However, the load $f(x, t)$ in equation (2.3) and equation (2.4) must be considered including its force and inertia effects, as is shown in Sect. 3.4.3.

This makes the solution of equations (2.3) and (2.4) much more difficult than the solution of equations (2.1) and (2.2), because in these equations it is possible, in accordance with initial assumptions, to neglect the inertia effects of the load; for details see [68].

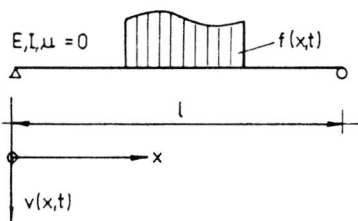

Fig. 2.2. Massless beam of span l.

If the beam stiffness in equation (2.1) is negligibly small, $I \to 0$, a string is obtained, whose carrying capacity is provided by the horizontal force N stretching the string. This gives rise to the equation

$$-N \frac{\partial^2 v(x,t)}{\partial x^2} + \mu \frac{\partial^2 v(x,t)}{\partial t^2} + 2\mu\omega_b \frac{\partial v(x,t)}{\partial t} = f(x,t) \qquad (2.5)$$

where N is positive, if it is in tension which is always so.

The idealization of railway bridges by equation (2.5) is not used, because railway bridges must always be sufficiently stiff (for details see [68], Chapter 14).

The theoretical model of suspension bridges can be derived from the equation

$$EI \frac{\partial^4 v(x,t)}{\partial x^4} - N \frac{\partial^2 v(x,t)}{\partial x^2} + \mu \frac{\partial^2 v(x,t)}{\partial t^2} + 2\mu\omega_b \frac{\partial v(x,t)}{\partial t} = f(x,t) \qquad (2.6)$$

which is a combination of equation (2.1) and equation (2.5) (for details see [68], Chapter 20). Suspension railway bridges are occasionally used in construction practice.

2.1.3 Continuous beams

The mathematical model of a continuous beam according to Fig. 2.3 is described by the same equations as those given in Sects 2.1.1 and 2.1.2 and the respective boundary conditions, Sect. 2.1.4. The continuity above intermediate supports is ensured by special conditions which form the core of the individual methods of analysis of continuous beams in structural mechanics, such as the slope-deflection method, the statical method, the method of initial parameters, the three-moment equation, the five moment equation, and others.

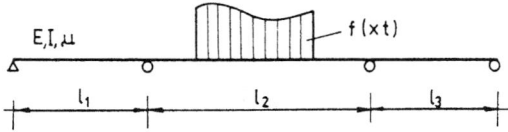

Fig. 2.3. Continuous beam.

Analogous procedures are applied in special cases of continuous beams, such as structures with suspended spans, cantilever ends, and so on.

2.1.4 Boundary and initial conditions

The differential equations (2.1), (2.2), (2.5) and (2.6) necessitate boundary conditions expressing mathematically the bearing of beam ends on supports according to Fig. 2.4. We can discern the following types of bearings of beams:

Hinged bearing (Fig. 2.4a); this has zero deflection and bending moment at the point $x = 0$ (or at the point $x = l$), so

$$v(x, t) = 0 \quad \text{and} \quad \frac{\partial^2 v(x, t)}{\partial x^2} = 0. \tag{2.7}$$

The hinged bearing (Fig. 2.4b) is described by the same equations as equation (2.7).

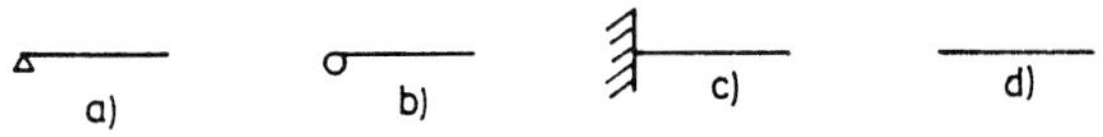

Fig. 2.4. Boundary conditions of beam in point $x = 0$: a) hinged bearing, b) sliding hinged bearing, c) clamped end, d) free end.

A *clamped* beam end (Fig. 2.4c) has zero deflection and rotation, so

$$v(x, t) = 0 \quad \text{and} \quad \frac{\partial v(x, t)}{\partial x} = 0. \tag{2.8}$$

The *free* end (Fig. 2.4d) has a zero bending moment and shear force, so

$$\frac{\partial^2 v(x, t)}{\partial x^2} = 0 \quad \text{and} \quad \frac{\partial^3 v(x, t)}{\partial x^3} = 0. \tag{2.9}$$

Continuous beams must obey conditions above the supports, i.e. the conditions of deflection, rotation, bending moment and the shear force limiting from the right and from the left hand sides.

The geometric conditions for $x = 0$ and for $x = l$

$$v(x, t), \quad \frac{\partial v(x, t)}{\partial x} \tag{2.10}$$

and the dynamic conditions

$$\frac{\partial^2 v(x, t)}{\partial x^2}, \quad \frac{\partial^3 v(x, t)}{\partial x^3} \tag{2.11}$$

are generally different.

In the integral-differential equations (2.2) or (2.4), the boundary conditions are already included in the influence functions $G(x, s)$ which must satisfy them.

2.1 Beams

The initial conditions express the initial deformation and velocity of the structure at the time $t = 0$ from which we begin the analysis. Naturally, they can also be the functions of spatial coordinates, such as for $t = 0$

$$v(x, t) = g_1(x) \quad \text{and} \quad \frac{\partial v(x, t)}{\partial t} = g_2(x). \tag{2.12}$$

2.2 Plates

Although reinforced and prestressed concrete railway bridges are often constructed as plates, they are not usually modelled in this way. Therefore, we give here only the differential equation of plates derived with analogous assumptions as equation (2.1):

$$D\left[\frac{\partial^4 w(x, y, t)}{\partial x^4} + 2\frac{\partial^4 w(x, y, t)}{\partial x^2 \partial y^2} + \frac{\partial^4 w(x, y, t)}{\partial y^4}\right]$$

$$+ \mu \frac{\partial^2 w(x, y, t)}{\partial t^2} = f(x, y, t) \tag{2.13}$$

where $w(x, y, t)$ — vertical deflection of the plate at the point x, y and at time t, see Fig 2.5,

$D = \dfrac{Eh^3}{12(1-v^2)}$ — bending stiffness of the plate,

E — modulus of elasticity of the plate,
h — plate thickness,
v — Poisson's number ($v < 1$),
μ — mass per unit area of the plate,
$f(x, y, t)$ — load per unit area of the plate.

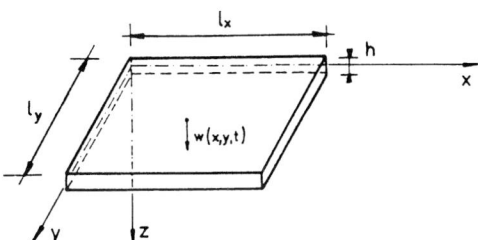

Fig. 2.5. Plate.

Orthotropic plates possess different mechanical properties in two mutually perpendicular directions. The orthotropic plate is often used as the superstructure of steel bridges with ballast. The differential equation of the plate is then

$$D_x \frac{\partial^4 w(x, y, t)}{\partial x^4} + 2D_{xy} \frac{\partial^4 w(x, y, t)}{\partial x^2 \partial y^2} + D_y \frac{\partial^4 w(x, y, t)}{\partial y^4}$$

$$+ \mu \frac{\partial^2 w(x, y, t)}{\partial t^2} = f(x, y, t) \tag{2.14}$$

where, in comparison with the homogeneous plate according to equation (2.13), its mechanical properties in the directions x and y are:

$$D_x = \frac{E_x I_x}{1 - v_x v_y},$$

$$D_y = \frac{E_y I_y}{1 - v_x v_y},$$

$$2D_{xy} = D_x v_y + D_y v_x + 2\left[1 - (v_x v_y)^{1/2}\right](D_x D_y)^{1/2}, \tag{2.15}$$

where E_x, E_y, v_x, v_y – moduli of elasticity and Poisson's ratios of normal deformation in the directions of x and y, respectively,

I_x, I_y – moment of inertia of plate unit width perpendicularly to x and y, respectively.

The expression of boundary conditions for plates is more difficult (for details see [120]). For instance, the boundary conditions of a rectangular plate (Fig. 2.5) for the edge $x = 0$ (i.e. for the axis y) are expressed as follows:

simple support $\quad w = 0, \quad \dfrac{\partial^2 w}{\partial x^2} + v\dfrac{\partial^2 w}{\partial y^2} = 0,$ \hfill (2.16)

clamped edge $\quad w = 0, \quad \dfrac{\partial w}{\partial x} = 0,$ \hfill (2.17)

free edge $\quad \dfrac{\partial^2 w}{\partial x^2} + v\dfrac{\partial^2 w}{\partial y^2} = 0; \quad \dfrac{\partial^3 w}{\partial x^3} + (2-v)\dfrac{\partial^3 w}{\partial x \partial y^2} = 0.$ (2.18)

2.3 Complex systems

More complex structural systems used in bridge engineering, such as trusses, frames, arches, etc., are analyzed in dynamics by the concentration of their mass in lumped masses, as shown in Sect. 2.4. In the present section, therefore, we shall deal only with certain simplifications and special cases.

2.3.1 Trusses

Lattice steel structures (trusses) are frequently found in medium and large span railway bridges.

For the purpose of dynamic analysis, trusses are often replaced by a mass beam of constant cross section so that equation (2.1) applies. The mass per unit length is determined from the total bridge weight G as

$$\mu = \frac{G}{gl} \qquad (2.19)$$

where g – acceleration due to gravity, and
l – span of the bridge.

In the case of variable depth of truss, the mass per unit length valid at midspan of the actual system may be used for μ as well.

The moment of inertia I of an equivalent beam is calculated so as to make the maximum static deflection v_{st} due to adequate load equal to the deflection of the actual lattice structure. As the static deflection of the structure due to an uniformly distributed load f_g (e.g. due to self-weight) is usually known, the computation is simple. For a simply supported beam, for instance, the deflection at midspan due to a uniformly distributed load f_g is

$$v(l/2) = \frac{5}{384} \frac{f_g l^4}{EI}, \qquad (2.20)$$

which can be made equal to the deflection of the equivalent beam $v_{st} = v(l/2)$. From this condition we obtain the moment of inertia

$$I = \frac{5}{384} \frac{f_g l^4}{E v_{st}}. \qquad (2.21)$$

The moment of inertia of the cross section I may also be obtained from the cross section area A_{sup} of the upper and A_{inf} of the lower chords of the truss of height h as follows:

$$I = \frac{1}{2} \left(A_{sup} + A_{inf} \right) h^2. \qquad (2.22)$$

The moment of inertia from equation (2.22) is usually higher than that from equation (2.21), particularly in the case of lattice structures of variable height.

2.3.2 Frames

The Koloušek's deformation (slope-deflection) method [120] is used most frequently in the dynamic analysis of frames for the investigation of natural frequencies and natural modes.

For the analysis of forced vibrations, the modal analysis as well as other methods of structural dynamics are suitable, [8], [120].

2.3.3 Curved bars

The equation of motion of a spatially curved bar of variable cross section can be expressed by a single vector equation

$$\mu(s) \frac{\partial^2 \mathbf{y}(s, t)}{\partial t^2} + 2\mu(s)\omega_b \frac{\partial \mathbf{y}(s, t)}{\partial t} - \mathbf{F}_r(s, t) = \mathbf{f}(s, t) \qquad (2.23)$$

derived by V. Koloušek [120]. Equation (2.23) is a vector sum of all forces applied to an element of the bar (Fig. 2.6) The independent variable s is measured along the bar centre line.

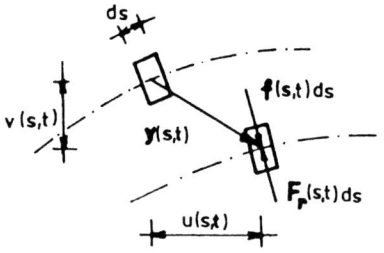

Fig. 2.6. Forces applied to a curved bar element.

In equation (2.23):

$\mathbf{y}(s, t)$ – vector of displacement with components $u(s, t)$, $v(s, t)$, $w(s, t)$,
$\mathbf{F}_r(s, t)$ – vector of reversible force, depending on the type of structure, with components $X(s, t)$, $Y(s, t)$, $Z(s, t)$,
$\mathbf{f}(s, t)$ – vector of load,
$\mu(s)$ – mass per unit length of the curved bar,
ω_b – circular frequency of viscous damping.

Equation (2.23) is usually solved by modal analysis. Equation (2.23) can be simplified for a planar circular arch with incompressible centre line, in which the fibre compression in the centre line and the influence of shear (in comparison with the influence of bending) may be neglected. If we introduce polar coordinates with the origin in the arch centre and denote the arch radius by r and the independent variable angle according to Fig. 2.7 by φ, we can adjust the equation of motion to the form (see [120] and [68]) of:

$$\frac{\partial^6 u(\varphi, t)}{\partial \varphi^6} + 2\frac{\partial^4 u(\varphi, t)}{\partial \varphi^4} + \frac{\partial^2 u(\varphi, t)}{\partial \varphi^2} + \frac{\mu r^4}{EI}\left[\frac{\partial^4 u(\varphi, t)}{\partial \varphi^2 \partial t^2} - \frac{\partial^2 u(\varphi, t)}{\partial t^2}\right]$$

$$= \frac{r^3}{EI}\left[r \frac{\partial f_n(\varphi, t)}{\partial \varphi} - r f_q(\varphi, t) + \frac{\partial^2 f_m(\varphi, t)}{\partial \varphi^2} + f_m(\varphi, t)\right] \qquad (2.24)$$

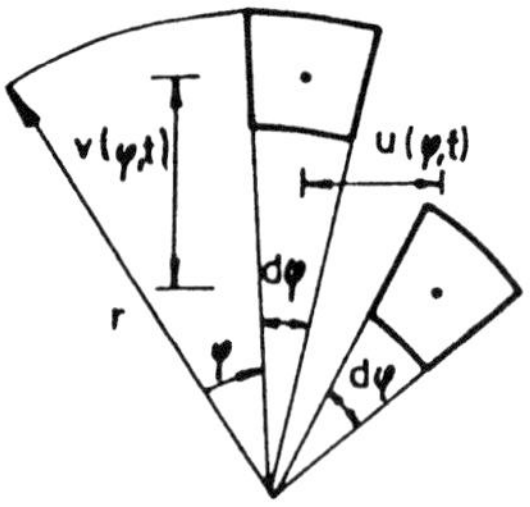

Fig. 2.7. Circular arch.

2.3 Complex systems 37

where $u(\varphi, t)$ – tangential displacement,
$v(\varphi, t)$ – radial displacement, $v(\varphi, t) = \partial u(\varphi, t)/\partial \varphi$,
$f_n(\varphi, t)$ – normal ⎫
$f_q(\varphi, t)$ – tangential ⎬ load per unit length of the arch,
$f_m(\varphi, t)$ – bending ⎭
μ – constant mass per unit length of the arch.

2.4 Lumped masses and models with continuously distributed mass

A typical theoretical model in structural dynamics is an element of volume which possesses six degrees of freedom in spatial vibrations (i.e. displacements in the directions x, y, z and rotations about the same axes).

This is used for the modelling of railway bridges by concentrating the masses of the individual structural elements at significant points of the structure, and by determining the respective stiffness in all directions and around all axes in each of these points. This procedure is illustrated in Fig. 2.8, where the whole lattice structure is concentrated in 44 lumped masses (see also [81]). The deflections of the individual structural elements between these points are usually linearly interpolated.

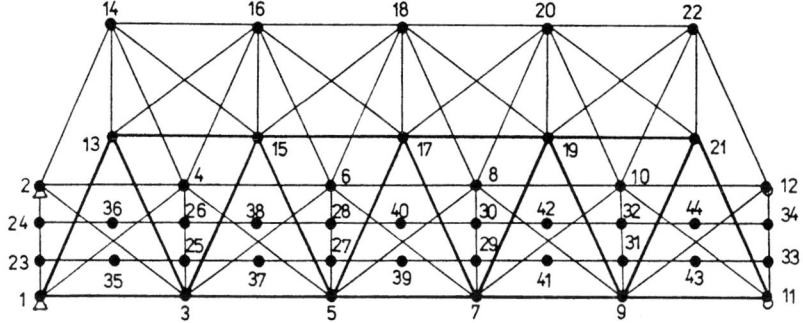

Fig. 2.8. Lattice bridge idealized by a system of 44 lumped masses: *1* to *22* – main girder including the lower and upper wind bracings, *23* to *35* – cross-beams, *35* to *44* – longitudinal girders.

In this procedure, it is an advantage to introduce global coordinates for the structure as a whole and local coordinates for the individual bars. This formulation makes it possible to consider also hinged or rigid joints of the individual bars and/or simplify the structure to a planar model by the selection of adequate input parameters, etc.

This method of idealization of a bridge results in equations of motion of a system of lumped masses which have a general form

$$[m]\{\ddot{q}\} + [b]\{\dot{q}\} + [k]\{q\} = \{F\}. \qquad (2.25)$$

Equation (2.25) is analogous to equation (1.1) which applies to the system with one degree of freedom. In equation (2.25), however, the individual symbols used have the following meanings:

- $[q]$ – generalized vector of displacement of lumped masses,
- $[\dot{q}]$ – generalized vector of velocities of lumped masses (the dot represents the derivative of the respective function with regard to time),
- $[\ddot{q}]$ – generalized vector of acceleration of lumped masses,
- $[m]$ – mass matrix,
- $[b]$ – damping matrix; the assumption $[b] = 2\beta\omega_1[m]$ results in significant simplification of computations; compare with equation (1.1),
- $[k]$ – stiffness matrix,
- $\{F\}$ – vector of modal forces and forces acting between the vehicle and the bridge structure.

Railway bridges may also be idealized by models with continuously distributed mass. The finite element method [92] is suitable for general forms; in the case of structures consisting of individual elementary members, such as beams, plates, and so on, also the folded plate method can be used, see Fig. 2.9.

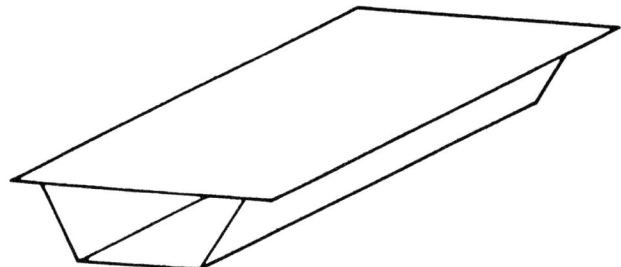

Fig. 2.9. Railway bridge model suitable for the application of the finite element method or the folded plate method.

2.5 Bridge deck modelling

2.5.1 Cross-beam effect

The cross-beam effect of steel railway bridges comprises the dynamic effects of the rolling stock (vehicles) passing along an open bridge deck of classical type, i.e. consisting of a system of longitudinal and transverse girders (Fig. 2.10). The stiffness of such a bridge deck is variable along its length: it is stiffer above the cross beams and less stiff between them (at midspan of longitudinal girders). The passage of a vehicle produces different deflections above the cross-beam (where the deflections are smaller) and at midspan of longitudinal girders (where they are larger).

Actually, the vehicle passes along a regularly undulating curved pathway. The ordinates of irregularities originating in this way, however, arise from the mutual

2.5 Bridge deck modelling

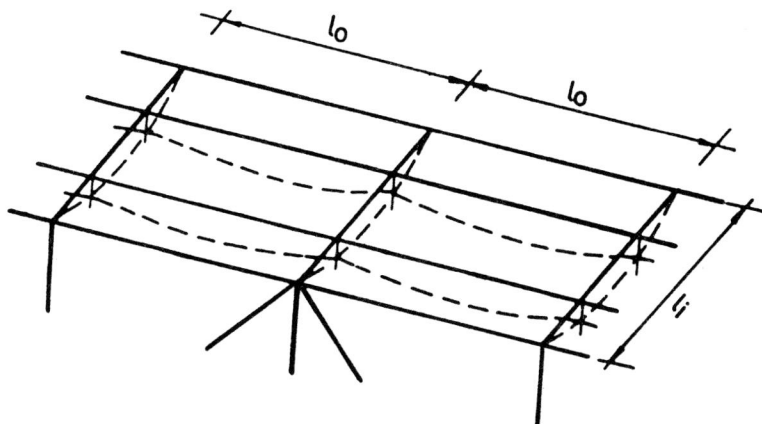

Fig. 2.10. Deflections of the deck grillage (longitudinal and transverse girders); the cross-beam effect.

interaction of the bridge and the vehicle as an integral dynamic system. When modelling the effects of the bridge deck upon the main girders, its influence may be characterized by a system of closely spaced elastic springs of variables stiffness along the bridge length (Fig. 2.11):

$$k(x) = k_1 + k_2 \cos 2\pi x / l_0 . \qquad (2.26)$$

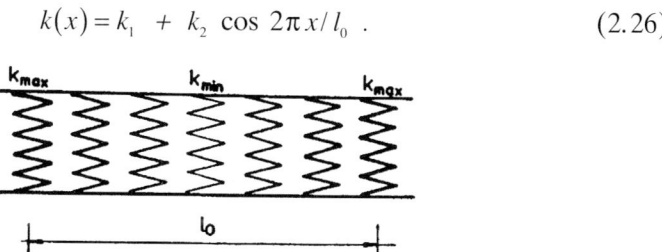

Fig. 2.11. Idealization of a bridge deck as an elastic layer of variable stiffness.

This characterization is made possible by the relatively low mass of the bridge deck when compared with the mass of the main girders and by the relatively small spans of longitudinal girders with respect to the bridge span.

If the distance l_0 equals the span of longitudinal girders, the resulting effect is called the *cross-beam effect*; if l_0 equals the spacing of sleepers, the effect is called the *sleeper effect*. If $k_2 = 0$, it is possible to include the influence of the rail bed.

The cross-beam effect appeared after the introduction of new locomotive types such as electric and diesel-electric [65]. These locomotives have bogies and their wheel base is often so arranged as to increase the cross-beam effect (Fig. 2.12). Locomotives with frames cannot be influenced by the bridge deck, because the irregularities of 2 m to 7 m length do not substantially affect them.

The cross-beam effect was investigated in detail in [65] and [68]. It has been found that it is characterized by the dimensionless parameter

$$\gamma = k_2 / k_1 , \qquad (2.27)$$

see Fig. 2.13. The values of k_1 and k_2 are computed from the maximum k_{max} and the minimum k_{min} stiffnesses of the bridge deck

$$k_1 = \frac{1}{2}\left(k_{max} + k_{min}\right),$$

$$k_2 = \frac{1}{2}\left(k_{max} - k_{min}\right). \qquad (2.28)$$

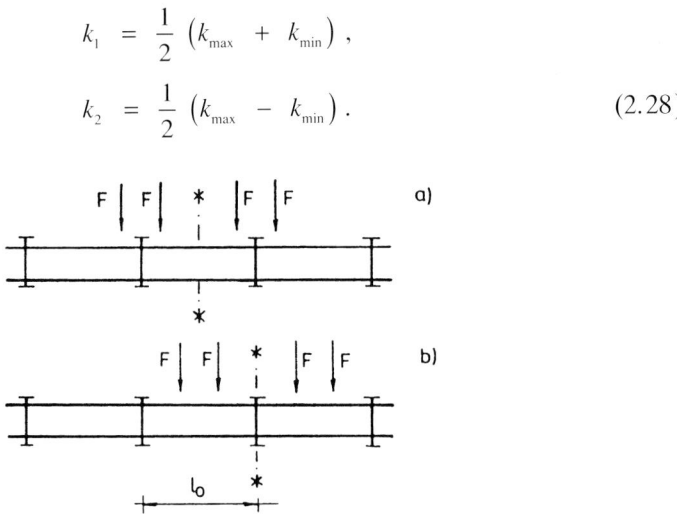

Fig. 2.12. Location of axle forces F with respect to the span l_0 of longitudinal beams; the cross-beam effect: a) the centre of the vehicle above the midspan point of the longitudinal girder, b) above the cross-beam.

The vehicle is placed in two positions such as to make the computed constant k_{max} maximum and k_{min} minimum; l_0 is double the distance of the centre of the vehicle in the two positions in which k_{max} and k_{min} were found.

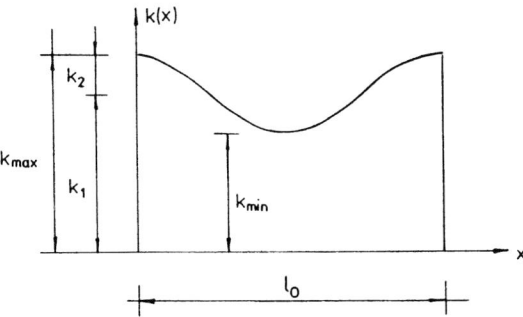

Fig. 2.13. Variable stiffnesses of an elastic layer.

The calculation is as follows: the centre of the vehicle is situated, for instance (Fig. 2.12a), above the midspan point of a longitudinal girder and the static deflection of the bridge deck y_{1i} is computed below every axle with an axle force F_i, where i is the successive number of vehicle axles and n their total number. These deflections are computed with regard to the interaction of longitudinal and transverse girders. The stiffness k_{max} is then

2.5 Bridge deck modelling

$$k_{max} = \sum_{i=1}^{n} \frac{F_i}{y_{1i}}, \quad (2.29)$$

Then, the centre of the vehicle is positioned above the cross-beam (e.g. according to Fig. 2.12b) and the deflections y_{2i} under every axle F_i are found, so that the stiffness is

$$k_{min} = \sum_{i=1}^{n} \frac{F_i}{y_{2i}}. \quad (2.30)$$

The value of the parameter γ, calculated from equation (2.27) with data obtained from equations (2.28) and (2.29), should be as low as possible. Experience has shown

$$\gamma < 0.3 \quad (2.31)$$

so that the dynamic effects are relatively small for the speeds currently attained on railways.

The practical consequence of the condition (2.31) is that the bridge deck stiffness should be as uniform as possible. An example of the computation of the parameter γ with necessary tables for deflections may be found in [65].

2.5.2 Ballast

For reasons of mechanized maintenance and noise reduction, modern railway bridges are built with a continuous rail bed (ballast). Its idealization is very difficult. Initially, the rail bed can be characterized as an elastic layer of constant stiffness according to Fig. 2.11 ($k_2 = 0$ in equation (2.26)). At high speeds, however, the sleeper effect shown in Fig. 2.14 can be important because the vertical deflection under the wheel is higher when it is between sleepers than it is when above a sleeper.

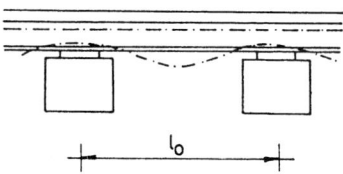

Fig. 2.14. The sleeper effect.

However, modern finite element methods can characterize the gravel bed, see C. S. Desai and A. M. Siriwardane [48]. According to Fig. 2.15, the rails, sleepers, ballast and possibly other parts of the permanent way are divided into finite elements; moreover, there are special boundary elements characterizing the behaviour of the transition between the substructure and the permanent way. The equations of motion of this system are similar to equations (2.25), but they contain numerous non-linearities because of the constitutive equations of soil mechanics. Such problems may be solved by the incremental iterative methods based on the Newton–Raphson method [123].

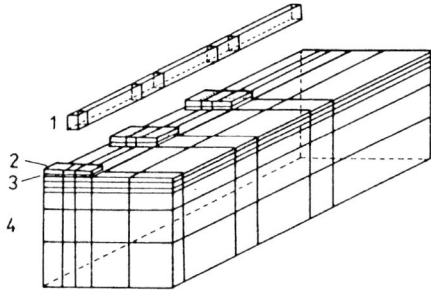

Fig. 2.15. Model of the permanent way and railway substructure according to C. S. Desai and A. M. Siriwardane [48]: *1* – rails, *2* – sleepers, *3* – boundary layer, *4* – gravel bed and substructure (not to scale).

2.6 Modelling other factors

2.6.1 Variable cross section

In the case of bars of variable cross sections both the mass per unit length $\mu(x)$ and the moment of inertia $I(x)$ depend on the longitudinal coordinate x. The differential equation for the deflection of a beam of variable cross section is analogous with equation (2.1):

$$\frac{\partial^2}{\partial x^2}\left[EI(x)\frac{\partial^2 v(x,\,t)}{\partial x^2}\right] + \mu(x)\frac{\partial^2 v(x,\,t)}{\partial t^2} = f(x,\,t)\,. \tag{2.32}$$

In this case the bending moment is found from the relation

$$M(x,\,t) = -EI(x)\frac{\partial^2 v(x,\,t)}{\partial x^2} \tag{2.33}$$

and the stress from equation

$$\sigma(x,\,t) = \frac{M(x,\,t)}{W(x)} \tag{2.34}$$

where $W(x)$ – cross section modulus, dependent on x.

Equation (2.32) with variable coefficients can be solved in a closed form for special cases only [120]. Therefore, one of the approximate methods (e.g. Ritz, Galerkin) is applied; it is also possible to proceed numerically by the division into finite elements according to Fig. 2.16. In this case, every element possesses a different mass; also the stiffness matrix in equation (2.25) is more complex.

2.6 Modelling other factors 43

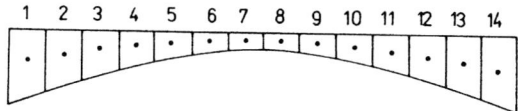

Fig. 2.16. Beam of variable cross section.

2.6.2 Prestressed concrete bridges

In prestressed concrete bridges, there are two principal cases: the prestressing tendons are either perfectly grouted or are entirely free.

In actual prestressed concrete railway bridges the reinforcement is bounded with concrete along the whole tendon length both in pre-tensioned and in post-tensioned beams. Thus the state approaches the first case. In this case the prestress in the tendons has no influence on the potential energy of the beam and, therefore, it does not cause any changes in its natural frequencies. The overall forces applied to the element of length do not vary, because the prestressing force is in equilibrium with the forces compressing concrete. Therefore, in the case of grouted tendons, we proceed with the dynamic analysis of the beam according to equation (2.1) as if the beam were not subjected to an axial force. We include the concrete cross section in full and the ideal cross section of the reinforcement into the cross section area of the beam and prestress. This procedure is applied, whether the beam is pre-tensioned or post-tensioned.

In the second case, when the beam is exposed to constant compression N at its ends only according to Fig. 2.17, the equation (2.6) is applied. The natural frequencies of such a beam are influenced by the axial force (for details see [68] Chapter 20). However, the tendons of such bridges would have to be free along their whole length, i.e. not embedded in concrete, situated on the neutral axis of the beam, and their stresses should not vary in the course of vehicle passage. Cases other than those mentioned above must be analyzed individually.

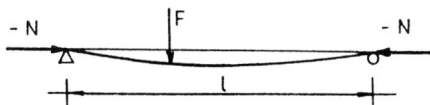

Fig. 2.17. Beam compressed by static axial force N and subjected to the transverse force F.

2.6.3 Influence of elastic foundation, shear and rotatory inertia

In Sect. 2.1 we have described the so-called Bernoulli–Euler beam model, which is most frequently used in theoretical modelling of bridge structures. This model, however, can be so generalized as to characterize other factors which can influence the stresses of the beam. These include the elastic foundation used for the modelling of the permanent way or even the substructure on the bridge, and the effect of shear stresses and rotatory inertia when the beam is deflected.

The last two effects come into consideration only for beams the depth-to-span ratio of which considerably exceeds 1:10, or for materials sensitive to shear stresses. In other cases, the influence of shear and rotatory inertia on the basic vibration modes is very small.

If we consider all forces and moments applied to an element of the beam of length dx as shown in Fig. 2.18, we can derive from the condition of equilibrium of vertical forces and moments differential equations describing the deformation of the beam (for details see [68], Chapter 23). Thus, we proceed from the simplest to more complex models (neglecting damping):

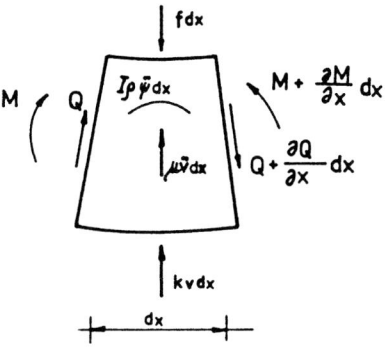

Fig. 2.18. Forces and moments applied to a length element of the beam on elastic foundation.

For the *Bernoulli–Euler model* of the beam on an elastic foundation (neglecting shear and rotatory inertia effects)

$$EI \frac{\partial^4 v(x,t)}{\partial x^4} + \mu \frac{\partial^2 v(x,t)}{\partial t^2} + k\, v(x,t) = f(x,t) \quad (2.35)$$

using the same notation as in equation (2.1), k is the stiffness of the elastic foundation (vertical force per unit length, resulting from unit deflection of the beam).

For the *Rayleigh model* of the beam on an elastic foundation (taking into account rotatory inertia, but neglecting shear effects)

$$EI \frac{\partial^4 v(x,t)}{\partial x^4} - \frac{EI}{c_1^2} \frac{\partial^4 v(x,t)}{\partial x^2 \partial t^2} + \mu \frac{\partial^2 v(x,t)}{\partial t^2} + k\, v(x,t) = f(x,t) \quad (2.36)$$

where

$$c_1^2 = E/\rho \quad (2.37)$$

and c_1 is the velocity of propagation of longitudinal waves along the beam, ρ being the beam mass per unit volume.

For the *shear model* of the beam on an elastic foundation (taking into account the influence of shear stresses on deformation, but neglecting rotatory inertia effects):

$$EI\frac{\partial^4 v(x,t)}{\partial x^4} - \frac{EI}{c_2^2}\frac{\partial^4 v(x,t)}{\partial x^2 \partial t^2} + \mu\frac{\partial^2 v(x,t)}{\partial t^2} + k\left[v(x,t) - i^2\frac{c_1^2}{c_2^2}\frac{\partial^2 v(x,t)}{\partial x^2}\right]$$

$$= f(x,t) - i^2\frac{c_1^2}{c_2^2}\frac{\partial^2 f(x,t)}{\partial x^2} \tag{2.38}$$

where

$$c_2^2 = \alpha_A G/\rho \tag{2.39}$$

and c_2 is the velocity of propagation of transverse waves along the beam,

α_A – constant dependent on cross section form,
G – modulus of elasticity in shear,
$i^2 = I/A$ – radius of gyration,
A – cross section area.

For the *Timoshenko model* of the beam on an elastic foundation (taking into account the influence of both shear stresses and rotatory inertia):

$$EI\frac{\partial^4 v(x,t)}{\partial x^4} - EI\left(\frac{1}{c_1^2} + \frac{1}{c_2^2}\right)\frac{\partial^4 v(x,t)}{\partial x^2 \partial t^2} + \frac{EI}{c_1^2 c_2^2}\frac{\partial^4 v(x,t)}{\partial t^4}$$

$$+ \mu\frac{\partial^2 v(x,t)}{\partial t^2} + k\left\{v(x,t) + i^2\frac{c_1^2}{c_2^2}\left[\frac{1}{c_1^2}\frac{\partial^2 v(x,t)}{\partial t^2} - \frac{\partial^2 v(x,t)}{\partial x^2}\right]\right\}$$

$$= f(x,t) + i^2\frac{c_1^2}{c_2^2}\left[\frac{1}{c_1^2}\frac{\partial^2 f(x,t)}{\partial t^2} - \frac{\partial^2 f(x,t)}{\partial x^2}\right]. \tag{2.40}$$

From equation (2.40) it is possible to obtain all simpler models of the beam: for $c_1 \to \infty$ we obtain equation (2.38), for $c_2 \to \infty$ we obtain equation (2.36), and for $c_1 \to \infty$ and $c_2 \to \infty$ equation (2.40) is simplified to equation (2.35).

Equations (2.35), (2.36), (2.38) and (2.40), are solved in detail in [68] for the case of a moving force.

2.7 Modelling of railway bridges

In the present-day era of advanced computer technology it would be possible to construct a general three-dimensional model of any railway bridge which would include all factors affecting its behaviour. The equations for the calculation of the bridge deflections would have the form of equations (2.25), and in addition would include a number of non-linearities and other factors. The solution would be lengthy and time consuming even if the most up-to-date computers were used.

Such a model would also require a large number of input parameters and data, some of which may not be known with sufficient accuracy. Hence, simplified models are suggested and usually used, which take into account only certain aspects. Such simplified models quite often satisfy engineering practice and requirements.

The choice of a representative, but approximate, model depends on the particular case, the purpose of calculations, the required accuracy and other factors. Further cases of modelling are described in Chapters 8, 9 and 14.

3. Modelling of railway vehicles

Railway vehicles are complex mechanical systems with many degrees of freedom, linear and non-linear springs and various types of damper. During their passage, they affect bridges by spatially oriented forces, i.e. vertical forces (wheel or axle forces), horizontal longitudinal forces (starting and braking forces) and horizontal transverses forces (centrifugal forces and lateral impacts) – see Fig. 3.1.

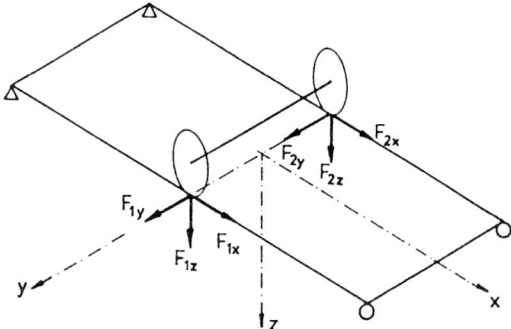

Fig. 3.1. Forces applied to railway bridges: F_{iz} – vertical forces, F_{ix} – horizontal longitudinal forces, F_{iy} – horizontal transverse forces (if only a plane model is considered, the vertical axis is denoted y).

According to Newton's law and the d'Alembert principle, the vehicles produce weight effects, i.e. vertical forces due to vehicle weight, and inertia effects, i.e. the effect of vehicle mass and its acceleration. Weight effects exist even if the vehicle does not move; therefore, they are the principal inputs for static analysis of bridges. The static forces act in a vertical direction. The inertia actions arise only when the vehicle is in motion and in all directions; therefore, they are the cause of dynamic effects.

Generally speaking, the loads on a railway bridge arise from the movement of vehicles over the bridge. It is a very complex problem and it is, therefore, often simplified in engineering practice. The simplification depends on the purpose of the analysis. For example, if our principal purpose is the dynamic analysis of a railway bridge, it is meaningless to consider all details of the contact forces between the wheel and the rail and similar factors, whose influence is only local and which do not affect more distant points, for example the main railway bridge beam, because of numerous filtrations or due to the Saint-Venant's local principle.

We shall deal now in greater detail with the simplifications most frequently used in the modelling of railway vehicles.

3.1 Moving vertical forces

3.1.1 Constant forces

If the inertia effects of moving vehicles are considerably less than their weight effects, the inertia forces can be entirely neglected. This can be permitted in the case of medium and large span bridges (over 30 m) the self-weight of which is considerably higher than the vehicle weight.

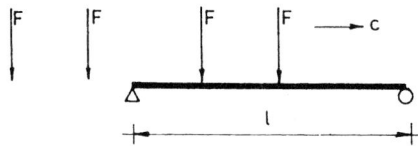

Fig. 3.2. Constant forces F moving at the velocity c along a mass beam of span l.

According to Fig. 3.2 the idealization thus originates by a system of moving constant forces F the magnitude of which equals the static axle or wheel forces. The inertia effects of the bridge beam are considered according to equation (2.1).

C. E. Inglis [106] and V. Koloušek [120] introduced a significiant simplification by separating the weight and the inertia effect of vehicles. The vehicle weight in the form of axle forces moves along the bridge, while the vehicle mass m is fixed at midspan (Fig. 3.3) or in some cases in the first third of the span.

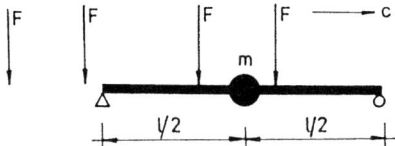

Fig. 3.3. Separation of weight and inertia effects of a vehicle with mass m during its passage along the bridge.

This assumption is suitable for large span bridges and considerably simplifies the theoretical analysis. The comparasion of theory with experiments is very good, when the natural frequency of the loaded bridge, which is variable in time due to vehicle motion (Fig. 4.15), approaches the natural frequency of the simplified case.

For the investigation of the action of the whole train, when the mean mass of the vehicle is uniformly distributed along the bridge, the above simplification is even more acceptable than in the case of a single vehicle, and considerably speeds the analysis even in computerized form.

A detailed solution of the movement of a constant force along a beam has been derived in [68], Chapter 1.

3.1.2 Harmonic variable force

The unbalanced counterweight on the driving wheels of steam locomotives produced an harmonic variable force which besides the force F characterized the action of steam locomotives, so

$$F(t) = F_0 \sin \Omega t ,\qquad (3.1)$$

where F_0 was the amplitude of force, dependent on the number of wheel revolutions per unit time. In the case of the CSD locomotive, series 524.1, the empirical amplitude of force F_0 in kN was

$$F_0 = (2.4 \div 3) \, n^2 \qquad (3.2)$$

where $n = c/u_r$ – number of revolutions of the driving wheels per second,
$\Omega = 2\pi n = c/r$ – circular frequency in equation (3.1),
r, u_r – radius or circumference of driving wheels, respectively,
c – velocity of motion.

The harmonic variable force (3.1) could also be used as the idealization of actions of regular sinusoidal irregularities on the rail head surface. (This is an approximaton, it assumes that the difference between the deformations of the vehicle springs and the bridge deflection under the vehicle is small in comparison with the depth of the irregularities.) The force amplitude and circular frequency would then be

$$F_0 = \frac{1}{2} ka ,$$

and

$$\Omega = 2\pi c/b \qquad (3.3)$$

where a, b – depth and length of sinusoidal irregularities, respectively,
k – spring stiffness.

The solution of the movement of the harmonic variable force along the beam can be found in [68], Chapter 2.

3.1.3 Continuous load

When the span of the bridge is large in comparision with the distance between axles, idealization by means of a continuous load can be used and is considered either as the load per unit length or per unit area according to Fig. 3.4. The system of wheel forces can be approximated to a uniformly distributed load f_{aL} in those cases where more than about four forces can be applied to the element of the bridge considered. In practice, a system of concentrated forces and uniformly distributed load is used, which models both local and global actions of the vehicle on small-span and large-span bridge elements.

Steady state bridge vibrations can be investigated by considering Fig. 3.4, provided the length of the loading strip (i.e. the train) is considerably larger than the bridge span, which is usually a fair assumption. The mass of the load is added to the bridge as in Section 3.1.1.

The solution of this problem may be found in [68], Chapter 3.

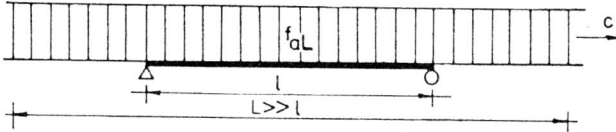

Fig. 3.4. Continuous moving load.

3.1.4 Random load

If the inertial action of vehicles is deadened by random actions of track irregularities, heterogenity of static axle forces or by other factors, the load can be expressed globally by a random load. The idealization of this approach gives rise to two extreme cases, see Figs 3.5 and 3.6.

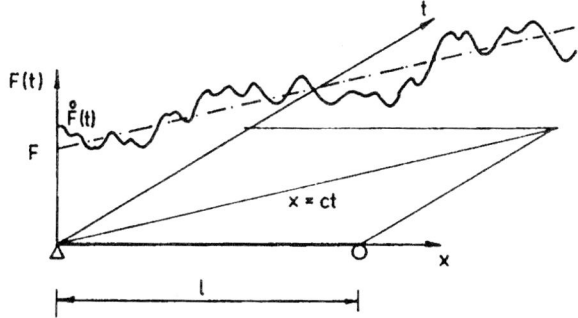

Fig. 3.5. Movement of a random force $F(t)$ along the beam.

Figure 3.5 shows the motion of a random variable force $F(t)$ along the beam. The force is a random (stationary and ergodic) function of time which can be resolved into a constant mean value and a random centred component

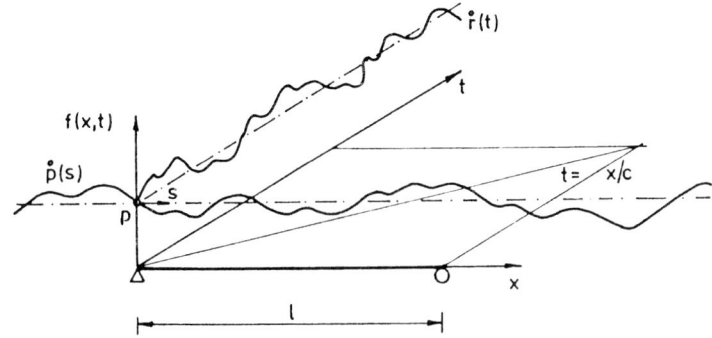

Fig. 3.6. Movement of an infinite random strip load $f(x, t)$ along the beam.

3.1 Moving vertical forces

$$F(t) = F + \varepsilon \overset{\circ}{F}(t), \quad (3.4)$$

where $F = E[F(t)]$ is the mean value (static axle or wheel force), $\overset{\circ}{F}(t)$ is the centred random component defined by equation (1.26), and $\varepsilon \ll 1$ is a small quantity if the random component of the force is small in comparison with the mean value.

The idealization of vehicle action according to (3.4) is adequate for bridges of very small spans, longitudinal girders, and so on, complying with the above assumptions. A detailed solution of this case may be found in [73]. It can be observed that the response of the beam is a non-stationary process even if the beam is loaded by a stationary force. This is due to the fact that the passage of the force along the beam is a transient process.

The other extreme case is shown in Fig. 3.6. The model represented in it is adequate for large span bridges, where track irregularities, imperfections of the gravel bed, the number of simultaneously acting vehicles with different axle distances and different axle forces, etc., appear. There is a great number of dynamic effects which, although often deterministic, act stochastically in the sum in accordance with the laws of probability.

According to Sect. 3.1.3 the bridge load can be considered as a continuous load which is a random variable both with regard to the coordinate x and to the time t and moves at a constant speed along the beam. The load was formulated in [74] as follows:

$$f(x, t) = [p + \varepsilon \overset{\circ}{p}(s)][1 + \overset{\circ}{r}(t)], \quad (3.5)$$

where $p = E[f(x,t)]$ is the constant mean value of the continuous load,

$\overset{\circ}{p}(s)$ — centred random component of the load depending on the moving coordinate s,

$\overset{\circ}{r}(t)$ — centred random component of the load depending on time t,

s — length coordinate with the origin moving at the speed c,

$\varepsilon \ll 1$.

The component $\overset{\circ}{p}(s)$ idealizes the composition of the traffic stream in railway traffic, while the component $\overset{\circ}{r}(t)$ characterizes the random character of the load in time, i.e. irregularities of the track, the springing of the vehicles, etc. For $\overset{\circ}{r}(t) = 0$ the load (3.5) characterizes the motion of a random strip load along the bridge. If $\overset{\circ}{p}(s) = 0$, the load $f(x,t)$ is random in time, but does not move.

A detailed solution is given in [74]. The response of the beam to such a load is a stationary process, as long as both random components $\overset{\circ}{p}(s)$ and $\overset{\circ}{r}(t)$ are stationary.

3.2 Mass elements

For the dynamic analysis of bridges it is sufficient to resolve the railway vehicles into individual mass elements, the simplest of which are lumped masses and rigid plates.

3.2.1 Lumped mass

A concentrated mass point adequately represents the axle and/or the wheel of railway vehicles (Fig. 3.7). The following forces are applied to the unsprung mass:

weight F of the mass m,

the inertia force $-m \dfrac{d^2 v(t)}{dt^2}$,

reversible forces from spring compression K,
damping forces of springs B,
track reaction $-R$,
and force directly applied to the mass m, e.g. an harmonic variable force $F(t)$ according to equation (3.1).

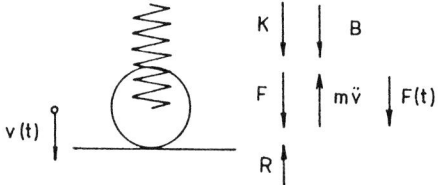

Fig. 3.7. Lumped mass and forces applied to it (positive force direction is downwards).

The choice of the state of equilibrium depends on whether we want to determine only the dynamic bridge response or the static and dynamic components. In the latter, more frequent case, it is necessary to add to the static force F also the static weight of the sprung mass affecting the wheel under consideration.

When considering the state of equilibrium we must distinguish between the cases, when the wheel is in constant contact with the rail, and the case where there is loss of contact. In the latter case it is more advantageous to consider the unloaded beam as the state of equilibrium.

The idealization of the whole vehicle by a sprung mass point according to Fig. 3.8 is acceptable when the bridge span is considerably larger than the vehicle axle base and when the unsprung masses are relatively small. With respect to the choice of the state of equilibrium the above notes apply to this case, too.

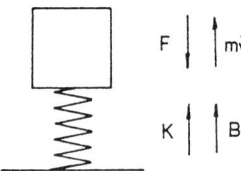

Fig. 3.8. Idealization of a vehicle by a sprung mass.

3.2 Mass elements   53

Besides the two basic cases according to Figs 3.7 and 3.8 there are various combinations of mass points possible, e.g. the concentration of all unsprung masses in one mass point and that of all sprung masses in another mass point, etc. Figure 3.9 shows this case which is adequate for large span bridges.

Every mass point possesses one degree of freedom if its motion is limited to a straight line. In Figs 3.7 to 3.9 the straight line has been represented by the vertical axis. On the other hand, an oscillating motion of a mass point in space provides three degrees of freedom, according to Table 3.1. The motion in the direction of x is considered uniform and independent of other components (in the majority of cases).

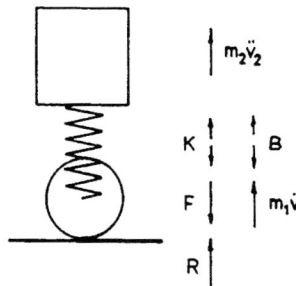

Fig. 3.9. Idealization of a vehicle by unsprung and sprung masses.

TABLE 3.1 Oscillating motion in the directions of coordinate axes according to Fig. 3.1

Motion in the direction of axis	Motion	Notation
x	longitudinal motion	$u(t)$
y	hunting	$v(t)$
z	bouncing	$w(t)$

3.2.2 Rigid plates and bodies

As a rule, the sprung elements of railway vehicles or their bogies are considerably stiffer than the bridge; therefore, they are idealized as rigid mass plates if their motion is considered to occur only in a vertical plane.

In this simplest case, the rigid plate possesses two degrees of freedom and can move in a vertical direction, $v(t)$, and rotate, $\eta(t)$, about its horizontal transverse axis passing through the centroid C_g of the vehicle according to Fig. 3.10. The following vertical forces are then applied to the plate:

the weight F of the plate with mass m (if the springs are not compressed in the state of equilibrium),
inertia force $-m\ddot{v}$,

reversible forces from spring compression K, and
damping forces in springs B.
In addition the following moments are applied:
moment of inertia $-I\ddot{\eta}$, where I is the moment of inertia of the mass plate with respect to the centroid C_g,
the reversible moments due to springs Kd, and
the damping moment due to springs Bd.

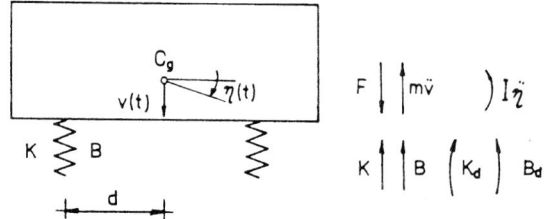

Fig. 3.10. Rigid plate and forces and moments applied to it (positive direction of forces is downwards, positive direction of moments is clockwise).

In a general three-dimensional case, the elements of railway vehicles can be idealized by mass rigid bodies which provide six degrees of freedom. The body can move in the directions of the coordinate axes according to Fig. 3.1 and Table 3.1 and rotate about these axes according to Table 3.2.

TABLE 3.2 Rotatory oscillating motion about coordinate axes according to Fig. 3.1

Rotatory motion about axis	Motion	Notation
x	rolling, sway motion	$\xi(t)$
y	pitching	$\eta(t)$
z	yawing	$\zeta(t)$

The individual elements of railway vehicles could be substituted also by elastic bodies possessing mass, such as beams, frames, trusses, walls, plates, etc. However, such accurate idealization is too detailed for bridges and would not bring any great advantage. In any case, the substitution always depends on the purpose of the analysis.

3.3 Springs and damping elements

The individual components of railway vehicles are interconnected by springs and damping elements. Their mathematical description is usually complicated, because they are non-linear in the majority of cases. For this reason they are often linearized.

3.3.1 Linear spring

The simplest springing element is the linear spring, in which the relation between the reversible force K and the relative displacement of both spring ends v is linear

$$K = kv = k(v_2 - v_1). \qquad (3.6)$$

In equation (3.6) k is the spring stiffness (see also equation (1.1) and Fig. 1.2). The relation (3.6) is presented graphically in Fig. 3.11.

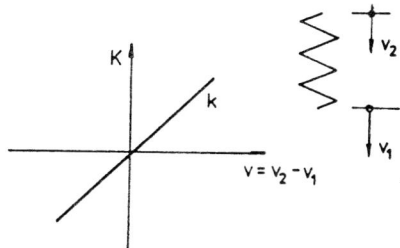

Fig. 3.11. Linear spring.

3.3.2 Non-linear spring

Non-linear springs occur frequently. A general equation between the force K and the relative displacement v is

$$K = k_1 v + k_2 v^2 + k_3 v^3 + \ldots . \qquad (3.7)$$

According to the magnitudes and signs of constants k_i special cases appear, for instance, a stiffening spring which occurs in rubber springing is represented by the equation (Fig. 3.12)

$$K = k_1 v + k_3 v^3 . \qquad (3.8)$$

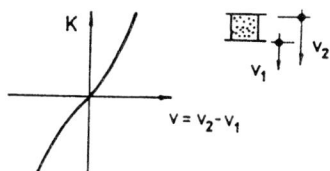

Fig. 3.12. Hardening spring.

The single-sided softening spring with the equation

$$K = k_1 v + k_2 v^2 \qquad (3.9)$$

occurs in the so-called C-springs (Fig. 3.13).

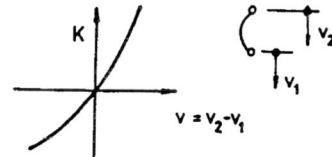

Fig. 3.13. Softening C-spring.

Bilinear springs may have a deviation in the origin (Fig. 3.14a) or in the point v_0 (Fig. 3.14b). In the former case (Fig. 3.14a) the following equation holds true

$$K = k_1 v \quad \text{for} \quad v < 0$$

and

$$K = k_2 v \quad \text{for} \quad v > 0. \tag{3.10}$$

In the latter case (Fig. 3.14b) the following equation is valid

$$K = k_1 v \quad \text{for} \quad v < v_0,$$

and

$$K = k_1 v_0 + k_2 (v - v_0) \quad \text{for} \quad v > v_0. \tag{3.11}$$

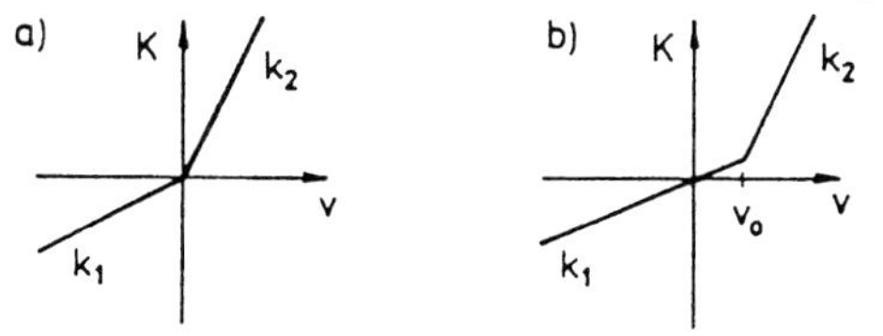

Fig. 3.14. Bilinear spring: a) point of deviation in the origin $v = 0$, b) deviation in the point $v = v_0$.

3.3.3 Stop

Stops are also used in vehicles to limit the displacement of the springs. Figure 3.15 shows the relation between the reversible force K and the relative displacement of springs, described by the equations

$$K = kv \quad \text{for} \quad v_1 < v < v_2,$$

$$K = kv + k_1(v - v_1) \quad \text{for} \quad v < v_1,$$

$$K = kv + k_2(v - v_2) \quad \text{for} \quad v > v_2. \tag{3.12}$$

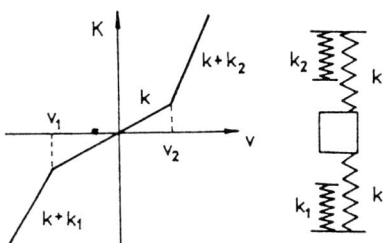

Fig. 3.15. Spring with stops.

A stiff stop would be represented by equation (3.12) by the introduction of high values of k_1 or k_2. Equation (3.12) can also be used for the linearization of a non-linear spring according to Fig. 3.12.

3.3.4 Viscous damping

The equations of motion must include the damping force, which, in the case of viscous damping, is

$$B = b\dot{v} \qquad (3.13)$$

so that it is proportional to the velocity of motion $\dot{v}$ (i.e. the velocity of relative motion of the spring ends), see also equation (1.1).

This assumption corresponds to the motion of a system immersed in a liquid, whose mass is subjected to viscous friction (Kelvin–Voight hypothesis). This damping is characterized by exponentially diminishing amplitudes of natural vibrations – see Fig. 1.3 (the ratio of two successive amplitudes is constant). A viscous damper is schematically represented in Fig. 3.16.

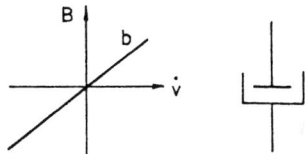

Fig. 3.16. Viscous damper.

3.3.5 Friction

In dry sliding friction (Coulomb hypothesis) a constant force acts in a direction opposing the motion, and it is proportional – through the friction coefficient μ – to the normal force N:
Hence

$$B = \mu N \quad \text{for} \quad \dot{v} > 0$$

and

$$B = -\mu N \quad \text{for} \quad \dot{v} < 0. \qquad (3.14)$$

A Coulomb damper is shown schematically in Fig. 3.17. This damping is characterized by the fact that the difference between two successive amplitudes of natural vibration is constant (the extreme positions of the displacement may be connected by a straight line).

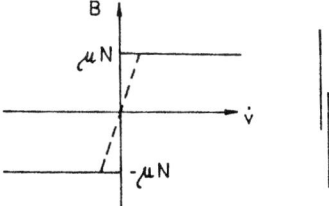

Fig. 3.17. Friction damper.

In numerical calculations the discontinuity at the point $\dot{v} = 0$ is sometimes replaced by a straight line, shown by a dashed line in Fig 3.17. The choice of width of this zone is important: a too wide zone introduces viscous damping into the system, while a too narrow zone may bring some numerical difficulties.

3.3.6 Elastic-plastic element

Railway vehicles may be provided with elastic-plastic springs which also possess damping. Fig. 3.18 shows the dependence of the force K on the displacement v which is first elastic and then plastic. When the load is removed the deformation is, once again, first elastic and then plastic. The energy used in every loading cycle is proportional to the area between these curves (the hysteresis loop) and represents the damping.

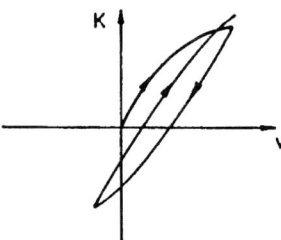

Fig. 3.18. Elastic-plastic element.

The behaviour of the elastic-plastic element described is often idealized by the model shown in Fig. 3.19, and depends on increasing ($\dot{v} > 0$) or decreasing ($\dot{v} < 0$) displacement.

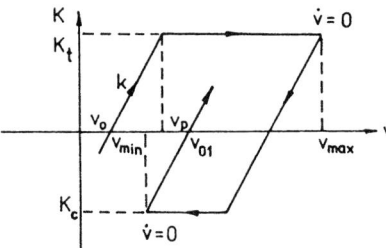

Fig. 3.19. Elastic-ideally plastic element.

Let us assume knowledge of the initial deformation v_0, the tensile yield limit K_t, the compressive yield limit K_c and the stiffness k (Fig. 3.19). Then we can find the deformation v_p at which plastic deformation occurs from

$$v_p = v_0 + K_t/k . \tag{3.15}$$

The reversible force K is then found as follows:
If $\dot{v} > 0$ elastic or plastic tensile deformation appears

$$K = K_t - k(v_p - v) \quad \text{for} \quad v < v_p ,$$
$$K = K_t \quad \text{for} \quad v > v_p . \tag{3.16}$$

This cycle terminates when $\dot{v} = 0$, for which we shall denominate the deformation as v_{max}. Then, the load removing cycle begins with $\dot{v} < 0$. In this process

$$K = K_t - k(v_{max} - v) \quad \text{for } v_{max} - (K_t - K_c)/k < v < v_{max},$$
$$K = K_c \quad \text{for } v < v_{max} - (K_t - K_c)/k. \quad (3.17)$$

The load removing cycle will terminate, once again, when $\dot{v} = 0$, for which v_{min} is defined. Then, again $\dot{v} > 0$, so

$$K = K_c + k(v - v_{min}) \quad \text{for } v < v_{min} - K_c/k. \quad (3.18)$$

When $v \geq v_{min} - K_c/k$, the cycle begins with a new value of

$$v_0 = v_{01} = v_{min} - K_c/k. \quad (3.19)$$

3.4 Modelling of a bridge and a running train

The equations for the calculation of the vibration of railway vehicles when crossing a bridge are derived on the basis of either the conditions of equilibrium of forces and moments, or by means of Lagrange equations of the second type. Recently, the component element method has also been used [132]. In this method railway vehicles are broken down into individual elements, i.e. lumped masses, plates, bodies, springs and damping elements according to Sects 3.2 and 3.3, where all relevant forces and moments are described. Then, the differential equations for the motion of the individual elements are written considering the conditions of equilibrium of forces and moments.

We shall demonstrate this method by an example of the movement of a number of railway vehicles (a train) across a multispan bridge according to Fig. 3.20 (plane model). Further models (including spatial models) and their equations may be found in [23], [81], [162], [165].

3.4.1 Initial assumptions

The bridge is idealized by a mass continuum according to Chapter 2. Fig 3.20 represents a continuous beam with 1, 2, ..., i, ..., I spans, of length l_i; vertical deflection is denoted by $v_i(x, t)$, moment of inertia $I_i(x)$ and mass per unit length $\mu_i(x)$ in the ith span. The coordinates x are measured from the left-hand end of the first span.

The damping of the bridge is proportional to the vibration velocity (constant ω_{bi}).

The boundary conditions are determined by the type of supports under consideration. The initial conditions are, generally,

$$v_i(x, 0) = v_{i0}(x) \quad \text{and} \quad \dot{v}_i(x, 0) = \dot{v}_{i0}(x). \quad (3.20)$$

The influence of the bridge deck will be replaced (Sect. 2.5) by an elastic layer of rigidity $k(x)$ which is generally a random variable depending on the length x.

Irregularities $r(x)$ which occur on the track surface, are random variables in the length x.

The train moves along the bridge with an irregular velocity; the motion of the first axle in time is determined by the function $u(t)$. The regular motion is described by the equation

$$u(t) = ct \tag{3.21}$$

where c is the velocity of motion and t is time.

The train consists of 1, 2, ..., q, ..., Q vehicles which possess 1, 2, ..., n, ..., N axles counted from right to left in Fig. 3.20. The static axle force of the q-th vehicle is F_q.

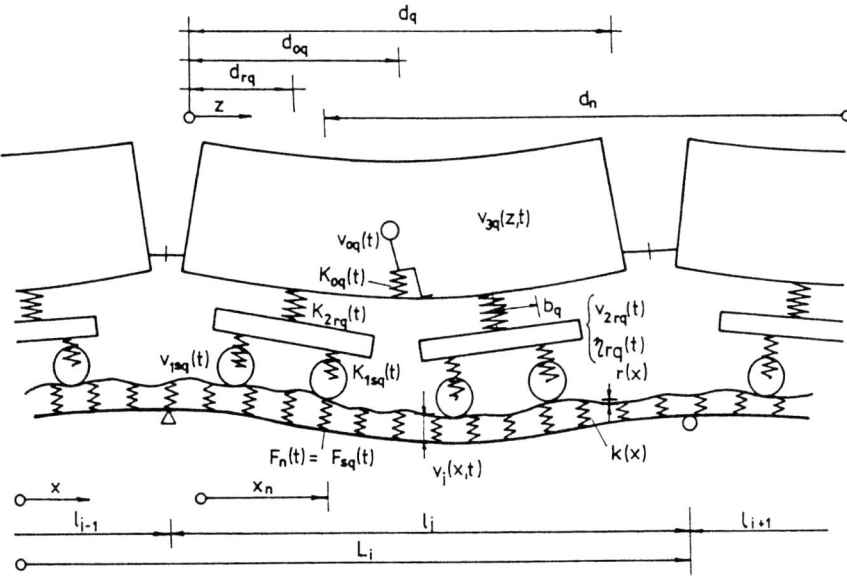

Fig. 3.20. Model of a train driving along a railway bridge.

The body of every vehicle is considered to be elastic, as shown in Fig. 3.20. For this reason, its motion is described by the Bernoulli–Euler equation for the beam with the notation of $v_{3q}(z, t)$ for vertical displacement, I_{3q} for the moment of inertia and μ_{3q} for the mass per unit length (both constant) and E_{3q}. The length coordinate z is measured from the left-hand side of every vehicle.

Secondary masses m_{2q} with moment of inertia I_{rq} (their number $r = 1, 2$ in every vehicle) are considered to be rigid plates which can move in the vertical direction $v_{2rq}(t)$ and rotate $\eta_{rq}(t)$.

Primary masses (wheels and axles) m_{1q} are idealized by lumped masses with vertical displacement $v_{1sq}(t)$, where the index s denotes the number of the axle in the qth vehicle.

Figure 3.20 (as well as the equations below) considers four-axle vehicles or locomotives which are the most frequent type of railway vehicles. However, the

3.4 Modelling of a bridge and a running train

formulation of the problem for two-axle or for six-axle vehicles using the index s would be similar.

We assume that in the body a passenger is seated whom we shall represent – until more adequate models are available – by a one degree of freedom system. In the qth vehicle it is the mass m_{0q} which can move in vertical direction $v_{0rq}(t)$.

The distance of the passenger from the left-hand end of the qth vehicle is d_{0q}, the distance of the rth bogie from the left-hand side is d_{rq}, half the axle base of the bogie is b_q and the distance of the nth axle from the first axle from the right-hand side is d_n.

The beginning of the movement is assumed at the moment of the entry of the first axle from the right-hand side on the left-hand edge of the first beam from the left. The train moves from left to right. The initial conditions are, generally (the vibrating train is assumed at $t = 0$):

$$v_{0q}(0) = v_{0q0}, \qquad \dot{v}_{0q}(0) = \dot{v}_{0q0},$$

$$v_{1sq}(0) = v_{1sq0}, \qquad \dot{v}_{1sq}(0) = \dot{v}_{1sq0},$$

$$v_{2rq}(0) = v_{2rq0}, \qquad \dot{v}_{2rq}(0) = \dot{v}_{2rq0},$$

$$\eta_{rq}(0) = v_{rq0}, \qquad \dot{\eta}_{rq}(0) = \dot{\eta}_{rq0},$$

$$v_{3q}(z, 0) = v_{3q0}(z), \qquad \dot{v}_{3q}(z, 0) = \dot{v}_{3q0}(z). \qquad (3.22)$$

The boundary conditions of the beam idealizing the body are those of the free end:

$$v''_{3q}(0, t) = 0, \qquad v''_{3q}(d_q, t) = 0,$$

$$v'''_{3q}(0, t) = 0, \qquad v'''_{3q}(d_q, t) = 0, \qquad (3.23)$$

where d_q is the length of the qth vehicle.

It is assumed that at the beginning of the movement all springs of the vehicle are compressed, while the beams of the bridge are not deformed (including the influence of the static action of axle forces).

3.4.2 Equations of motion

With these assumptions it is possible to put together the respective differential equations of motion for the vertical displacement of all parts of the model shown in Fig. 3.20. The positive direction of motion is downwards, the tensile force is positive.

For the motion of the passenger,

$$- m_{0q} \ddot{v}_{0q}(t) - K_{0q}(t) = 0, \qquad q = 1, 2, \ldots, Q, \qquad (3.24)$$

where the force in the spring

$$K_{0q}(t) = k_{0q}\left[v_{0q}(t) - v_{3q}(d_{0q}, t)\right] + B_{0q} \qquad (3.25)$$

has the first (elastic) component proportional to the difference of the respective displacements, and the second component B_{0q} denotes damping or non-linear action.

A similar system of notation for forces is used for other springs.

Flexural vibrations of the vehicle body are described by the equations

$$E_{3q}I_{3q}v_{3q}^{IV}(z,t) + \mu_{3q}\ddot{v}_{3q}(z,t) + 2\mu_{3q}\omega_{b3q}\dot{v}_{3q}(t)$$
$$= \delta(z - d_{0q})K_{0q}(t) - \sum_{r=1}^{2}\delta(z - d_{rq})K_{2rq}(t), \quad q = 1, 2, ..., Q, \quad (3.26)$$

where $\delta(x)$ is the Dirac function (see Section 3.4.3), and

$$K_{2rq}(t) = k_{2q}\left[v_{3q}(d_{rq},t) - v_{2rq}(t)\right] + B_q \quad (3.27)$$

is the force in the spring between the body and the secondary mass.

Vertical motion of the secondary mass is determined by the equation

$$-m_{2q}\ddot{v}_{2rq}(t) + K_{2rq}(t) - \sum_{s}K_{1sq}(t) = 0,$$

$$r = \begin{cases}1\\2\end{cases} \quad \text{for} \quad s = \begin{cases}1, 2\\3, 4,\end{cases} \quad q = 1, 2, ..., Q, \quad (3.28)$$

where the force between the primary and the secondary masses is denoted by

$$K_{1sq}(t) = k_{1q}\left[v_{2rq}(t) - (-1)^s b_q \eta_{rq}(t) - v_{1sq}(t)\right] + B_{1q}. \quad (3.29)$$

For rotatory motion of secondary mass

$$-I_{rq}\ddot{\eta}_{rq}(t) + b_q\sum_{s}(-1)^s K_{1sq}(t) = 0$$

$$\text{where} \quad r = \begin{cases}1\\2\end{cases} \quad \text{for} \quad s = \begin{cases}1, 2\\3, 4,\end{cases} \quad q = 1, 2, ..., Q. \quad (3.30)$$

Vertical motion of the primary mass is described by the equation

$$-m_{1q}\ddot{v}_{1sq}(t) + K_{1sq}(t) - F_{sq}(t) = 0, \quad s = 1, 2, 3, 4, \quad q = 1, 2, ..., Q \quad (3.31)$$

where the dynamic component of the nth axle force is

$$F_{sq}(t) = F_n(t) = k(x_n)\left[v_{1sq}(t) - \varepsilon_{in}v_i(x_n,t) - r(x_n)\right] + B \quad (3.32)$$

and it characterizes the interactions between the vehicle and the bridge with an elastic layer and irregularities; B expresses the damping force in the elastic layer.

The Bernoulli–Euler differential equation applied to bridge beams gives

$$E_i\frac{\partial^2}{\partial x^2}\left[I_i(x)v_i''(x,t)\right] + \mu_i(x)\ddot{v}_i(x,t) + 2\mu_i(x)\omega_{bi}\dot{v}_i(x,t)$$
$$= \sum_{n}\varepsilon_{in}\delta(x - x_n)\left[F_q + F_n(t)\right], \quad i = 1, 2, ..., I. \quad (3.33)$$

The deflection of the ith span $v_i(x,t)$ is valid over the interval of $L_{i-1} \leq x \leq L_i$.

3.4 Modelling of a bridge and a running train

In equation (3.32) and equation (3.34) the notation

$$x_n = u(t) - d_n, \quad n = 1, 2 \ldots, N, \quad d_1 = 0 \qquad (3.34)$$

is used for the coordinate of the nth axle, and the symbol

$$\varepsilon_{in} = \begin{cases} 1 \\ 2 \end{cases} \text{ for } \begin{cases} L_{i-1} \leq x_n \leq L_i \\ x_n < L_{i-1}, \, x_n > L_i \end{cases} \qquad (3.35)$$

characterizes whether the nth axle is loading the ith girder or not.

The condition of the contact between the wheel and the rail requires

$$F_q + F_n(t) \geq 0. \qquad (3.36)$$

When this condition has not been complied with, the wheel loses contact with the rail and

$$F_q + F_n(t) = 0. \qquad (3.37)$$

is substituted in equation (3.33) until the contact has been regained. This produces an impact.

The system of differential equations (3.24) to (3.33) may be computed by any appropriate method of numerical mathematics. So far only some special cases have been analyzed (see [23], [191]) while the general formulation proposed here would necessitate a fairly large computer. To illustrate this point, a train with 100 axles and a 3-span bridge ($N = 100$, $Q = 25$, $I = 3$), which is a current case would result in the following number of differential equations:

Equation	Number of equations
(3.24)	25
(3.26)	25
(3.28)	50
(3.30)	50
(3.31)	100
(3.33)	3
Together	253

Moreover, equations (3.26) and (3.33) are partial differential equations for which the solution by resolution into natural modes would result in an even greater number of ordinary differential equations. For these reasons only some special cases are solved, usually according to the analysis requirements.

3.4.3 Movement of the vehicle along the bridge

As has already been suggested in equation (3.33), the movement of the vehicle along the bridge is often expressed by means of the Dirac function $\delta(x)$ which in mechanics characterizes the action of unit force concentrated in point $x = 0$.

In mathematics, the Dirac function is considered a generalized function which has the following properties:

$$\int_{-\infty}^{\infty} \delta(x)\,dx = 1, \qquad (3.38)$$

$$\int_{-\infty}^{\infty} \delta(x-a)\,f(x)\,dx = f(a), \qquad (3.39)$$

or

$$\int_{a}^{b} \delta(x-\xi)\,f(x)\,dx = \begin{cases} 0 & \text{for } \xi < a < b \\ f(\xi) & \text{for } a < \xi < b \\ 0 & \text{for } a < b < \xi \end{cases} \qquad (3.40)$$

where $f(x)$ is a continuous function within the interval of $\langle a, b \rangle$ and a, b, ξ are constants. For further details see [68] Chapter 1.1.

Therefore, if a concentrated force F moves along a beam type structure, the load per unit length, which appears on the right-hand side of the differential equation (2.1), has the form of

$$f(x, t) = \delta[x - u(t)]\,F, \qquad (3.41)$$

where $u(t)$ expresses the law of the force movement. For uniform motion it is the function (3.21).

The equation (3.33) formulated above did not reveal the difficulties which are encountered when analysing the movement of a load along a structure. Consider the simplified example shown in Fig. 3.21.

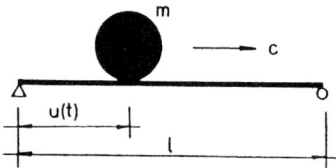

Fig. 3.21. Movement of mass m along a beam obeying the mass.

The mass m of the weight F moves along a mass beam. If neither the mass of the beam nor the moving mass may be neglected, it is necessary to consider the forces of inertia of the moving mass at the point of contact. The position, naturally, changes in the course of the movement of the load, and is a function of time $u(t)$. Then, the load on the right-hand side of equation (2.1) is, according to d'Alembert's theorem,

$$f(x, t) = \delta[x - u(t)]\left\{F - m\,\frac{d^2 v[u(t), t]}{dt^2}\right\}. \qquad (3.42)$$

Consequently, the load depends on the response of the beam in the place of contact $v[u(t), t]$, which introduces complications into both theoretical and

3.4 Modelling of a bridge and a running train

numerical solutions. The derivative in (3.42) must be calculated as follows:

$$\frac{d^2 v[u(t),\ t]}{dt^2} = \frac{\partial^2 v}{\partial u^2}\left(\frac{du}{dt}\right)^2 + 2\frac{\partial^2 v}{\partial u \partial t}\frac{du}{dt} + \frac{\partial v}{\partial u}\frac{d^2 u}{dt^2} + \frac{\partial^2 v}{\partial t^2}. \tag{3.43}$$

In the case of uniform movement at velocity c it is

$$x = u(t) = ct, \qquad \frac{du}{dt} = c \quad \text{and} \quad \frac{d^2 u}{dt^2} = 0. \tag{3.44}$$

The derivative of (3.43) is

$$\frac{d^2 v(ct,\ t)}{dt^2} = \left[c^2\frac{\partial^2 v(x,\ t)}{\partial x^2} + 2c\frac{\partial^2 v(x,\ t)}{\partial x \partial t} + \frac{\partial^2 v(x,\ t)}{\partial t^2}\right]_{x=ct}. \tag{3.45}$$

The first term on the right-hand side of equation (3.45) expresses the influence of track curvature, the second term the influence of Coriolis acceleration, and the third term the influence of the acceleration of the moving load in a vertical direction. The action of the first and particularly of the second term is usually small in comparison with the action of the third term.

These considerations can be extended to cover the movement of the load in a plane, e.g. on a plate, see equations (2.13), (2.14) and Fig. 2.5. In that case the load $f(x, y, t)$ of the mass $\mu(x, y, t)$ per unit area possesses the form

$$f(x,\ y,\ t) - \mu(x,\ y,\ t)\frac{d^2 w}{dt^2}, \tag{3.46}$$

while the movement of this load is determined by the parametric functions of time

$$x = u(t), \qquad y = v(t). \tag{3.47}$$

The derivative in (3.46) in this case is:

$$\frac{d^2 w}{dt^2} = \left[\frac{\partial^2 w}{\partial x^2}\left(\frac{du}{dt}\right)^2 + \frac{\partial^2 w}{\partial y^2}\left(\frac{dv}{dt}\right)^2 + \frac{\partial^2 w}{\partial t^2} + 2\frac{\partial^2 w}{\partial x \partial y}\frac{du}{dt}\frac{dv}{dt}\right.$$

$$\left. + 2\frac{\partial^2 w}{\partial x \partial t}\frac{du}{dt} + 2\frac{\partial^2 w}{\partial y \partial t}\frac{dv}{dt} + \frac{\partial w}{\partial x}\frac{d^2 u}{dt^2} + \frac{\partial w}{\partial y}\frac{d^2 v}{dt^2}\right]_{x=u(t),\ y=v(t)}. \tag{3.48}$$

Equation (3.43) or equation (3.48) reveals that the consideration of vehicle mass considerably complicates the analysis, as noted in Sect. 2.1.2.

4. Natural frequencies of railway bridges

The most important dynamic characteristics of railway bridges are their natural frequencies which actually characterize the extent to which the bridge is sensitive to dynamic loads. They are measured by the number of vibrations per unit time. The unit of the frequency is Hertz (Hz), which is the number of cycles executed per second.

Mechanical systems with continuously distributed mass have an infinite number of natural frequencies, only the lowest of which have any practical application. If excitation forces are applied to a system over a wide spectrum of frequencies, the structure selects only the frequencies near its own natural frequencies and reacts to these. This is the reason for the great importance attached to natural frequencies.

The notation for natural frequencies is f_j and the subscript $j = 1, 2, 3, \ldots$ indicates their sequence. Apart from f_j the natural circular frequency is ω_j where

$$\omega_j = 2\pi f_j. \tag{4.1}$$

The unit of circular frequency is s^{-1}.

Natural frequency is also connected to the period of vibration by

$$T_j = 1/f_j. \tag{4.2}$$

T_j is measured in seconds and expresses the duration of one cycle.

4.1 Calculation of natural frequencies

The calculation of the natural frequencies of an undamped mechanical system is based on equation (2.25) in which $[b] = 0$ and $\{F\} = 0$ are substituted:

$$[m]\{\ddot{q}\} + [k]\{q\} = \{0\}. \tag{4.3}$$

In equation (4.3) the modified form is used, where $[m]$ is a diagonal mass matrix and $[k]$ a square symmetrical stiffness matrix.

If

$$\{q\} = \{q_j\} e^{i\omega_j t}. \tag{4.4}$$

is the jth natural mode of harmonic vibrations with a circular frequency ω_j then after substitution in equation (4.3) we obtain

$$\left([k] - \omega_j^2[m]\right)\{q_j\} = \{0\}. \tag{4.5}$$

4.1 Calculation of natural frequencies

This equation is soluble only if its determinant equals zero, i.e.

$$\left\| [k] - \omega_j^2 [m] \right\| = 0 . \quad (4.6)$$

The circular natural frequencies ω_j are found from equation (4.6) and also the quantities f_j or T_j according to equations (4.1) or (4.2). There are special methods for the solution of equations of the type (4.6) which are described in all text-books on dynamics or numerical mathematics, e.g. [8], [120], [174]. Some of these methods calculate only the lowest or approximate values of natural frequencies.

The roots ω_j of equation (4.6) pertain to characteristic vectors $\{q_j\}$ which we obtain from equation (4.5), with significant orthogonal characteristics. Consequently, for a system of n degrees of freedom there are n natural frequencies and n modes of natural vibration, $j = 1, 2, 3, ..., n$.

4.1.1 Beams

The calculation of the frequencies of natural vibrations of an undamped beam of constant cross section is based on equation (2.1), in which the loading and damping terms are considered equal to zero. Hence,

$$EI \frac{\partial^4 v(x, t)}{\partial x^4} + \mu \frac{\partial^2 v(x, t)}{\partial t^2} = 0 . \quad (4.7)$$

In harmonic vibrations with frequencies ω_j the solution of equation (4.7) is assumed to be of the form

$$v(x, t) = \sum_{j=1}^{\infty} v_j(x) \sin \omega_j t , \quad (4.8)$$

where $v_j(x)$ are the modes of natural vibration.

After substituting (4.8) in equation (4.7) we obtain

$$EI \sum_{j=1}^{\infty} \frac{d^4 v_j(x)}{dx^4} \sin \omega_j t - \mu \sum_{j=1}^{\infty} \omega_j^2 v_j(x) \sin \omega_j t = 0 . \quad (4.9)$$

Equation (4.9) must be obeyed by every one of the j natural modes, so that it can be simplified into a set of independent ordinary differential equations

$$\frac{d^4 v_j(x)}{dx^4} - \frac{\lambda_j^4}{l^4} v_j(x) = 0 , \quad j = 1, 2, 3 ..., \quad (4.10)$$

where

$$\lambda_j^4 = l^4 \frac{\mu \omega_j^2}{EI} . \quad (4.11)$$

The general solution of the homogenous differential equation (4.10) is

$$v_j(x) = A_1 \sin \frac{\lambda_j x}{l} + A_2 \cos \frac{\lambda_j x}{l} + A_3 \sinh \frac{\lambda_j x}{l} + A_4 \cosh \frac{\lambda_j x}{l} \quad (4.12)$$

where the coefficients A_i, $i = 1, 2, 3, 4$, are integration constants determined from the boundary conditions of the beam.

For a simply supported beam the boundary conditions (2.7) apply, i.e.

$$\text{for } x = 0 \quad v_j(x) = 0, \quad \frac{d^2 v_j(x, t)}{dx^2} = 0$$

and

$$\text{for } x = l \quad v_j(x) = 0, \quad \frac{d^2 v_j(x)}{dx^2} = 0. \tag{4.13}$$

If we substitute the solution (4.12) into the four conditions (4.13), we obtain four equations for the unknown constants A_i. For a simply supported beam it follows that

$$v_j(0) = A_2 + A_4 = 0,$$

$$v_j''(0) = \frac{\lambda_j^2}{l^2}(-A_2 + A_4) = 0,$$

$$v_j(l) = A_1 \sin \lambda_j + A_3 \sinh \lambda_j = 0.$$

and

$$v_j''(l) = \frac{\lambda_j^2}{l^2}(-A_1 \sin \lambda_j + A_3 \sinh \lambda_j) = 0. \tag{4.14}$$

Equations (4.14) are satisfied only when

$$A_2 = A_3 = A_4 = 0 \tag{4.15}$$

and

$$\sin \lambda_j = 0. \tag{4.16}$$

The constant A_1 can be arbitrary; for a natural mode the notation

$$A_1 = q_j \tag{4.17}$$

is used. Note that natural vibration may occur with any amplitude A_j or q_j which ought to be taken into account in Table 4.1 and in Figs 4.1 to 4.5 and 4.9 to 4.12.

Equation (4.16) is the frequency equation, which indicates that for a simply supported beam

$$\lambda_j = j\pi, \quad j = 1, 2, 3, \ldots, \tag{4.18}$$

λ_j being a dimensionless frequency parameter.

The natural circular frequency follows from equation (4.11):

$$\omega_j^2 = \frac{\lambda_j^4}{l^4} \frac{EI}{\mu} \tag{4.19}$$

so that

$$f_j = \frac{\lambda_j^2}{2\pi l^2} \left(\frac{EI}{\mu}\right)^{1/2}. \tag{4.20}$$

4.1 Calculation of natural frequencies

TABLE 4.1. Characteristics of natural vibrations of single span beams of constant cross section (natural frequency (4.20) and vibration modes (4.12) according to [14])

Beam, Figure	Boundary conditions	No. of natural mode j	λ_j	Frequency equation	Natural vibration mode $v_j(x)$	A_j	Equation for A_j
Simply supported, Fig. 4.1	$v_j(0) = 0$ $v_j''(0) = 0$ $v_j(l) = 0$ $v_j''(l) = 0$	1 2 3 4 5 >5	3.142 6.283 9.425 12.566 15.708 $j\pi$	$\sin \lambda_j = 0$	$\sin \dfrac{\lambda_j x}{l}$	– – – – – –	–
Both ends clamped, Fig. 4.2	$v_j(0) = 0$ $v_j'(0) = 0$ $v_j(l) = 0$ $v_j'(l) = 0$	1 2 3 4 5 >5	4.730 7.853 10.996 14.137 17.279 $(2j+1)\dfrac{\pi}{2}$	$\cos \lambda_j \cosh \lambda_j = 1$	$\cosh \dfrac{\lambda_j x}{l} -$ $- \cos \dfrac{\lambda_j x}{l} -$ $- A_j \left(\sinh \dfrac{\lambda_j x}{l} -\right.$ $\left. - \sin \dfrac{\lambda_j x}{l} \right)$	0.983 1.001 1 1 1 1	$(\cosh \lambda_j - \cos \lambda_j)/$ $/(\sinh \lambda_j - \sin \lambda_j)$
Both ends free, Fig. 4.3	$v_j''(0) = 0$ $v_j'''(0) = 0$ $v_j''(l) = 0$ $v_j'''(l) = 0$		the same as for beam with both ends clamped		$\cosh \dfrac{\lambda_j x}{l} +$ $+ \cos \dfrac{\lambda_j x}{l} -$ $- A_j \left(\sinh \dfrac{\lambda_j x}{l} +\right.$ $\left. + \sin \dfrac{\lambda_j x}{l} \right)$	the same as for beam with both ends clamped	
Cantilever Fig. 4.4	$v_j(0) = 0$ $v_j'(0) = 0$ $v_j''(l) = 0$ $v_j'''(l) = 0$	1 2 3 4 5 >5	1.875 4.694 7.855 10.996 14.137 $(2j-1)\dfrac{\pi}{2}$	$\cos \lambda_j \cosh \lambda_j + $ $+ 1 = 0$	$\cosh \dfrac{\lambda_j x}{l} -$ $- \cos \dfrac{\lambda_j x}{l} -$ $- A_j \left(\sinh \dfrac{\lambda_j x}{l} -\right.$ $\left. - \sin \dfrac{\lambda_j x}{l} \right)$	0.734 1.018 0.999 1 1 1	$(\sinh \lambda_j - \sin \lambda_j)/$ $/(\cosh \lambda_j + \cos \lambda_j)/$
One end clamped, one end simply supported Fig. 4.5	$v_j(0) = 0$ $v_j'(0) = 0$ $v_j(l) = 0$ $v_j''(l) = 0$	1 2 3 4 5 >5	3.927 7.069 10.210 13.352 16.493 $(4j+1)\dfrac{\pi}{4}$	$\operatorname{tg} \lambda_j = \operatorname{tgh} \lambda_j$	$\cosh \dfrac{\lambda_j x}{l} -$ $- \cos \dfrac{\lambda_j x}{l} -$ $- A_j \left(\sinh \dfrac{\lambda_j x}{l} -\right.$ $\left. - \sin \dfrac{\lambda_j x}{l} \right)$	1.001 1 1 1 1 1	$\cotg \lambda_j$

Therefore, the natural frequency of beams and, consequently, of the majority of railway bridges depends on the following quantities:

l – span,
λ_j – boundary conditions,
E – modulus of elasticity of the beam or bridge material,
I – moment of inertia of the cross section of the beam or bridge,
μ – mass of the beam or bridge per unit length.

The roots λ_j of the frequency equations and the natural modes $v_j(x)$ of single span beams with different types of support together with further characteristics of natural vibrations, are presented in Table 4.1 and Figs 4.1 to 4.5.

According to equation (4.8) a beam can vibrate with simple harmonic motion in an infinite number of modes. The simplest natural mode is the sine curve

$$v_j(x) = q_j \sin \frac{j \pi x}{l}, \qquad (4.21)$$

which can be obtained from (4.12), (4.15) to (4.18) for a simply supported beam.

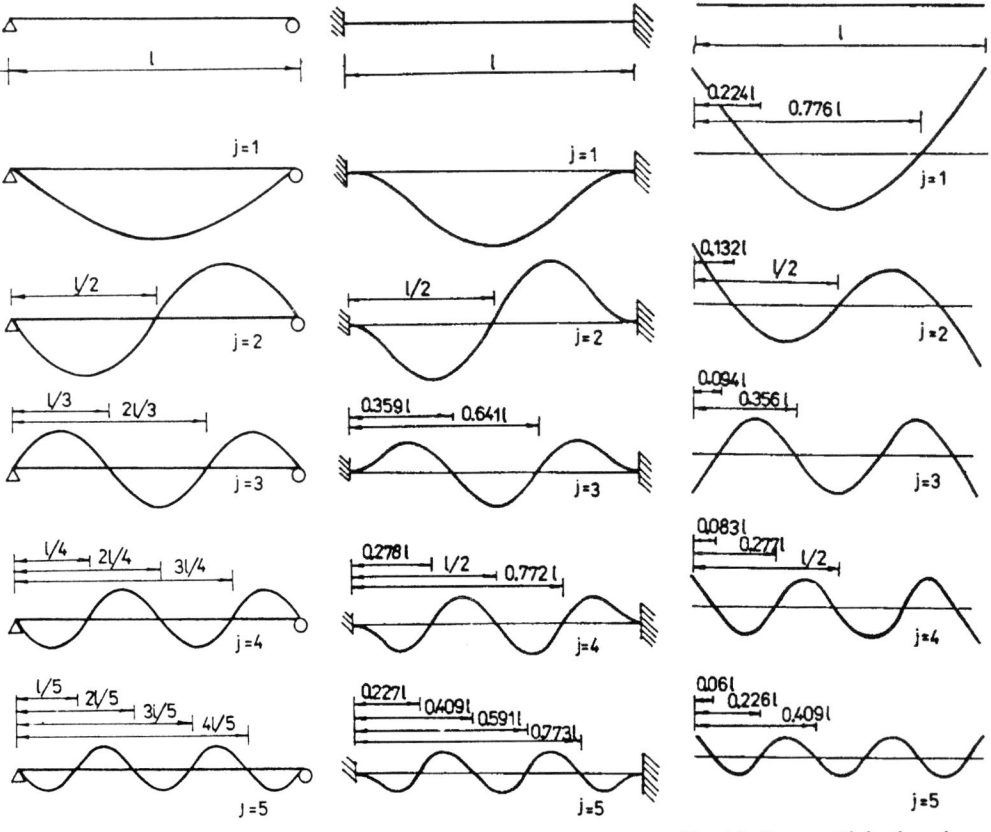

Fig. 4.1. Simply supported beam: modes of natural vibration.

Fig. 4.2. Beam with both ends clamped: modes of natural vibration.

Fig. 4.3. Beam with both ends free: modes of natural vibration.

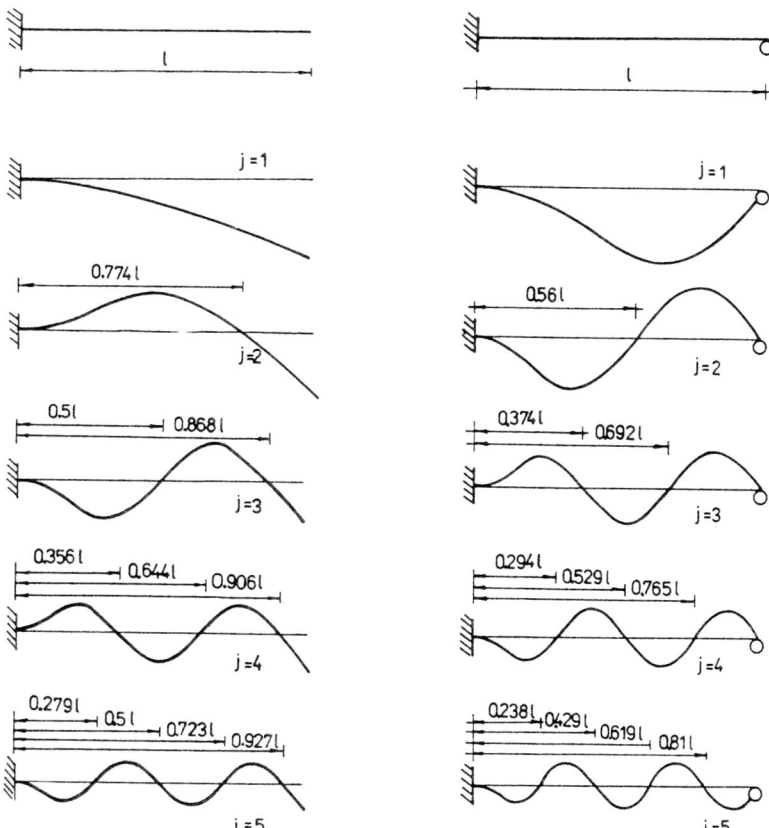

Fig. 4.4. Cantilever beam: modes of natural vibration.

Fig. 4.5. Beam with left end clamped and right end simply supported: modes of natural vibration.

However, vibrations without external forces need not be simple harmonic motion; they can consist of a combination of simple harmonic motions each of which has a different frequency and their phases may be mutually displaced. For this reason natural vibrations are sometimes distinguished from free vibrations. A complete solution of equation (4.7) for free vibrations of a simply supported beam is

$$v(x, t) = \sum_{j=1}^{\infty} q_j \sin \frac{j \pi x}{l} \sin (\omega_j t + \varphi_j), \qquad (4.22)$$

where φ_j is the phase of the jth natural mode.

The time-history of free vibrations depends on the mode or modes in which the beam has been deflected at the beginning of the motion and on its velocity.

This phenomenon complicates the evaluation of natural frequencies and logarithmic decrement of damping from records of deflections or stresses, which occurred after a vehicle has left the bridge. If before that moment the

TABLE 4.2. Roots λ_i of frequency equation for the computation of natural frequency (4.20) of continuous beam of constant cross section of n equal spans according to [33]

Continuous beam of n spans, Figure	No. of spans n	Roots λ_i of frequency equation — No. of natural mode j				
		1	2	3	4	5
Both ends hinged, Figs. 4.6, 4.9 to 4.12	1	3.142	6.283	9.425	12.566	15.708
	2	3.142	3.927	6.283	7.069	9.425
	3	3.142	3.550	4.304	6.283	6.692
	4	3.142	3.393	3.927	4.461	6.283
	5	3.142	3.299	3.707	4.147	4.555
	6	3.142	3.267	3.550	3.927	4.304
	7	3.142	3.236	3.456	3.770	4.084
	8	3.142	3.205	3.393	3.644	3.927
	9	3.142	3.205	3.330	3.550	3.801
	10	3.142	3.205	3.299	3.487	3.707
	11	3.142	3.173	3.267	3.424	3.613
	12	3.142	3.173	3.267	3.393	3.550
Both ends clamped, Fig. 4.7	1	4.730	7.853	10.996	14.137	17.279
	2	3.927	4.744	7.069	7.855	10.210
	3	3.550	4.304	4.744	6.692	7.446
	4	3.393	3.927	4.461	4.744	6.535
	5	3.299	3.707	4.147	4.555	4.744
	6	3.267	3.550	3.927	4.304	4.587
	8	3.205	3.393	3.644	3.927	4.210
	10	3.205	3.299	3.487	3.707	3.927
	12	3.173	3.267	3.393	3.550	3.739
Left end clamped, right end hinged, Fig. 4.8	1	3.927	7.069	10.210	13.352	16.493
	2	3.393	4.461	6.535	7.603	9.677
	3	3.267	3.927	4.587	6.409	7.069
	4	3.205	3.644	4.210	4.650	6.347
	5	3.205	3.487	3.927	4.367	4.681
	6	3.173	3.393	3.739	4.116	4.461
	8	3.173	3.299	3.393	3.770	4.084
	10	3.142	3.236	3.456	3.582	3.801
	12	3.142	3.236	3.380	3.487	3.644

4.1.2 Continuous beams

The calculation of natural frequencies of continuous beams is analogous with the calculation of single-span beams (Sect. 4.1.1). Equation (4.7) is applied to beams of constant cross section in every span and the natural vibration modes can be found from equation (4.12). Naturally, the integration constants A_i are different in every span. Thus, an n span continuous beam possesses $4n$ integration constants. The integration constants are determined by the solution of a system of linear algebraic equations following out of the boundary conditions of the beam and of the conditions for deflection, rotation, bending moment and shear forces above internal and external supports of the continuous beam. This system of equations is homogeneous; therefore, its non-trivial solution necessitates that the determinant of the system should equal zero. In this way we obtain the frequency equation from which λ_j are found.

The roots λ_j of the frequency equations for three types of continuous beams according to Figs 4.6 to 4.8 of n equal spans are given in Table 4.2; the natural modes are represented in Figs 4.9 to 4.12. From the data of Table 4.2, the natural frequency of a continuous beam of constant cross section can by found using equation (4.20).

Natural frequencies of a continuous beam of constant cross section have concentrated zones. Every zone contains as many natural frequencies as there are spans of the continuous beam. The frequencies in every dense zone are near one another. In Table 4.2 they are boxed by thick lines.

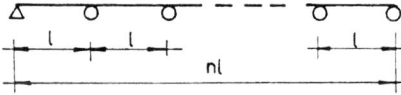

Fig. 4.6. Continuous beam of n equal spans hinged at both ends.

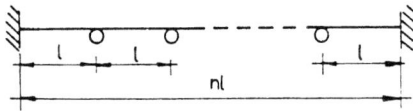

Fig. 4.7. Continuous beam of n spans clamped at both ends.

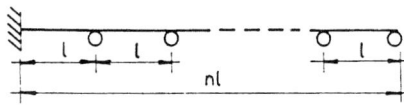

Fig. 4.8. Continuous beam of n equal spans, left end clamped and right end simply supported.

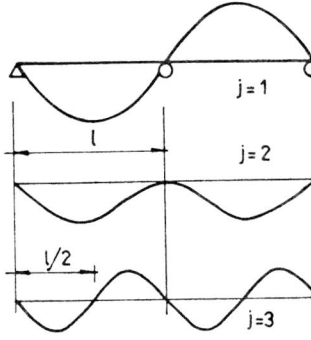

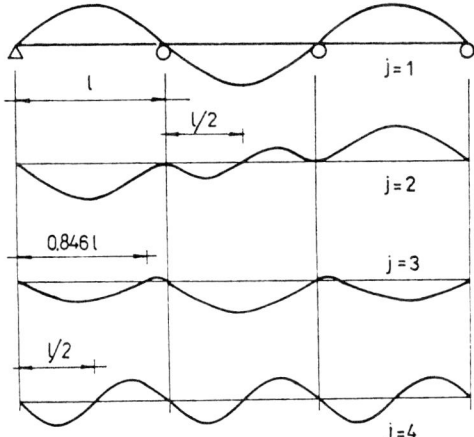

Fig. 4.9. Continuous beam of two equal spans: modes of natural vibration. The first dense region of natural frequencies $j = 1$ and 2.

Fig. 4.10. Continuous beam of three equal spans: modes of natural vibration. The first dense region of natural frequencies $j = 1$ to 3.

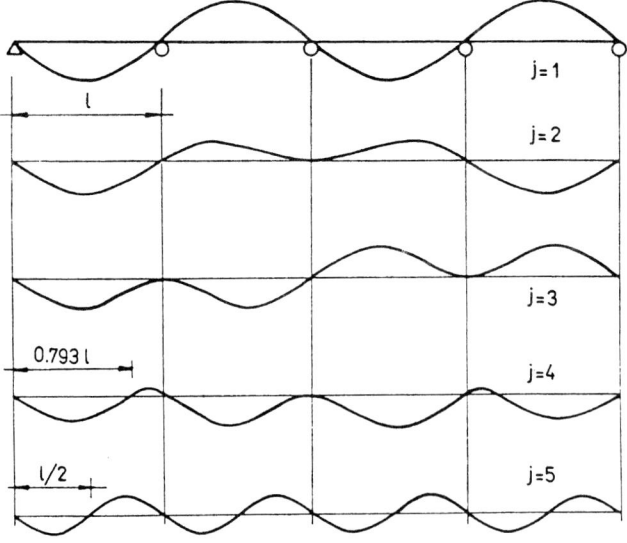

Fig. 4.11. Continuous beam of four equal spans: modes of natural vibration. The first dense region of natural frequencies $j = 1$ to 4.

4.1 Calculation of natural frequencies

Figure 4.13 and Table 4.3 show the dependence of the roots λ_1 and λ_2 of frequency equations on the ratio of lengths l_1/l_2 of a continuous beam of constant cross section with two spans of different lengths l_1 and l_2. The length l_2 is substituted for l in equation (4.20).

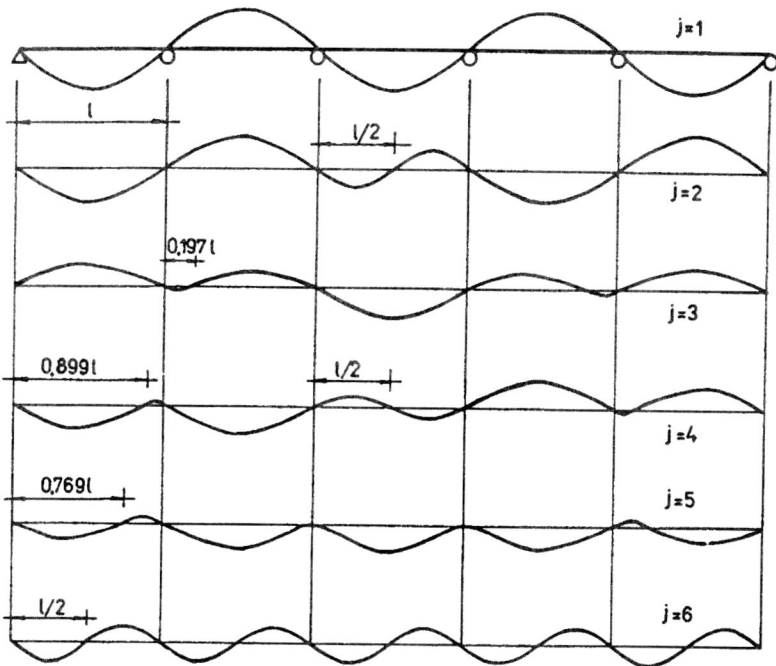

Fig. 4.12. Continuous beam of 5 equal spans: modes of natural vibration. The first dense region of natural frequencies $j = 1$ to 5.

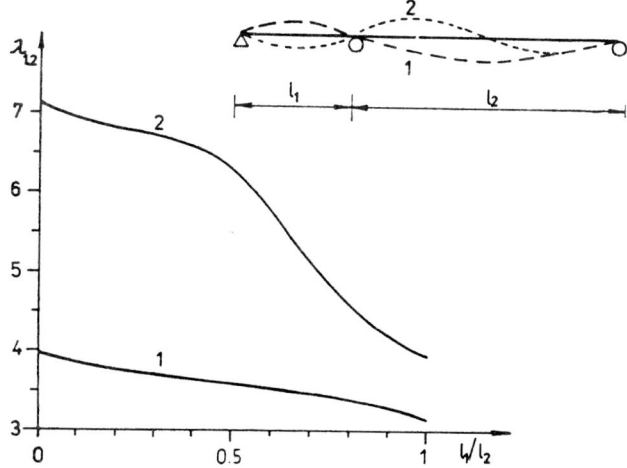

Fig. 4.13. Continuous beam of constant cross section with 2 unequal spans and its first two natural modes 1, 2. Roots of frequency equation λ_1 and λ_2 plotted against ratio of spans l_1/l_2.

TABLE 4.3. Roots λ_1 and λ_2 of frequency equation of continuous beam of two unequal spans l_1 and l_2 (Fig. 4.13)

Ratio of spans, Fig. 4.13	Frequency parameter (l in Eq. (4.20) is substituted by l_2)	
l_1/l_2	λ_1	λ_2
0	3.926 60	7.068 58
0.1	3.814 27	6.883 86
0.2	3.729 77	6.763 02
0.3	3.663 01	6.664 49
0.4	3.607 04	6.544 34
0.5	3.556 41	6.283 19
0.6	3.505 99	5.698 28
0.7	3.449 66	5.059 73
0.8	3.378 48	4.549 93
0.9	3.279 48	4.176 79
1	3.141 59	3.926 60

TABLE 4.4. Roots λ of frequency equation of three-span symmetric girder (Fig. 4.14) (l in Eq. (4.20) is substituted by l_2)

Ratio of spans, Fig. 4.14	Frequency parameter of the first symmetric (S) and antisymmetric (A) vibrations	
l_1/l_2	λ_{1S}	λ_{1A}
0	4.730 04	7.853 20
0.1	4.471 37	7.459 54
0.2	4.289 39	7.214 08
0.3	4.150 17	7.011 98
0.4	4.033 91	6.756 97
0.5	3.926 60	6.283 19
0.6	3.815 95	5.582 16
0.7	3.689 17	4.915 23
0.8	3.534 26	4.365 93
0.9	3.347 73	3.920 80
1	3.141 59	3.556 41
1.1	2.934 63	3.253 71
1.2	2.739 81	2.998 64
1.3	2.562 34	2.780 91
1.4	2.402 93	2.592 91
1.5	2.260 34	2.428 96

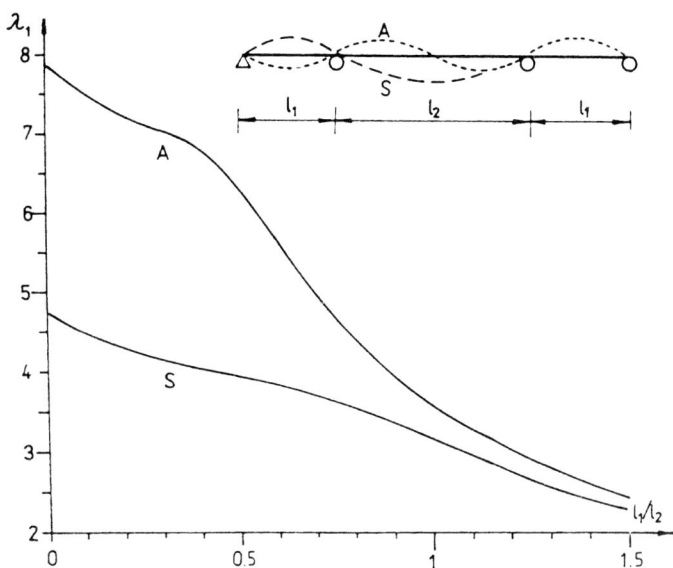

Fig. 4.14. Three span continuous symmetric beam of constant cross section and its first natural mode for symmetric (S) and antisymmetric (A) vibrations. The roots of frequency equation λ_1 plotted against ratio of spans l_1/l_2.

The continuous symmetric three span bridge of $l_1 + l_2 + l_1$, the lengths of the outside spans of which are equal, occurs frequently in practice. The first roots λ_1 of the frequency equation for the symmetric and antisymmetric vibrations of this continuous beam of constant cross section are plotted in Fig. 4.14 against the ratio of lenghts l_1/l_2 – see also Table 4.4. Once again, the length l_2 is substituted for l in equation (4.20).

In both cases the frequency parameter λ was determined using the Koloušek deformation (slope-deflection) method according to [120].

4.1.3 Plates

Rectangular plates frequently occur in railway bridges. Their natural frequency can be characterized by a form similar to equation (4.20), i.e.

$$f_{ij} = \frac{\lambda_{ij}^2}{2\pi l_x^2} \left(\frac{D}{\mu}\right)^{1/2} \quad (4.23)$$

where according to Sect. 2.2:

l_x – plate span in the direction of x (Fig. 2.5),

$D = \dfrac{Eh^3}{12(1-v^2)}$ – flexural rigidity of the plate of thickness h and Poisson's number v,

μ – plate mass per unit area.

The lowest values of the roots λ_{ij}^2 of the frequency equations for several ratios of plate dimensions l_x/l_y in the directions x and y, respectively, are given in Table 4.5.

For a rectangular plate, simply supported on all four sides, a procedure similar to that described in Sect. 4.1.1 can be used for the derivation of expressions for natural modes and frequency parameters λ_{ij} (natural frequency is found, once again, from equation (4.23)):

$$w_{ij}(x, y) = q_{ij} \sin \frac{i\pi x}{l_x} \sin \frac{j\pi y}{l_y}, \quad (4.24)$$

$$\lambda_{ij}^2 = \pi^2 \left(i^2 + j^2 \frac{l_x^2}{l_y^2}\right), \quad i, j = 1, 2, 3, \ldots . \quad (4.25)$$

There is a number of methods for the calculation of natural frequencies of more complex railway bridges, which have been described in references [8], [120]. In recent years, the finite element method has also been applied which leads to equations of the type (4.6).

4.1.4 Natural frequencies of loaded bridges

If the mass of the unloaded bridge μ per unit length is increased to the value of $\bar{\mu}$ (e.g. by loading the bridge with a whole train), the natural frequencies of

TABLE 4.5. Roots λ_{ij}^2 of frequency equation for rectangular plates (according to [95])

Boundary conditions 1 – simply supported ---- 2 – clamped edge ==== 3 – free edge _____	Ratio of spans l_y/l_x	Frequency parameters λ_{ij}^2 for Eq. (4.23)		
		Lowest frequency	Second frequency	Third frequency
All edges simply supported (1,1,1,1)	1	19.74	see eq. (4.25)	
All edges free (3,3,3,3)	1	14.1	20.5	23.9
Two adjacent clamped, two free (2,3,3,2)	1	6.96	24.1	26.8
Two opposite clamped, two free (2,3,2,3)	2 1 0.5	3.51 3.49 3.47	5.37 8.55 14.9	22.0 21.4 21.6
Two opposite simply supported, one clamped, one free (1,2,1,3)	2 1 0.5	17.3 23.7 51.7	51.7	58.7
Two opposite simply supported, two clamped (1,2,1,2)	2 1 0.5	23.8 29.0 54.8	54.8	69.3
All edges clamped (2,2,2,2)	∞ 3 2 1	22.4 23.2 24.6 36.0	73.4	108.3

4.1 Calculation of natural frequencies

the loaded bridge $\bar{f}_j$ can be found from the natural frequencies of the unloaded bridge f_j (equation (4.20)) by writing

$$\bar{f}_j = f_j \left(\frac{\mu}{\bar{\mu}} \right)^{1/2}. \qquad (4.26)$$

When a bridge of a large span is loaded only by a concentrated load of weight F (e.g. by a vehicle), such a load is resolved into a Fourier series and added to the self-weight of the bridge. For instance, for the simply supported beam the mass per unit length of the bridge subjected to such a load is (see [68], Sect. 1.4.3):

$$\bar{\mu} = \mu \left(1 + \frac{2F}{G} \sin^2 \frac{j \pi x_0}{l} \right), \qquad (4.27)$$

where F – weight of the vehicle,
G – weight of the bridge,
x_0 – coordinate of the vehicle position.

For the first natural frequency of the loaded simply supported beam loaded by the force F at midspan $x_0 = l/2$ we obtain, from equations (4.24) and (4.23),

$$\bar{f}_1 = f_1 (1 + 2F/G)^{-1/2}. \qquad (4.28)$$

The change of natural frequency of the loaded simply supported beam as a function of the position x_0 calculated from equations (4.28) and (4.27) is shown in Fig. 4.15; this clearly illustrates how the position of the load influences the natural frequency of the loaded structure. If the vehicle is near the middle of the span, however, this effect is relatively small. Consequently, the frequency characteristics of the bridge/vehicle system depend on the position of the travelling vehicle and, consequently, on time.

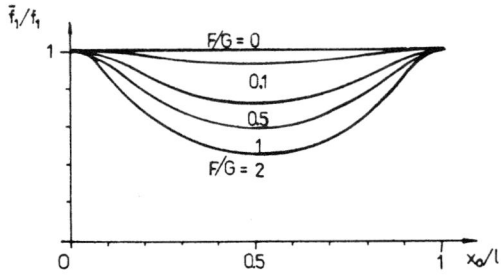

Fig. 4.15. The first natural frequency $\bar{f}_1$ of loaded simply supported beam plotted against the position x_0 of load of weight F; the weight of the beam is G.

Equations (4.26) and (4.27) may also be used to estimate the natural frequency of an unloaded bridge from the measured natural frequency of the bridge under load and the knowledge of the position and magnitude of the load:

$$f_j = \bar{f}_j \left(\frac{\bar{\mu}}{\mu} \right)^{1/2}. \qquad (4.29)$$

4.2 Experimental results

In this Section we summarize the results of measurements of the natural frequencies on 113 railway bridges. The tests were carried out on bridges of the former Czechoslovak State Railways by several Czech and Slovak institutes, of seven European railways within the framework of international research programs ORE D 23, D 128, D 154 and D 160 of the Office for Research and Experiments (ORE) of the International Union of Railways (UIC), [159] to [165], and on bridges of the German Federal Railways (DB), [25].

With respect to natural frequencies the bridges were divided into five groups:
 1. Steel truss bridges with and without ballast;
 2. Steel plate girder bridges with ballast (this group includes also bridges with orthotropic bridge deck, box girder bridges and Langer beams stiffened by an arch);
 3. Steel plate girder without ballast (including with open decks);
 4. Concrete bridges (reinforced and prestressed concrete bridges and bridges with steel girders embedded in concrete) with ballast;
 5. Concrete bridges without ballast.

It should be noted that the tests mostly determine only the first natural frequency of vibrations of the main beam. The vibrations of various other elements of the bridge, such as cross-beams, longitudinal beams (stringers), orthotropic deck slabs, etc., are influenced considerably by the vibrations of the main bearing system, so that their natural frequencies cannot be found without special equipment.

4.2.1 Dynamic bridge stiffness

Railway bridges are often idealized by a beam, the natural frequency of which, according to (4.20), can be found from

$$f_j = \frac{\lambda_j^2}{2\pi l^2} B, \qquad (4.30)$$

where l – span of the bridge,
 λ_j – solution of frequency equations, which depend on the structure of the bridge, i.e. on boundary conditions of the bridge span, see Tables 4.1 to 4.5.

The constant B for beams, characterizing their dynamic stiffness, is – according to equation (4.20) –

$$B = \left(\frac{EI}{\mu}\right)^{1/2}. \qquad (4.31)$$

Note that for plates equation (4.23) has to be applied.

B characterizes the material of the bridge structure (E), the stiffness of the cross section (I) and the mass of the bridge (μ). Its unit is $m^2 s^{-1}$, if l is measured in metres and f_j in Hertz in equation (4.30).

With respect to the dynamic stiffness B the bridges were divided into the groups mentioned above: truss bridges are stiff and usually of large spans, so that the existence of the ballast does not play any decisive role. Plate girder bridges, on the other hand, are usually of medium and small spans and possess a lower moment of inertia; consequently the presence or absence of the ballast is of some significance. Concrete bridges are considerably heavier than steel bridges, but possess a lower modulus of elasticity E.

The dynamic stiffness of bridges may be found from measurements of first natural frequencies and the span:

$$B = \frac{2}{\pi} f_1 l^2 \qquad (4.32)$$

assuming that $\lambda_1 = \pi$. This applies to simply supported beams or continuous beams with any number of equal spans of constant cross section. It may also be used approximately in other cases, if the variations of cross section or the differences in spans are not too great.

The dependence of the dynamic bridge striffness B on the span l for the five groups of bridges was analyzed by regression analysis. From the four basic regression types (linear, exponential, logarithmic and power regression) linear regression

$$B = a + bl \qquad (4.33)$$

has proved the best one, which means that the dynamic stiffness B grows approximately linearly with the span l (see Figs 4.16 to 4.20, showing also the measured values).

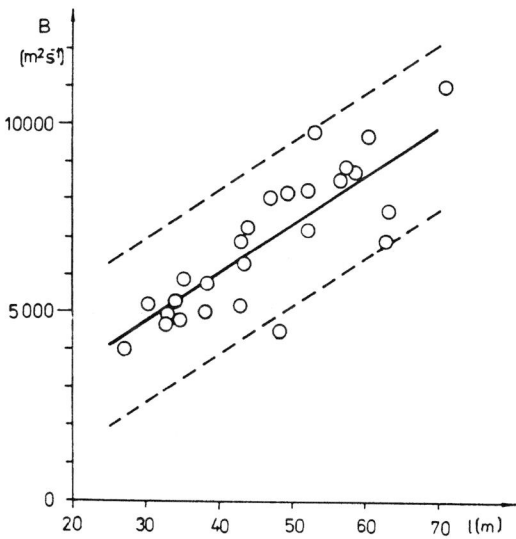

Fig. 4.16. Dynamic stiffness B plotted against span l for steel truss bridges (regression coefficients are in Table 4.6).

82   4. Natural frequencies of railway bridges

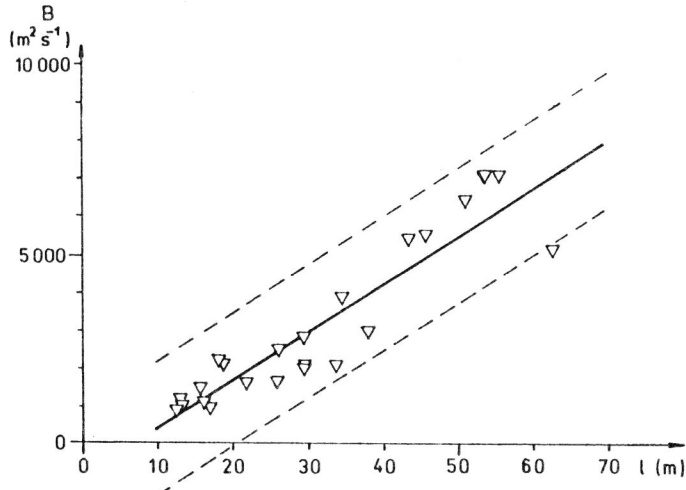

Fig. 4.17. Dynamic stiffness B plotted against span l for steel plate girder bridges with ballast (regression coefficients are in Table 4.6).

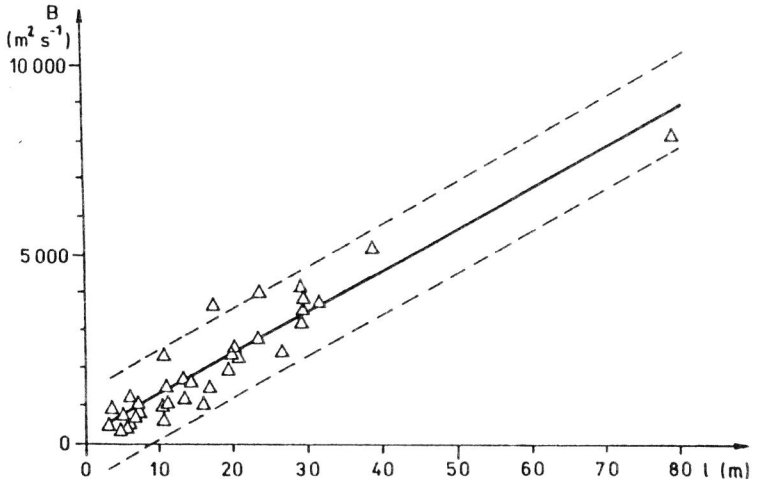

Fig. 4.18. Dynamic stiffness B plotted against span l for steel plate girder bridges without ballast (regression coefficients are in Table 4.6).

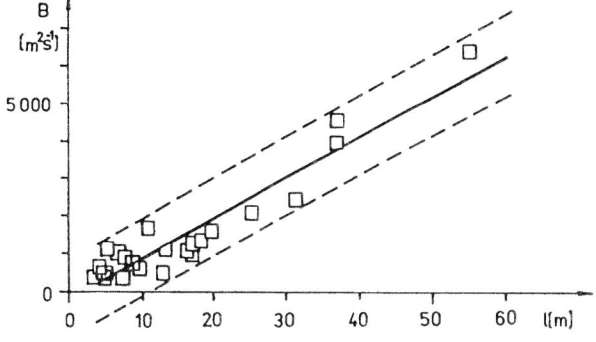

Fig. 4.19. Dynamic stiffness B plotted against span l for concrete bridges with ballast (regression coefficients are in Table 4.6).

4.2 Experimental results

The regression coefficients a, b in equation (4.33), found by the least square method, are tabulated in Table 4.6, showing also the regions of validity (from l_{min} to l_{max}) and the correlation coefficient r, which is very high in all cases ($r \geqslant 0.92$).

The range of reliability s for observations at point x for the general linear regression

$$y = a + bx \tag{4.34}$$

was calculated according to [181] as follows:

$$s = t_{n-2;P} s_{y.x} \left[1 + \frac{1}{n} + \frac{\left(x - \frac{1}{n} \sum_i x_i \right)^2}{Q_x} \right]^{1/2}, \tag{4.35}$$

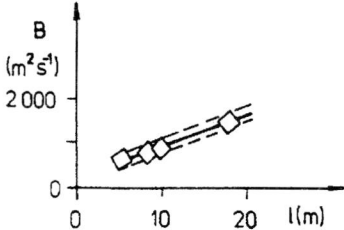

Fig. 4.20. Dynamic stiffness B plotted against span l for concrete bridges without ballast (regression coefficients are in Table 4.6).

where $t_{n-2;P}$ is the value of Student's distribution (for tables see [181]) for $n - 2$ degrees of freedom (n being the number of measurements) and the probability of error P. The computations considered the two-sided test for $P = 0.05$, which means that the results assure a 95% reliability. Further notations:

$$s_{y.x} = \left(\frac{Q_y - b Q_{xy}}{n - 2} \right)^{1/2},$$

$$Q_x = \sum x_i^2 - \frac{1}{n} \left(\sum x_i \right)^2,$$

$$Q_y = \sum y_i^2 - \frac{1}{n} \left(\sum y_i \right)^2,$$

$$Q_{xy} = \sum x_i y_i - \frac{1}{n} \left(\sum x_i \right) \left(\sum y_i \right). \tag{4.36}$$

This means that for the regression relation (4.34) the mean value of the quantity y at point x is

$$y = a + bx \tag{4.37}$$

with lower limit $\quad y = a + bx - s$
and upper limit $\quad y = a + bx + s$
for $(1 - P)\,100\%$ reliability.

TABLE 4.6. Dynamic stiffness B according to eq. (4.31), linear regression $B = a + bl$

Railway bridges		Ballast: yes + no −	No of measurements n	Regression coefficients		Range of reliability	Correlation coefficient	Validity region	
				a	b	s	r	l_{min}	l_{max}
				$m^2 s^{-1}$	$m\ s^{-1}$	$m^2 s^{-1}$	1	m	
Steel	Truss girder	+/−	27	870.647	129.635	2159.848	0.973	25	200
	Plate girder	+	23	−878.385	126.598	1768.429	0.920	10	70
		−	36	214.987	110.678	1172.399	0.941	3	80
Concrete		+	23	−186.774	108.934	1043.584	0.947	4	60
		−	4	135.425	77.699	111.937	0.996	5	20

The data of Table 4.6 and Figs 4.16 to 4.20 can be used for the computation of natural frequencies of bridges from equation (4.30). The mean value of the dynamic stiffness of bridges from Figs 4.16 to 4.20 is usually considered. If the stiffness is higher, the values nearer the upper limit may be chosen; if the bridge has a limited structural height or if it is heavier, the values nearer the lower limit should be used. The mean value is shown by a solid line in Figs 4.16 to 4.20, while the lower and upper limits for 95% reliability are represented by dashed lines.

4.2.2 Statistical evaluation of natural frequencies

The measured natural frequencies of 113 railway bridges are plotted against span in Figs 4.21 to 4.25 for the individual groups and in Fig. 4.26 regardless of their structure. From the four types of regressions (linear, exponential, logarithmic and power regression) the power regression

$$f_1 = al^b \tag{4.38}$$

has proved the best.

The regression coefficients a, b as well as other data are given in Table 4.7. The correlation coefficients ($r \geq 0{,}83$), show, once again, that the proposed regression model (4.38) is adequate enough.

The power regression of the type

$$y = ax^b \tag{4.39}$$

can be converted into linear regression (4.34) by taking logs. The computation then proceeds in accordance with the equations (4.34) to (4.37). The mean value at x is (4.39), while the lower limit is

$$y = ax^b s^{-1}$$

and the upper limit

$$y = ax^b s^{+1} \tag{4.40}$$

for the $(1 - P)\,100\%$ reliability.

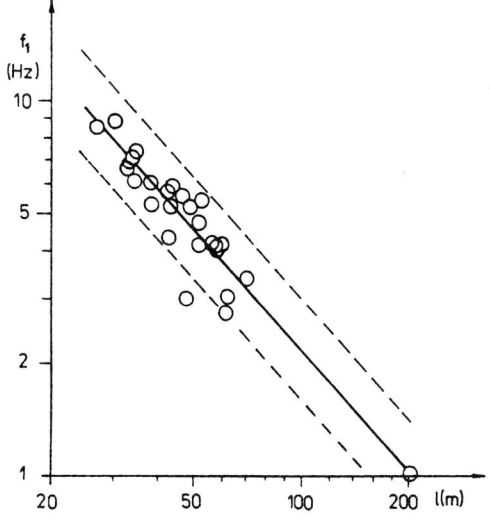

Fig. 4.21. First natural frequency f_1 plotted against span l for steel truss bridges (regression coefficients are in Table 4.7).

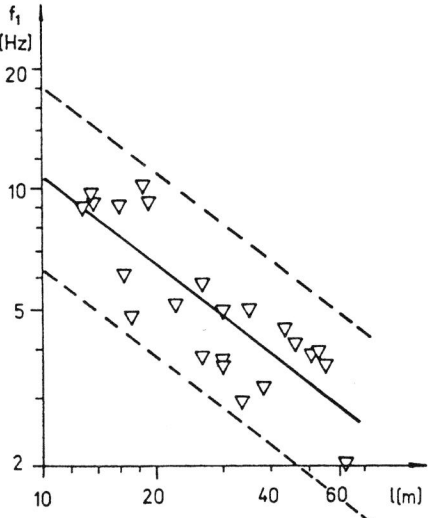

Fig. 4.22. First natural frequency f_1 plotted against span l for steel plate girder bridges with ballast (regression coefficients are in Table 4.7).

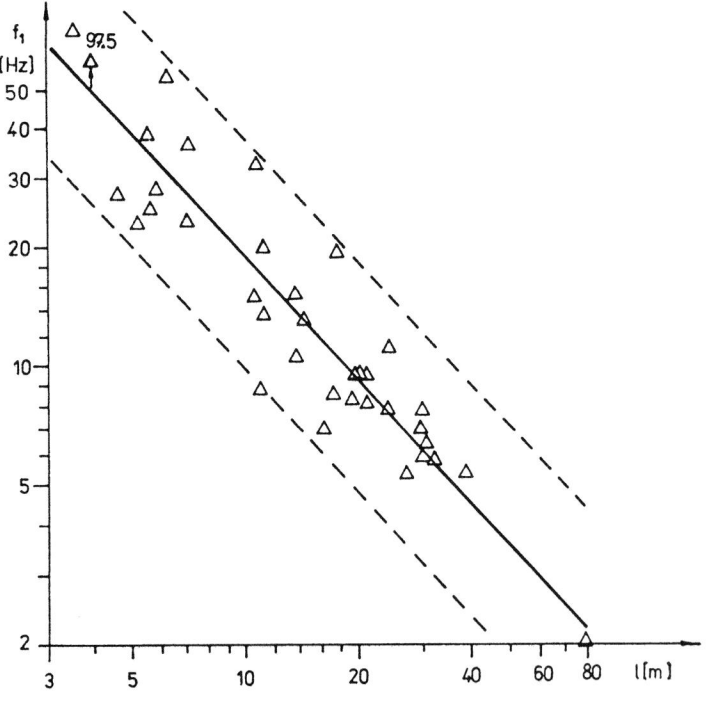

Fig. 4.23. First natural frequency f_1 plotted against span l for steel plate girder bridges without ballast (regression coefficients are in Table 4.7).

4.2 Experimental results

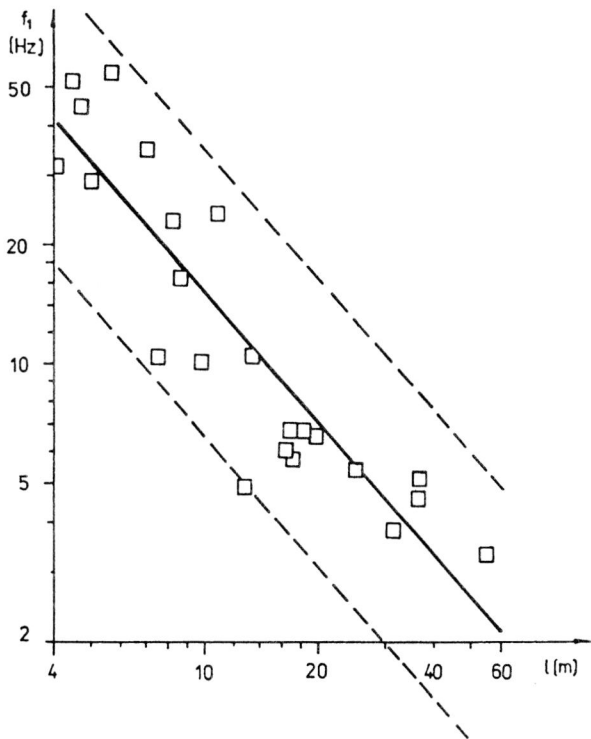

Fig. 4.24. First natural frequency f_1 plotted against span l for concrete bridges with ballast (regression coefficients are in Table 4.7).

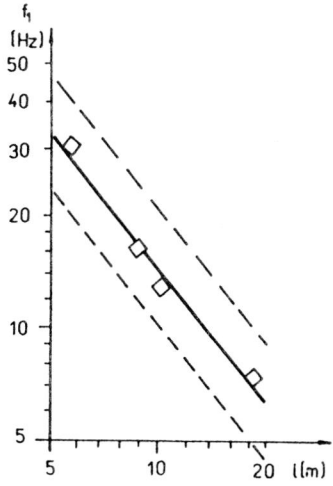

Fig. 4.25. First natural frequency f_1 plotted against span l for concrete bridges without ballast (regression coefficients are in Table 4.7).

Consequently, the dependence of the first natural frequencies of railway bridges on their span can be expressed by equation (4.38); this can be shown graphically by a descending straight line (see Figs. 4.21 to 4.26). The mean value is indicated by a solid line, the lower and upper limits for 95% reliability are represented by dashed lines.

Figure 4.21 reveals the high reliability of determination of the first natural frequency of steel truss bridges, as they are of a relatively uniform type. The group of plate girder bridges, on the other hand, contains bridges of various types, so that the range is wider. In the group of concrete bridges without ballast only the results of four structures were available, which is a very small sample. It is of interest, however, that the regression coefficients for concrete bridges with and without ballast are very close to each other, which justifies the conclusion that the ballast has only a very small influence on the first natural frequency of concrete bridges (the ratio of mass of the ballast and that of the concrete bridge being low).

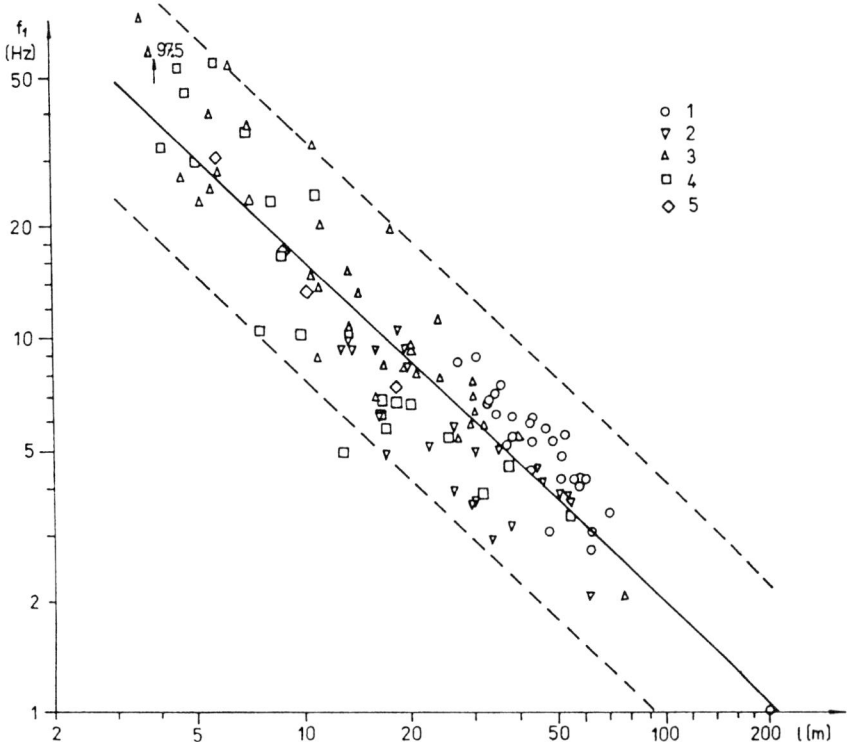

Fig. 4.26. First natural frequency f_1 of railway bridges plotted against span l (regression coefficients are in Table 4.7):
1 – steel truss bridges, 2 – steel plate girder bridges with ballast, 3 – steel plate girder bridges without ballast, 4 – concrete bridges with ballast, 5 – concrete bridges without ballast,
—— mean value for all bridges, – – – lower and upper limits of 95% reliability for all bridges.

4.2 Experimental results

TABLE 4.7. First natural frequency of railway bridges, power regression $f_1 = al^b$

Railway bridges		Ballast: yes + no –	No of measurements n	Regression coefficients		Range of reliability	Correlation coefficient	Validity region	
				a	b	s	r	l_{min}	l_{max}
				m^{-b}s^{-1}	1	1	1	m	
Steel	Truss girder	+/–	27	306.754	-1.073	1.363	0.944	25	200
	Plate girder	+	23	59.477	-0.743	1.717	0.829	10	70
		–	36	208.039	-1.036	1.939	0.923	3	80
Concrete		+	23	190.415	-1.102	2.312	0.906	4	60
		–	4	225.353	-1.191	1.424	0.995	5	20
Together		+/–	113	133.006	-0.911	2.080	0.899	3	200

Figure 4.26 represents the measured first natural frequencies of all 113 railway bridges. The figure shows that practically all experimental data may be found within the zone of 95% reliability around the mean value. Consequently, Fig. 4.26 can be used for the estimation of the mean value and of the lower and the upper limit of the first natural frequency of railway bridges of any type, material and structural system.

4.2.3 Empirical formulae

On the basis of statistical evaluation of measured natural frequencies it is possible, after the rounding-off of regression coefficients from Table 4.7, to propose empirical formulae for the estimation of the first natural frequencies.

For railway bridges of all types, materials and structural systems it follows that

$$f_1 = 133 l^{-0.9}, \quad s = 2.1 \qquad (4.41)$$

where f_1 is in Hertz, and l in metres. These units are also used in all formulae which follow. The reliability constant s gives, after multiplication or division by the mean value, the upper and lower value, respectively, for the 95% reliability according to equation (4.40). This empirical dependence is represented in Figs 4.26 and 4.27, lines *1* and *2*.

For steel truss bridges the formula

$$f_1 = 307 l^{-1.1}, \quad s = 1.4 \qquad (4.42)$$

can be recommended, see Fig. 4.21. Line *3* in Fig. 4.27 represents the mean value only, the same applies also to lines *4* to *7*.

For steel plate girder bridges with ballast the relation

$$f_1 = 59 l^{-0.7}, \quad s = 1.7 \qquad (4.43)$$

is satisfactory; see Fig. 4.22 and line *4* in Fig. 4.27.

For steel plate girder bridges without ballast

$$f_1 = 208 l^{-1}, \quad s = 1.9, \qquad (4.44)$$

see Fig. 4.23 and line *5* in Fig. 4.27.

For concrete bridges with ballast the relation

$$f_1 = 190 l^{-1.1}, \quad s = 2.3 \qquad (4.45)$$

forms an optimum, see Fig. 4.24 and line *6* in Fig. 4.27, while for concrete bridges without ballast

$$f_1 = 225 l^{-1.2}, \quad s = 1.4, \qquad (4.46)$$

see Fig. 4.25 and line *7* in Fig. 4.27.

The regions of validity of all the above formulae are given in Table 4.7 and presented in Figs 4.21 to 4.26.

The International Union of Railways [212] recommends the region within

4.2 Experimental results

which the first natural frequencies of unloaded railway bridges of all types and materials should lie. The lower limit of this region is

and
$$f_1 = 80/l \quad \text{for } 4 \leqslant l \leqslant 20 \text{ m}$$

$$f_1 = 23.58 \, l^{-0.592} \quad \text{for } 20 \leqslant l \leqslant 100 \text{ m}, \quad (4.47)$$

while the upper limit is
$$f_1 = 94.76 \, l^{-0.748} \quad \text{for } 4 \leqslant l \leqslant 100 \text{ m}. \quad (4.48)$$

The graphical representation of the functions (4.47) and (4.48) is shown in Fig. 4.27 as lines *8* and *9*.

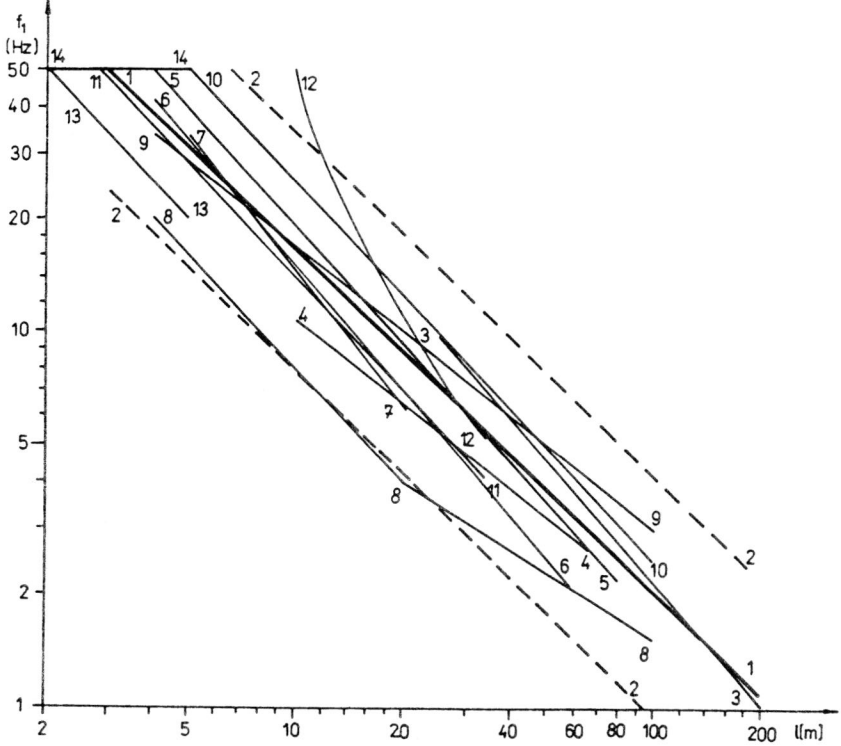

Fig. 4.27. First natural frequency f_1 of railway bridges plotted against span l:
1 – mean value obtained by statistical evaluation of all bridges according to equation (4.41), *2* – lower and upper limits of 95% reliability for all bridges, *3* – mean value for steel truss bridges (4.12), *4* – mean value for steel plate girder bridges with ballast (4.43), *5* – mean value for steel plate girder bridges without ballast (4.44), *6* – mean value for concrete bridges with ballast (4.45), *7* – mean value for concrete bridges without ballast (4.46), *8* – lower limit according to the recommendations of the International Union of Railways (4.47), *9* – upper limit acording to the recommendations of the International Union of Railways (4.48), *10* – steel bridges (4.49), *11* – lower limit for reinforced concrete bridges (4.50) and mean value for prestressed concrete bridges (4.52), *12* – upper limit for reinforced concrete bridges (4.50), *13* – lower limit for reinforced concrete bridges of very small spans (4.51), *14* – upper limit for reinforced concrete bridges of very small spans (4.51).

The author presented the following simple empirical formula in [64] and [68]:

$$f_1 = 250/l,\qquad(4.49)$$

this was based on earlier measurements of natural frequencies of steel bridges. It is line *10* in Fig. 4.27.

According to [22] for the spans up to 23 m (up to approximately 33.5 m), the first natural frequencies of reinforced concrete bridges vary within the limits of

$$1/(0.0072\,l) \leq f_1 \leq 1/(0.0072\,l - 0.05),\qquad(4.50)$$

which are represented by lines *11* and *12* in Fig. 4.27.

For spans below 5 m, the relation [22]

$$100/l \leq f_1 \leq 50,\qquad(4.51)$$

is better, see lines *13* and *14* in Fig. 4.27.

The first natural frequencies of prestressed concrete railway bridges usually are lower than those of reinforced concrete bridges [22] and follow approximately the lower limit of (4.50) (this applies to the spans up to 33.5 m, see line *11* in Fig. 4.27)

$$f_1 = 1/(0.0072\,l).\qquad(4.52)$$

Figure 4.27 shows quite well that the empirical frequency versus span relations lie within the zone of 95% reliability, i.e. between the lines denoted by *2*. In fact, line *1* represents the mean value. From the other formulae, the lower limit *8* of the International Union of Railways recommendation is in good agreement with the lower statistical limit *2*. The upper limit *9*, however, is too low in comparison with the upper statistical limit *2* and does not cover some steel bridges.

In all cases, however, the first natural frequency can be found from the formula

$$f_1 = 17.753\, v_{st}^{-1/2},\qquad(4.53)$$

whenever the midspan deflection v_{st} (in mm) of the bridge due to self-weight can be determined; f_1 is in Hertz. Equation (4.53) follows from the formula for the midspan deflection of a simply supported beam loaded by a uniformly distributed load μg

$$v_{st} = \frac{5}{384}\frac{\mu g l^4}{EI},\qquad(4.54)$$

which is substituted in equation (4.20). It yields for a simply supported beam, $\lambda_1 = \pi$

$$f_1 = \frac{\pi}{2}\left(\frac{5}{384}g\right)^{1/2} v_{st}^{-1/2}.\qquad(4.55)$$

4.2 Experimental results

In equations (4.54) and (4.55), $g = 9.81$ m s^{-2}. The numerical evaluation of the constants in equation (4.55) yields equation (4.53), which is frequently used for bridges.

Otherwise, it is possible to estimate natural frequencies by means of equation (4.30) and the data from Table 4.6 or equation (4.20) with the moment of inertia according to either equation (2.21) or (2.22). The choice depends on which quantity (E, I, μ, B or v_{st}) is known or can be determined easily. In all cases we assume that the bridge span l is known.

Further empirical relations can be found in [22]. They are based on permissible stresses used in the design of bridges according to the theories of permissible stresses or limit states.

5. Damping of railway bridges

Damping is a desirable property of building materials and structures which, in majority of cases, reduces the dynamic response and causes the bridge to reach its state of equilibrium soon after the passage of vehicles or other excitation.

The physical causes of damping are very complex. During vibration one form of energy changes into another (potential energy into kinetic energy and vice versa) and part of the energy is lost by plastic deformations of the material or is changed into other forms of energy, e.g. thermal, acoustic, etc. In this way the energy supplied by the passage of vehicles is irreversibly dissipated into the environment.

The sources of damping of bridge structures are both internal and external. The internal sources of damping includes viscous internal friction of building materials experienced during their deformation, their non-homogeneous properties, cracks, and so on. The external sources of bridge damping includes friction in supports and bearings, friction in the permanent way and particularly in the ballast, friction in the joints of the structure, aerodynamic resistance of the structure (which is very small with regard to the great rigidity of railway bridges), viscoelastic properties of soils and rocks below or beyond the bridge piers and abutments, and so on.

It is obvious that the number of sources of damping of vibrations of railway bridges is high and that it is almost impossible to take them all into account in engineering calculations. Damping depends on the material (steel, reinforced concrete, prestressed concrete) and on the state (presence of cracks, ballast, etc.) of the structure. The magnitude of damping also depends on the amplitude of vibrations [25], [122]; in this respect, however, the influence of the forced vibrations has not yet been fully investigated. The component of forced vibrations is usually substantially higher in bridges than the amplitude of free vibrations after the departure of the vehicle from the bridge. However, the damping depends very little on vibration frequency in the region of low frequencies up to 50 Hz, and this is the range within which railway bridges vibrate most frequently.

5.1 Damped vibrations of a beam during the passage of a force

There are a number of damping hypotheses, most of which are described in [197]. For this reason, we will limit our considerations to the three most important damping theories and to the assessment of their influence on the vibration of a simply supported beam during the passage of a constant force F according to Fig. 5.1, i.e. to the basic case as in Chapter 3. We shall base our considerations on the assumption that at the time $0 \leq t \leq l/c$ the greatest static and dynamic deflections of the beam take place which are important in its design. Free vibrations which occur after the departure of the force from the beam possess considerably lower amplitudes and, therefore, this second phase of the phenomenon is of lower significance in practice (only being used for the evaluation of natural frequencies and damping characteristics and for assessment of fatigue).

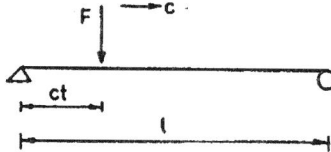

Fig. 5.1. Motion of a constant force F along a simply supported beam of span l at velocity c.

5.1.1 Viscous damping proportional to the velocity of vibration

The Kelvin–Voigt theory of viscous damping assumes that damping is proportional to the velocity of vibration in every element of the beam. This hypothesis actually expresses the external damping of a body immersed in a liquid. Because of its mathematical simplicity, however, this hypothesis has found by far the widest application even though it assumes that damping depends on vibrations frequency, which is at variance with some laboratory experiments [197]. However, global results in the case of complex and big structures, such as railway bridges, usually are in satisfactory agreement with experiments.

A partial differential equation may be derived from the condition of equilibrium of vertical force and bending moments in a beam element (Fig. 5.2)

$$EI \frac{\partial^4 v(x,t)}{\partial x^4} + \mu \frac{\partial^2 v(x,t)}{\partial t^2} + 2\mu\omega_b \frac{\partial v(x,t)}{\partial t} = \delta(x - ct)F \quad (5.1)$$

for the motion of constant force F at velocity c along the beam (the symbols having the same meaning as in equation (2.1), see Sects 2.1.1, 3.1.1 and 3.4.3).

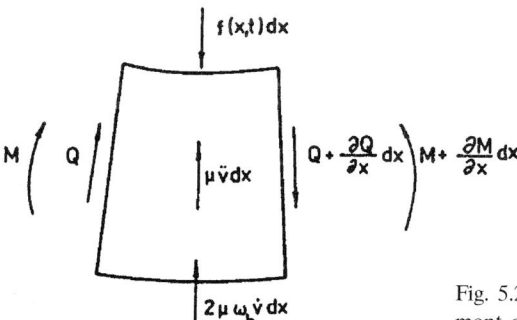

Fig. 5.2. Forces and moments applied to an element of the beam with damping proportional to vibration velocity.

The complete solution of equation (5.1) for all cases of viscous damping and velocities has been derived in [68], Chapter 1. Therefore, only the result is given here:

$$v(x, t) = v_0 \sum_{j=1}^{\infty} \frac{1}{j^2 \left[j^2 (j^2 - \alpha^2)^2 + 4\alpha^2 \beta^2 \right]}$$

$$\cdot \left[j^2 (j^2 - \alpha^2) \sin j\omega t - \frac{j\alpha \left[j^2 (j^2 - \alpha^2) - 2\beta^2 \right]}{(j^4 - \beta^2)^{1/2}} e^{-\omega_b t} \right.$$

$$\left. \cdot \sin \omega'_j t - 2j\alpha\beta \left(\cos j\omega t - e^{-\omega_b t} \cos \omega'_j t \right) \right] \sin \frac{j\pi x}{l}, \qquad (5.2)$$

which applies to the period of passage of the force along the simply supported beam, $0 \leq t \leq l/c$, for subcritical damping $\omega_b < \omega_j$, and for $\alpha \neq j$.

The following notation has been used in equation (5.2):
– static deflection of the beam at midspan due to force F applied at the same point

$$v_0 \approx \frac{2F}{\mu l \omega_1^2} = \frac{2Fl^3}{\pi^4 EI}, \qquad (5.3)$$

– dimensionless velocity parameter

$$\alpha = \frac{\omega}{\omega_1} = \frac{c}{2 f_1 l}, \qquad (5.4)$$

– dimensionless damping parameter (damping ratio)

$$\beta = \frac{\omega_b}{\omega_1} = \frac{\vartheta}{2\pi}, \qquad (5.5)$$

– circular frequency of force passage

$$\omega = \frac{\pi c}{l}, \qquad (5.6)$$

– circular frequency of damped vibrations of unloaded beam at subcritical damping ($\omega_b < \omega_j$)

$$\omega'^2_j = \omega_j^2 - \omega_b^2, \qquad (5.7)$$

– circular frequency of undamped vibrations of unloaded simply supported beam according to equation (4.18) and equation (4.19)

$$\omega_j^2 = \frac{j^4 \pi^4}{l^4} \frac{EI}{\mu}, \qquad f_j = \omega_j / (2\pi), \qquad (5.8)$$

and the logarithmic decrement of damping ϑ according to equation (1.8) in Chapter 1.

Figure 5.3 shows the time-history of midspan deflection of the beam for various velocities α of force movement and for different damping β, calculated from equations (1.30) to (1.39) in [68]. The figure shows that for usual velocities

5.1 Damped vibrations of a beam during the passage of a force

($\alpha \ll 1$) and low damping ($\beta \ll 1$) the greatest dynamic deflection at midspan occurs when the force is above midspan of the beam or immediately afterwards.

For current cases ($\alpha \ll 1$, $\beta \ll 1$) the first term $j = 1$ of equation (5.2) yields satisfactory results, so that equation (5.2) can be simplified to

$$v(x, t) = \frac{v_0}{1 - \alpha^2} (\sin \omega t - \alpha e^{-\omega_b t} \sin \omega_1 t) \sin \pi x/l , \qquad (5.9)$$

where the second term in brackets represents natural damped vibrations and is substantially less than the first term, which expresses the forced motion, i.e. a quasistatic case.

a)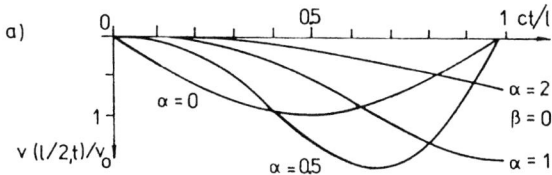

b)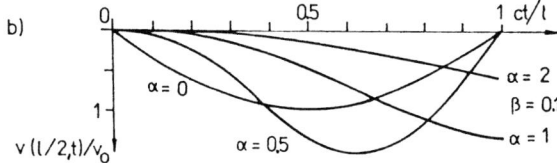

c)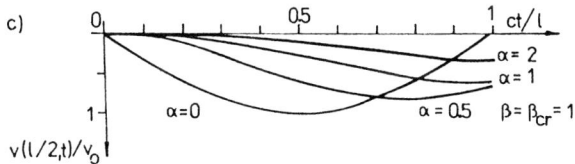

d)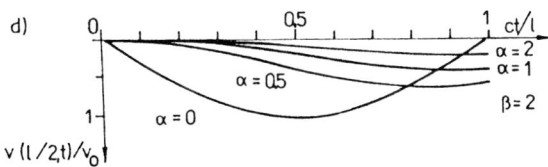

Fig. 5.3. Dynamic deflection at midspan of a simply supported beam $v(l/2, t)/v_0$ for various dimensionless velocities α and damping β:
a) $\beta = 0$; $\alpha = 0;, 0.5; 1; 2$;
b) $\beta = 0.1$; $\alpha = 0; 0.5; 1; 2$;
c) $\beta = \beta_{cr} = 1$; $\alpha = 0; 0.5; 1; 2$;
d) $\beta = 2$; $\alpha = 0; 0.5; 1; 2$.

It is often maintained that *high vibration modes* have *higher damping* than the basic mode. If we denote the circular frequency of damping of the jth mode as ω_{bj} then equation (5.1) will contain the expression ω_{bj} instead of ω_b and the solution of equation (5.2) will be identical except for the notation of the damping terms:

$$\omega'^2_j = \omega^2_j - \omega^2_{bj},$$
$$\beta_j = \omega_{bj}/\omega_1 = \vartheta_j/(2\pi). \qquad (5.10)$$

Although this does not yield any new quality it does give rise to practical difficulties concerning the evaluation of the deflection or stress-time records to obtain the respective damping coefficients ω_{bj} or ϑ_j, when the pure higher vibration modes could rarely be obtained.

Another hypothesis from viscous damping is that *damping is proportional to the rate of stress variations*. It expresses internal damping as a function of stress. According to this theory, it is assumed that the stress σ depends on the velocity of strain ε

$$\sigma = bE \frac{\partial \varepsilon}{\partial t}, \qquad (5.11)$$

where b is the damping coefficient with the unit (s).

As the strain at point y (which is the distance from the neutral axis) of the cross section A is

$$\varepsilon = v''y, \qquad (5.12)$$

the bending moment, which damps the vibration of the beam, has the form

$$M = \int \sigma y \, dA = \int bE\dot{v}''y^2 \, dA = bEI\dot{v}''. \qquad (5.13)$$

It is assumed (according to Fig. 5.4a) that the stresses are linearly distributed along the cross section.

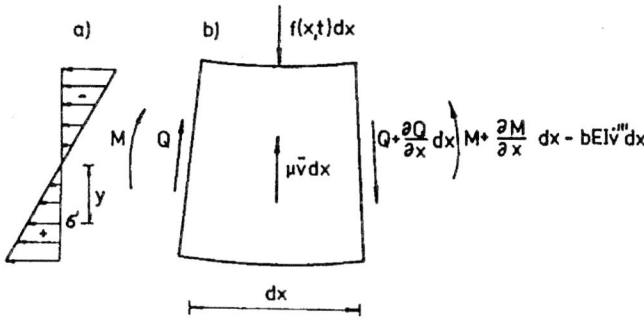

Fig. 5.4. Damping proportional to the rate of stress variation:
a) stress distribution along the cross section,
b) forces and moments applied to an element of the beam.

5.1 Damped vibrations of a beam during the passage of a force

From the conditions of equilibrium of vertical forces and bending moments in an element of the beam as shown in Fig. 5.4b the differential equation for the deformation of the beam is given by

$$EI \frac{\partial^4 v(x,t)}{\partial x^4} + bEI \frac{\partial^5 v(x,t)}{\partial x^4 \partial t} + \mu \frac{\partial^2 v(x,t)}{\partial t^2} = \delta(x - ct)F. \quad (5.14)$$

When

$$b = 2\omega_b / \omega_j^2, \quad (5.15)$$

the solution of equation (5.14) is represented by equation (5.2) using the notation $\omega_b = b\omega_j^2/2$.

The assumption of damping proportional to stress variation rate, therefore, leads to the same results as the assumption of damping proportional to the vibration velocity, although both theories have been based on different approaches, see Fig. 5.2 and Fig. 5.4b.

5.1.2 Dry friction

The hypothesis of Coulomb dry friction assumes that a constant force acts on every element of the beam against the direction of its movement. The partial differential equation follows from the conditions of equilibrium of vertical forces and moments in Fig. 5.5

$$EI \frac{\partial^4 v(x,t)}{\partial x^4} + \mu \frac{\partial^2 v(x,t)}{\partial t^2} \pm b = \delta(x - ct)F, \quad (5.16)$$

where the upper sign of b holds true for $\dot{v}(x,t) > 0$ and the lower sign for $\dot{v}(x,t) < 0$.

The constant b has the dimension of force per unit length and, for railway bridges, is about 1 to 3 kN m^{-1}.

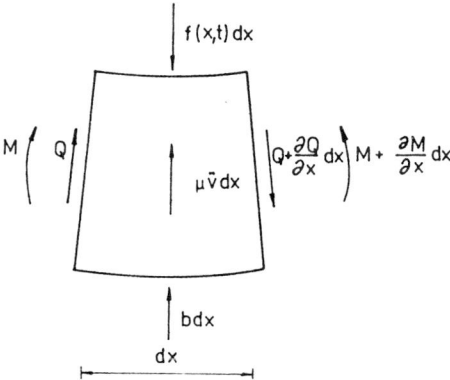

Fig. 5.5. Friction; forces and moments applied to an element of the beam at $\dot{v}(x,t) > 0$.

For simplificaton assume that the solution of equation (5.16) for the simply supported beam has the form

$$v(x, t) = v_0 q(t) \sin \pi x / l ,\qquad (5.17)$$

where v_0 is given by equation (5.3). Alternatively, a numerical solution of equation (5.16) or (5.18) can be derived.

The Galerkin method with the basic function (5.17) provides instead of equation (5.16) an ordinary differential equation

$$\frac{d^2 q(t)}{dt^2} + \omega_1^2 q(t) \pm \omega_1^2 \Phi = \omega_1^2 \sin \omega t, \quad \text{for} \quad \dot{q}(t) \gtrless 0 , \qquad (5.18)$$

with the notatiton (5.8)

$$\omega_1^2 = \frac{\pi^4 EI}{l^4 \mu} \qquad (5.19)$$

for the first circular natural frequency of the simply supported beam, equation (5.6) for ω and the dimensionless friction force

$$\Phi = \frac{2bl}{\pi F} . \qquad (5.20)$$

The function $q(t)$ represents the dimensionless midspan deflection of the bridge with initial conditions

$$q(t) = \dot{q}(t) = 0 \quad \text{for} \quad t = 0 . \qquad (5.21)$$

The force on the beam (Fig. 5.1) starts at time $t = 0$.

The motion of the force along the beam damped by internal Coulomb friction can be divided into three phases:

(1) In the first phase, the deflection $q(t)$ will be zero, until the friction forces have been overcome by beam deformations. It holds that

$$\ddot{q}(t) + \omega_1^2 q(t) = \omega_1^2 (\sin \omega t - \Phi) = 0 , \qquad (5.22)$$

from which it follows (assuming equation (5.21)) that

$$q(t) = 0 \qquad (5.23)$$

and that it will hold true until

$$\sin \omega t_1 = \Phi . \qquad (5.24)$$

It follows from equation (5.24) that equation (5.23) is valid during the interval

$$0 \leq t \leq t_1 \qquad (5.25)$$

where

$$t_1 = \frac{1}{\omega} \arcsin \Phi \approx \frac{\Phi}{\omega} = \frac{2bl^2}{\pi^2 cF}$$

is the time at the end of the first phase.

(2) For the second phase, we shall introduce a new independent variable

$$\tau = t - t_1 , \qquad (5.26)$$

which has its beginning at t_1 in absolute time, and whose initial conditions for $q(\tau)$ equal zero, since it follows the preceeding phase.

According to (5.18) the equation

$$\ddot{q}(\tau) + \omega_1^2 q(\tau) = \omega_1^2 \left[\sin \omega(t_1 + \tau) - \Phi \right] \qquad (5.27)$$

now holds true, as long as $\dot{q}(\tau) > 0$.

We shall derive the solution of the differential equation (5.27) with zero initial conditions by means of equations (27.17), (27.18), (27.31) and (27.32) from [68] in the following form:

$$q(\tau) = \frac{\cos \omega t_1}{1 - \alpha^2} (\sin \omega \tau - \alpha \sin \omega_1 \tau)$$

$$- \Phi \left[1 - \frac{1}{1 - \alpha^2} (\cos \omega \tau - \alpha^2 \cos \omega_1 \tau) \right] \qquad (5.28)$$

and its derivative with respect to time is

$$\dot{q}(\tau) = \frac{\omega_1 \alpha}{1 - \alpha^2} \left[\cos \omega t_1 (\cos \omega \tau - \cos \omega_1 \tau) - \Phi(\sin \omega \tau - \alpha \sin \omega_1 \tau) \right]. \qquad (5.29)$$

Both equations hold true while $\dot{q}(\tau) > 0$.

For small parameters α (5.4) and Φ (5.20), equation (5.24) and $\cos \omega t_1 \approx 1$, it follows from equations (5.28) and (5.29) approximately

$$q(\tau) = \frac{1}{1 - \alpha^2} (\sin \omega \tau - \alpha \sin \omega_1 \tau) - \Phi \left(1 - \frac{1}{1 - \alpha^2} \cos \omega \tau \right) \qquad (5.30)$$

and

$$\dot{q}(\tau) = \frac{\omega_1 \alpha}{1 - \alpha^2} (\cos \omega \tau - \cos \omega_1 \tau - \Phi \sin \omega \tau) . \qquad (5.31)$$

The condition $\dot{q}(\tau_2) = 0$ is complied with according to equation (5.31) approximately at the instant when the force traverses the midspan point

$$\tau_2 \approx t_2 = \frac{l}{2c} . \qquad (5.32)$$

At that moment, according to equations (5.30) and (5.31), the dimensionless deflection and its derivative with respect to time are approximately

$$q(t_2) \approx \frac{1}{1 - \alpha^2} - \Phi ,$$

$$\dot{q}(t_2) \approx 0 . \qquad (5.33)$$

Phase 2 terminates at time t_2 and equation (5.33) represents the initial conditions for the third phase.

(3) In the third phase, the dimensionless deflection according to equation (5.18) is described by the differential equation

$$\ddot{q}(\tau) + \omega_1^2 q(\tau) = \omega_1^2 \left[\sin \omega(t_2 + \tau) + \Phi\right] \tag{5.34}$$

which holds true while $\dot{q}(\tau) < 0$. An independent variable

$$\tau = t - t_2 \tag{5.35}$$

has been introduced in equation (5.34).

The solution of equation (5.34) with initial conditions (5.33) is, according to equations (27.19), (27.33), and (27.17) from [68],

$$q(\tau) = \frac{1}{1 - \alpha^2} \cos \omega\tau + \Phi(1 - 2 \cos \omega_1\tau),$$

$$\dot{q}(\tau) = \left(\frac{-\alpha}{1 - \alpha^2} \sin \omega\tau + 2\Phi \sin \omega_1\tau\right)\omega_1 \tag{5.36}$$

for $\dot{q}(\tau) < 0$.

The force leaves the beam when $t_3 = l/c$, i.e. according to equations (5.35) and (5.32) when $\tau_3 = t_3 - t_2 = l/(2c)$. At this instant, the dimensionless deflection at beam midspan according to equations (5.36), (5.4) and (5.6) is approximately

$$q(\tau_3) \approx \Phi\left(1 - 2 \cos \frac{\pi}{2\alpha}\right) \tag{5.37}$$

and the deflection of the beam is usually stopped by dry friction.

The derived expression can be used for the evaluation of the dry friction coefficient b. From the recorded time-history of midspan deflection the residual deflection v_r at the moment of vehicle departure from the bridge is evaluated. Then according to equations (5.17) and (5.18)

$$v(l/2, l/c) = v_r = v_0 \Phi[1 - 2 \cos \pi/(2\alpha)] \tag{5.38}$$

and, using equation (5.20), we obtain

$$b = \frac{\pi F}{2l[1 - 2 \cos \pi/(2\alpha)]} \frac{v_r}{v_0}. \tag{5.39}$$

All the values in equation (5.39) can easily be found from tests. The term in square brackets depends on the phase of bridge motion, because it can acquire the values from -1 to $+3$. The residual deflection v_r is easily recognizable in test records, as the record before the arrival of the vehicle on the bridge sometimes does not coincide with the record after the departure of the vehicle. This indicates the presence of dry friction.

Dry friction theory describes well the influence of ballast on railway bridges of minor spans.

5.1 Damped vibrations of a beam during the passage of a force

In case of poor bearing maintenance, the bearings are clogged with impurities and resist motion. This case can be modelled according to Fig. 2.17, by considering that the beam is subjected to the vertical force F and a *horizontal longitudinal friction force* from the bearings. That is, a force

$$N(t) = \pm f\left(\frac{G}{2} + F\frac{ct}{l}\right) \tag{5.40}$$

applied to its ends, which is proportional to the reaction of the beam at the supports below the moving bearings. In this equation
 f – friction coefficient and
 G – self-weight of the bridge.

The upper sign of the force $N(t)$ applies to the motion of the rolling bearing towards the left, the lower sign to the motion towards the right, see Fig. 2.17 or Fig. 5.1. This movement is characterized approximately by the deflection rate of the beam $\dot{v}(x, t)$ which means that the upper sign applies to $\dot{v}(x, t) > 0$ and the lower sign to $\dot{v}(x, t) < 0$.

With these assumptions, it is possible to derive the differential equation (2.6) using the conditions of equilibrium of forces and moments, which in this particular case is of the form

$$EI\frac{\partial^4 v(x, t)}{\partial x^4} + \mu\frac{\partial^2 v(x, t)}{\partial t^2} - N(t)\frac{\partial^2 v(x, t)}{\partial x^2} = \delta(x - ct)F. \tag{5.41}$$

The solution of equation (5.41) with (5.40) leads to Bessel function, see [113], equations (2.10), (2.14) and (2.162). An approximate solution in the form of equation (5.17) by the Galerkin and perturbation methods for $\dot{q}(t) > 0$ yields

$$q(t) = q_0(t) + q_1(t), \tag{5.42}$$

where

$$q_0(t) = \frac{1}{1 - \alpha^2}(\sin\omega t - \alpha\sin\omega_1 t), \tag{5.43}$$

$$\begin{aligned}q_1(t) = &-\frac{f}{N_{cr}(1 - \alpha^2)}\left\{\frac{G}{2}\left[\frac{1}{1 - \alpha^2}(\sin\omega t - \alpha\sin\omega_1 t)\right.\right.\\
&\left.- \frac{\alpha}{2}(\sin\omega_1 t - \omega_1 t\cos\omega_1 t)\right]\\
&+ \frac{F}{\pi}\left[\frac{2}{(1 - \alpha^2)^2}\left(\frac{1 - \alpha^2}{2}\omega t\sin\omega t - \alpha^2\cos\omega t + \alpha^2\cos\omega_1 t\right)\right.\\
&\left.\left.- \frac{\alpha\omega t}{4}(\sin\omega_1 t - \omega_1 t\cos\omega_1 t)\right]\right\}. \end{aligned} \tag{5.44}$$

In the above case, the relations (5.3) to (5.8) hold true; apart from that, the critical axial force is denoted by

$$N_{cr} = \pi^2 EI/l^2. \tag{5.45}$$

$q_0(t)$ represents the motion due to a force F (compare with equations (5.9)), while $q_1(t)$ represents the motion of the beam due to the force $N(t)$. The dimensionless deflections (5.42) to (5.44) hold approximately true for $t < l/(2c)$. The direction of the axial force $N(t)$ influences the natural frequency of the beam in equation (20.4) and (20.6) from [113]. In the first phase of the motion $0 < t < l/(2c)$, ω_1 is higher, while in the second phase, $l/(2c) < t < l/c$, ω_1 is lower in comparsion with equation (5.19).

For low values of f and α the expression $q_1(t)$ is considerably lower than $q_0(t)$.

5.1.3 Complex theory of internal damping

Sorokin's theory of internal damping [197] is based on the relation between stress and strain in the complex form

$$\sigma^* = (1 + i\gamma)E\varepsilon^* , \qquad (5.46)$$

where γ is the coefficient of internal damping which is generally dependent on the instantaneous vibration displacement. The quantities σ^* and ε^* are complex, the other quantities being real. The expression (5.46) characterizes the event that the non-reversible deformation $\gamma\varepsilon$ lags behind the elastic deformation ε by a certain phase angle and that damping does not depend on the frequency of vibration.

The mean value of the coefficient of internal damping is

	according to	[122]	and [25]	
for steel bridges	$\gamma =$	0.027	0.031	
for reinforced concrete bridges	$\gamma =$	0.1	0.072	(5.47)

and its relation to the logarithmic decrement of damping (1.8) is

$$\gamma = \vartheta/\pi . \qquad (5.48)$$

From equation (5.46), the partial differential equation for damped vibrations of a beam with complex load $f^*(x, t)$ is

$$(\beta_1 + i\beta_2)EI \frac{\partial^4 v^*(x, t)}{\partial x^4} + \mu \frac{\partial^2 v^*(x, t)}{\partial t^2} = f^*(x, t) \qquad (5.49)$$

where $v^*(x, t)$ is the deflection of the beam in complex form, and the coefficients β_1 and β_2 depend on γ:

$$\beta_1 = \frac{1 - \gamma^2/4}{1 + \gamma^2/4} ,$$

$$\beta_2 = \frac{\gamma}{1 + \gamma^2/4} . \qquad (5.50)$$

According to Sorokin [197], the real load

$$f(x, t) = \delta(x - ct)F \qquad (5.51)$$

is matched to the complex load with (5.51) as the real component. In our particular case, it is approximately the load

$$f^*(x, t) = \frac{2}{l} F e^{i(\omega t - \pi/2)} \sin(\pi x/l) = \frac{2}{l} F(\sin \omega t - i \cos \omega t) \sin(\pi x/l), \quad (5.52)$$

the real component of which, after development into Fourier series and for $j = 1$, corresponds to equation (5.51).

The solution of equation (5.49) with $f^*(x, t)$ from equation (5.52) is assumed to be of the form

$$v^*(x, t) = v_0 q^*(t) \sin(\pi x/l) \quad (5.53)$$

for a simply supported beam.

After the substitution of (5.53) and (5.52) in (5.49) we obtain an ordinary differential equation

$$\ddot{q}^*(t) + \omega_1^{*2} q^*(t) = \omega_1^2 e^{i(\omega t - \pi/2)}, \quad (5.54)$$

where the complex natural frequency is

$$\omega_1^{*2} = (\beta_1 + i\beta_2)\omega_1^2 = (1 + i\gamma/2)^2 \omega_1'^2,$$
$$\omega_1' = \omega_1 (1 + \gamma^2/4)^{-1/2}, \quad (5.55)$$

and ω_1 is determined by either equation (5.8) or (5.19).

The solution of equation (5.54) covers two complex initial conditions, the real components of which are assumed to equal zero. The solution also includes the terms with growing amplitudes which, however, are considered equal to zero. The mathematical aspect of Sorokin's theory has been criticized because of its ambiguity and improved by V. Koloušek and I. Babuška [120]. Nevertheless, Sorokin's hypothesis is often in good agreement with experiments and is frequently referred to in the literature.

The procedure described in Sect. 19 of [197] yields the real part of the solution of equation (5.54)

$$q(t) = \frac{1}{A^2 + \gamma^2} \left\{ A_1 \sin \omega t - (\alpha A_1 B - A_2 \gamma^2/2) \exp\left(\frac{-\gamma}{2} \omega_1' t\right) \right.$$
$$\left. \times \sin \omega_1' t - A_2 \gamma \left[\cos \omega t - \exp\left(-\frac{\gamma}{2} \omega_1' t\right) \cos \omega_1' t \right] \right\}, \quad (5.56)$$

where

$$A = 1 - \gamma^2/4 - \alpha^2 B^2,$$
$$B^2 = 1 + \gamma^2/4,$$
$$A_1 = (1 - \gamma^2/4)A + \gamma^2,$$
$$A_2 = 1 - \gamma^2/4 - A = \alpha^2 B^2. \quad (5.57)$$

The dimensionless deflection (5.56) at midspan of the beam is analogous with equation (5.2), and for $\gamma = 0$ and $j = 1$ both expressions are identical. As the coefficient of internal damping γ for bridges is small (see equation (5.47)), the Sorokin's theory does not yield very different results in comparison with the Kelvin–Voigt theory (for the case solved above). This is because the forced vibration prevails over the natural vibration in the case of a beam loaded by a moving force and the excitation lasts for a limited time.

Sorokin's theory also permits the consideration of a *non-linear internal damping* coefficient which depends on the vibration amplitude a_0 in accordance with

$$\gamma(a_0) = \gamma \left(\frac{a_0}{\alpha_0 + a_0} + \beta_0 a_0^k \right), \qquad (5.58)$$

where α_0, β_0 and k are empirical coefficients [197].

It the case of non-stationary vibrations the coefficient $\gamma(a_0)$ is approximately constant over the time of a half cycle, which corresponds to the travel of the force along the beam. For this reason, equation (5.54) is linear over this period, with constant coefficients, and as such it can be integrated. The non-linearity according to equation (5.58) can be introduced only into the results. Naturally, the procedure is only approximate.

In experiments, such as those described in [25], it is often observed that the *damping is directly proportional to displacement*, i.e. according to equation (5.58)

$$\gamma(a_0) = \gamma_0 a_0, \qquad (5.59)$$

where $\alpha_0 \to \infty$, $k = 1$, $\gamma_0 = \gamma \beta_0$.

The maximum displacement in equation (5.56) takes place approximately at the moment $t = l/(2c)$ and its magnitude is

$$a_0 = q(l/(2c)) = \frac{A_1}{A^2 + \gamma^2} \qquad (5.60)$$

for small α and γ. When neglecting $\gamma^2/4$ in comparison with unity we obtain, from equations (5.57) and (5.59), and for $A = 1 - \alpha^2$, $A_1 = 1 - \alpha^2 + \gamma^2$ and $B = 1$, a cubic equation for the amplitude a_0

$$a_0 = \frac{1 - \alpha^2 + \gamma_0^2 a_0^2}{(1 - \alpha^2)^2 + \gamma_0^2 a_0^2}. \qquad (5.61)$$

Its approximate solution is

$$a_0 \approx \frac{1 + \gamma_0^2/(1 - \alpha^2)^3}{1 - \alpha^2 + \gamma_0^2/(1 - \alpha^2)^3}. \qquad (5.62)$$

which is, for small γ_0 very near the maximum displacement $1/(1 - \alpha^2)$ in equations (5.9), (5.30), (5.43) and (5.56).

From the various theories of damping it is possible to conclude that damping influences free vibrations after the departure of the vehicle from the bridge, and that the theories do not yield too different results during the passage of the vehicle along the bridge, when the static together with dynamic action is a maximum. The loading of railway bridges is a transient process of a relatively short duration, in which the application of simple hypotheses is permitted (e.g. the Kelvin–Voigt hypothesis). For a more accurate analysis it would be necessary to introduce the hypotheses and procedures described in Sects 5.1.1, 5.1.2 and 5.1.3.

5.2 Experimental results

The coefficients which characterize damping cannot be determined by theoretical calculations. They must be obtained from experiments on bridges.

Usually, the deflection- and stress-time records are used for the evaluation of the logarithmic decrement of damping ϑ by the procedure described in Fig. 1.3 from two or more successive amplitudes of free vibrations (equation (1.8)). The basic theory is explained in all text books of dynamics and has been briefly described in Sect. 1.2.

5.2.1 Statistical evaluation of logarithmic decrement of damping

Our own measurements, the experiments carried out by other Czech and Slovak institutes, the tests of the Office for Research and Experiments (ORE) and the measurements carried out on the bridges of the German Federal Railways (DB) [25], have furnished the logarithmic decrements of damping ϑ for 73 railway bridges, which are plotted against span l in Figs 5.6 and 5.7. The figures differentiate steel truss bridges, steel plate girder bridges with ballast, steel plate girder bridges without ballast, concrete bridges with ballast and concrete bridges without ballast.

Statistical analysis of the logarithmic decrements of damping has yielded the following results:
- a great variation in results depending on the methods and gauges used for measurements,
- very poor correlation between ϑ and span l and natural freqency f_1 of the bridges,
- poor correlation with the occurrence of ballast for spans greater than 20 m,
- in minor spans ($l < 20$ m), the damping due to dry friction is observed according to the theory described in Sect. 5.1.2. The smaller the span, the higher the dependence on friction in bearings and on the influence of continuous gravel bed,
- logarithmic decrements of damping of concrete bridges are higher than those of steel bridges,
- the differences among the individual bridge types in the groups of steel bridges and in the group of concrete bridges are statistically insignificiant.

For these reasons, the experimental results were divided into two sets only, concerned with steel and concrete bridges. The results for the bridges with spans over 20 m yield logarithmic decrements of damping characterizing two principal structural materials of railway bridges – steel and concrete. In bridges with spans less than 20 m, this phenomenon is made indistinct by the influence of bearings and continuous gravel bed. This is due to the mixed influence of viscous damping and dry friction which results in increasing ϑ with decreasing span l.

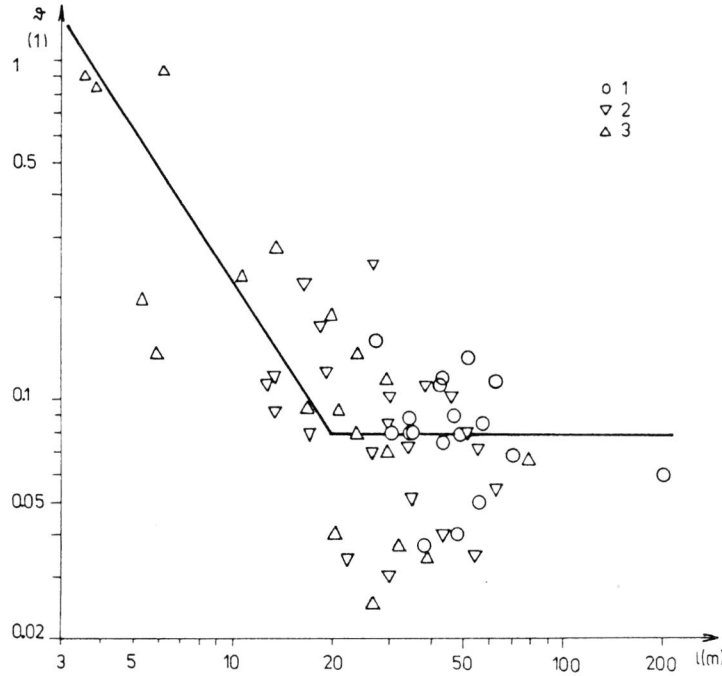

Fig. 5.6. Logarithmic decrement of damping ϑ of steel railway bridges for various spans:
1 – steel truss bridges, *2* – steel plate girder bridges with ballast, *3* – steel plate girder bridges without ballast,
────── mean values (5.64), (5.65).

For these reasons, Figs 5.6 and 5.7 contain, apart from test results, also the curves of mean values within the range of $l > 20$ m and within the range of $l < 20$ m. The empirical relation

$$\vartheta = \vartheta_0 \left(\frac{l_0}{l}\right)^a, \tag{5.63}$$

evaluates by the least square method the results obtained by tests on small spans. In equation (5.63) ϑ_0 is the mean value of logarithmic decrement of damping for $l > 20$ m, $l_0 = 20$ m, and the exponent a is given in Table 5.1.

5.2 Experimental results

Table 5.1 gives, in addition to the mean values of ϑ, their standard deviation s, the 1.96 s multiple which, with the assumption of Gauss distribution, includes 95% of results, and the reliability region $t_{n-1;p} s n^{-1/2}$, where $t_{n-1;p}$ is the value of Student distribution for the number n of measured data for $P\%$ of two-sided reliability.

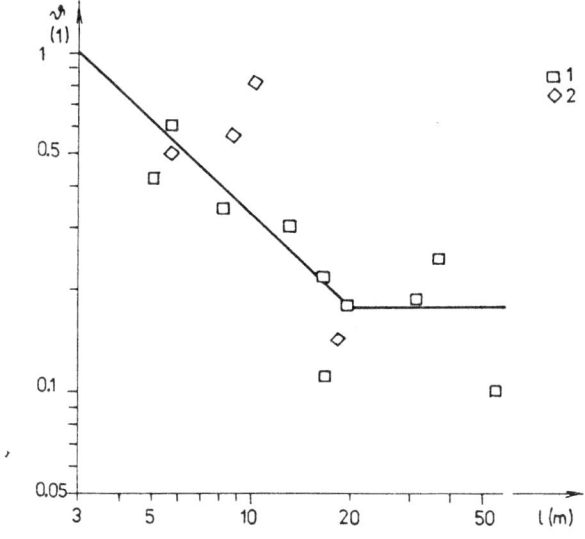

Fig. 5.7. Logarithmic decrement of damping ϑ of concrete railway bridges for various spans l:
1 – concrete bridges with ballast, 2 – concrete bridges without ballast,
─── mean values (5.66), (5.67).

The analysis of the damping of railway bridges also shows that it is essential to record the magnitude of the amplitude at which data are measured, because of the non-linear dependence of damping on amplitude, and the necessity of evaluating also the residual deformation v_r for the purpose of ascertaining the dry friction according to equation (5.39).

TABLE 5.1. Logarithmic decrement of damping ϑ of railway bridges

Quantity	Notation	Reliability region	Railway bridges	
			Steel	Concrete
Number of measurements	n		43	3
Mean value	ϑ		0.080	0.177
Standard deviation	s	$l > 20$ m	0.041	0.073
Reliability (for Gauss distribution)	1.96 s		0.080	0.143
Reliability region	$t_{n-1;p} s n^{-1/2}$		0.013	0.182
Number of measurements	n	$l < 20$ m	16	11
Exponent in equation (5.63)	a		1.494	0.918

Similar statistical analysis of the logarithmic decrements of damping of 27 railway bridges of the German Federal Railways (DB) was carried out by W. Braune [25], who came to the following conclusions:
- the dependence of ϑ on span l, stiffness EI, deflection due to self-weight of the bridge v_g, dynamic stiffness $(EI/\mu)^{1/2}$ and the first natural frequency f_1 is statistically insignificiant,
- ϑ depends on the displacement in natural vibrations at least linearly. In practice, however, this phenomenon will find hardly any application because of limitations of deflection prescribed for bridges,
- with poor bearings, ϑ increases to as much as 0.4 to 0.8,
- there is no great difference in values of ϑ between the bridges with and without ballast,
- the set of results was classified into six groups according to bridge types and materials; it has come to light, however, that it could be classified into three groups only. The mean values and reliability limits are given in Table 5.2. The conclusions of [25] are in good agreement with our results.

TABLE 5.2. Logarithmic decrement of damping ϑ of DB railway bridges according to W. Braune [25]

Quantity	Notation	Railway bridges		
		Steel	Steel box	Concrete
Number of measurements	n	9	8	10
Mean value	ϑ	0.117	0.079	0.228
Reliability region	$t_{n-1;p} s n^{-1/2}$	0.028	0.028	0.068

5.2.2 Empirical formulae

On the basis of statistical evaluation of logarithmic decrements of damping in railway bridges, it is possible to recommend the following empirical values (originated by the rounding-off the data of Table 5.1):
- for steel bridges $l > 20$ m

$$\vartheta = 0.08, \quad s = 0.04, \tag{5.64}$$

- for steel bridges $l < 20$ m

$$\vartheta = 0.08 (20/l)^{1.5}, \tag{5.65}$$

- for concrete bridges $l > 20$ m

$$\vartheta = 0.18, \quad s = 0.07, \tag{5.66}$$

- for concrete bridges $l < 20$ m

$$\vartheta = 0.18 (20/l)^{0.9}. \tag{5.67}$$

The span l substituted in equations (5.65) and (5.67) is in metres.

5.2 Experimental results

The author's measurements of older data [64] yielded the empirical formula

$$\vartheta = (0.3l - 1.2 \times 10^{-3} l^2)^{-1} , \tag{5.68}$$

where l is the span in metres.

In the former USSR, according to [23], the relation

$$\vartheta = \frac{f_1}{10(1 + 10/f_1)} \tag{5.69}$$

is recommended, where f_1 is the first natural frequency in Hertz.

6. Influence of vehicle speed on dynamic stresses of bridges

Vehicle speed is the most important parameter influencing the dynamic stresses in railway bridges. In general, the dynamic stresses in bridges increase with increasing speed. We will show, that they depend also on the bridge and vehicle dynamic system, track irregularities and other parameters.

Considering the general tendency to increasing speed, the dynamic stresses in bridges at higher vehicle speeds have been given considerable attention. For example, the Office for Research and Experiments (ORE) of the International Union of Railways (UIC) concentrated several research programs in this field to elucidate the problem with respect to stresses [159], fatigue [162], noise [161] and the effect on man [165]. Also Japanese National Railways tested their bridges at high vehicle speeds [141]. Up-to-date field measurements have verified dynamic stresses in railway bridges at vehicle speeds up to 250 km h^{-1} (Fig. 6.6), while theoretical results have been extrapolated up to 500 km h^{-1} (Fig. 6.3).

Theoretical analyses consider constant speed of motion of vehicles along the bridge, which is usually so in practice. The actual speed depends on the horizontal and vertical track alignment on the bridge. Even in the cases where the bridges are in poor condition or are temporary the speed remains constant even though it is reduced.

However, in those cases when the vehicles or the train start or brake on bridges the speed is variable. These speed considerations form the sections of this chapter.

6.1 Constant velocity of motion

The simplest case, i.e., the motion of a concentrated force F along a simply supported beam at constant velocity, was solved in Sect. 5.1.1, and the response of the beam to such an excitation is determined by equation (5.2), which shows that the deflection of the beam $v(x, t)$ is a function of a dimensionless velocity parameter α, defined by equation (5.4).

From equation (5.2) the maximum midspan deflection of the beam max $v(l/2, t)/v_0$ was found, where v_0 is determined by equation (5.3), and plotted in Fig. 6.1 against the velocity parameter α (5.4) for various damping β (5.5). Fig. 6.1 shows that the dynamic actions increase with increasing velocity to about $\alpha \approx 0.5$ to 0.7 at subcritical damping. For greater α the midspan deflection of the beam drops, while for very small α the dynamic deflection approaches the static deflection.

6.1 Constant velocity of motion

It follows from equation (5.4) that the velocity parameter α is less than 1 for contemporary railway speeds and usual natural frequencies and bridge spans. This explains why the dynamic response of bridges increases with increasing speed.

The phenomenon has also been verified on more complex models than that of a force moving along a simply supported beam. For instance, Fig. 6.2 reproduces (according to [68]) a theoretical model of a multiaxle vehicle of the Czech Railways (CD) travelling along a beam with an elastic layer and track irregularities.

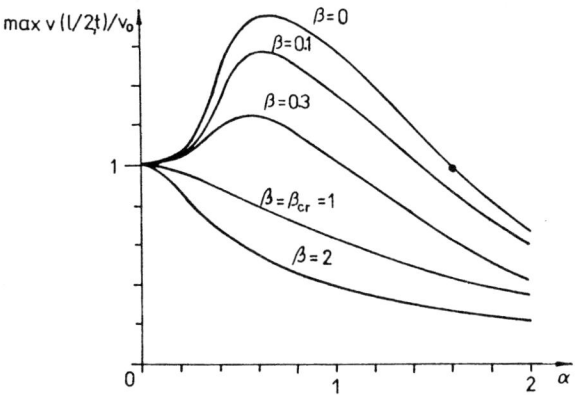

Fig. 6.1. Maximum dynamic deflection at midspan of a simply supported beam, max $v(l/2, t)/v_0$ as a function of velocity parameter α and damping β.

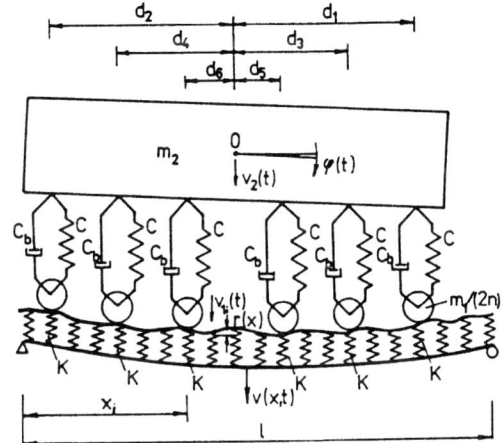

Fig. 6.2. Theoretical CD model according to [68]: beam with an elastic layer K and track irregularities $r(x)$ loaded by a multiaxle vehicle.

According to this model the dependence of the dynamic coefficient δ on velocity during the passage of an electic four-axle locomotive of type E 10, weight 850 kN, along a prestressed concrete bridge of span 10 m, Fig. 6.3, was calculated with the observance of the sleeper effect for various depth of track irregularities $a = 0$; 0.25; 0.5 mm. Fig. 6.3 shows the dynamic coefficients calculated for the stresses.

Fig. 6.3 confirms the (roughly) rising tendency of the dynamic coefficient with increasing speed. There are also various local maxima, the position and magnitude of which depend on the dynamic bridge/vehicle system. Dahlberg [43] and Melcer [142] have supplemented this explanation with the statement that the highest effects of a multiaxle system is attained at various velocities in various positions of the moving axles. The consequence of this phenomenon is the theoretically stepped-up dependence of δ on c. In field tests of bridges, δ has a considerable variation which conceals the above facts.

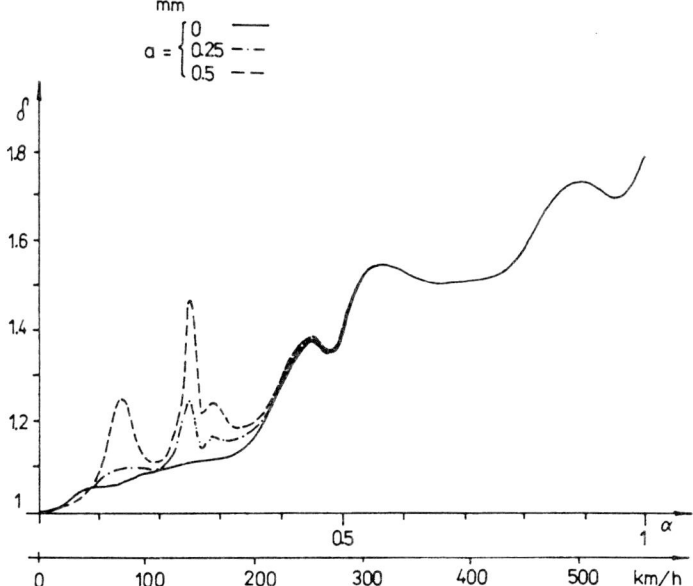

Fig. 6.3. Dynamic coefficient δ plotted against velocity c for a prestressed concrete bridge of span $l = 10$ m, loaded by an electric four-axle locomotive of DB, type E 10, weight 850 kN. The influence of track irregularities of a depth $a = 0$; 0.25 and 0.5 mm, sleeper effect [68].

Great attention was paid to the effect of high speeds (over 200 km h^{-1}) on the dynamic behaviour of railway bridges in experimental bridge tests. Fig. 6.4 compares the stress-time history found from the measurements carried out by the DB on the bridge across the river Paar (steel plate girder bridge of span 19.6 m) during the passage of an electric four-axle locomotive of type E 10, weight 850 kN, at a speed of 200 km h^{-1} with the calculations of the CD according to the theoretical model shown in Fig. 6.2.

Fig. 6.5 presents a comparison of theoretical and measured stresses on the SNCF bridge near Angerville (composite bridge of span 26.4 m) during the passage of an RTG train drawn by an electric six-axle locomotive type CoCo 6 500, weight 1 140 kN, at the velocity of 241 km h⁻¹ [162]. Fig. 6.6 give the calculated and measured (E = 48 000 MPa) bending moment at midspan of a prestresed concrete slab of the DB bridge at Rheda Nonnenstrasse, span 16.5 m, during the passage of a test train at a velocity of c = 251 km h⁻¹, six-axle electric locomotive, type 103 118-6, weight 1 148 kN, and three four-axle measuring cars [162].

The measurements on the bridges of Japanese National Railways also achieved good agreement between calculated and measured stress-time or deflection histories during the passage of trains or car sets at high speeds [141].

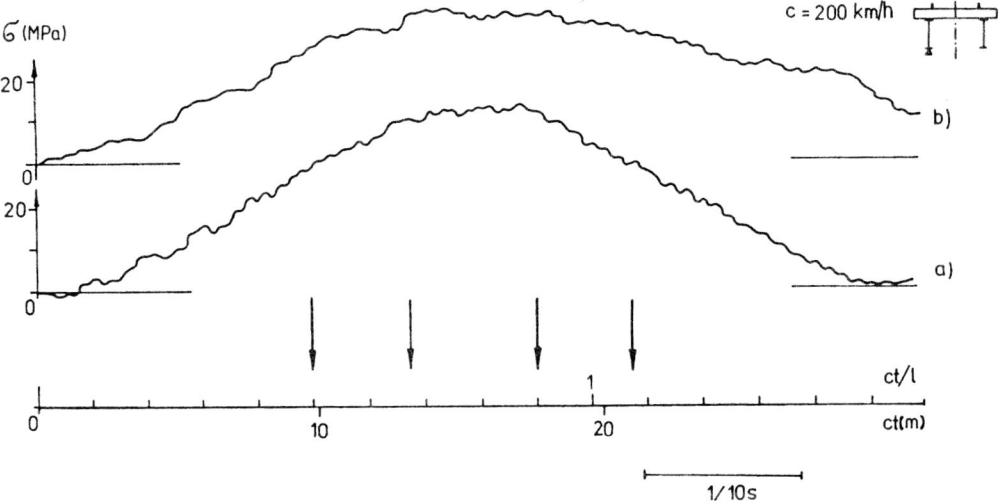

Fig. 6.4 a) Theoretical, b) experimental stress-time history at midspan of a steel plate girder bridge of the DB of span l = 19.6 m during the passage of an electric locomotive, type E 10, at speed c = 200 km⁻¹, [159].

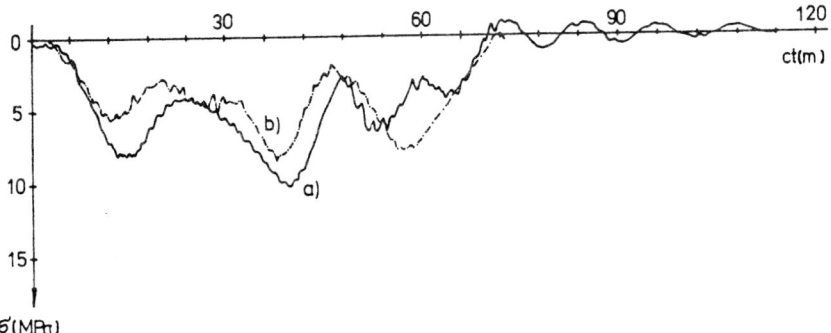

Fig. 6.5. a) Theoretical, b) experimental stress-time history at midspan of a composite (steel-concrete) bridge of the SNCF near Angerville, l = 26.4 m, during the passage of an RTG set at speed c = 241 km h⁻¹, [162].

In general, the mathematical models from Chapters 2 and 3 are adequate for the computations of dynamic stresses and dynamic coefficients even for speeds over 200 km h^{-1}. Dynamic coefficients roughly increase with increasing vehicle speed; however, in measurements they show a relatively large variation. This is contributed to the fact that complex models are described by many parameters, the accuracy of which is difficult to estimate in many cases. Therefore, it is recommended to calculate more cases for a number of parameters.

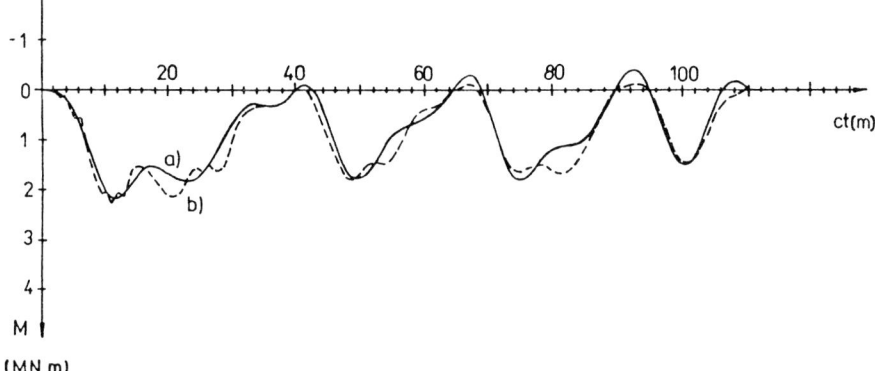

Fig. 6.6. a) Theoretical, b) experimental bending moment at midspan of a prestressed concrete slab of a DB bridge in Rheda Nonnenstrasse, span l = 16.5 m, during the passage of a test set at speed c = 251 km h^{-1}, [162]

6.2 Variable velocity of motion

Vehicles on bridges need not always travel at a constant speed, for example when starting or braking the train. In these cases, the time coordinates of wheel contact points are described by the function of time $u(t)$.

The simplest model, i.e. the simply supported beam during the passage of constant force F at variable speed, was analyzed in [69]. This case is described by the differential equation according to Fig. 6.7 and equation (3.41)

$$EI \frac{\partial^4 v(x, t)}{\partial x^4} + \mu \frac{\partial^2 v(x, t)}{\partial t^2} = f(x, t) = \delta[x - u(t)]F \ . \qquad (6.1)$$

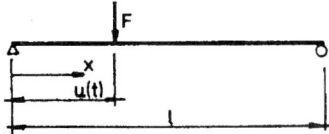

Fig. 6.7. Constant force F moving along a simply supported beam; the motion of the contact point is described by the function of time $u(t)$.

6.2 Variable velocity of motion

The general solution of equation (6.1) is

$$v(x, t) = \sum_{j=1}^{\infty} \left\{ \frac{v_j(x)}{M_j \omega_j} \int_0^{t} \left[\int_0^{l} f(x, \tau) v_j(x) \, dx \right] \sin \omega_j(t - \tau) \, d\tau \right.$$

$$\left. + \frac{\mu}{M_j} G_1(j) v_j(x) \cos \omega_j t + \frac{\mu}{M_j \omega_j} G_2(j) v_j(x) \cos \omega_j t \right\} \quad (6.2)$$

where (other symbols being identical with those of equation (2.1)):
- $v_j(x)$ — jth natural vibration mode of a beam of length l,
- M_j — generalized mass,
- ω_j — natural circular frequency of the beam,
- $g_1(x) = v(x, 0)$ — initial deflection of the beam at $t = 0$,
- $g_2(x) = \dot{v}(x, 0)$ — initial velocity of the beam at $t = 0$,

$$G_1(j) = \int_0^{l} g_1(x) v_j(x) \, dx,$$

$$G_2(j) = \int_0^{l} g_2(x) v_j(x) \, dx.$$

If we substitute the right-hand side of equation (6.1) into equation (6.2) and use (3.39), the first component of the right-hand side of equation (6.2) acquires the form of

$$v(x, t) = \sum_{j=1}^{\infty} \frac{F v_j(x)}{M_j \omega_j} \int_0^{t} v_j[u(\tau)] \sin \omega_j(t - \tau) \, d\tau. \quad (6.3)$$

Equation (6.3), together with the remainder of equation (6.2), expressing the influence of initial conditions, describes the deflection of the beam during the motion of a force F characterized by the function $u(t)$. Equation (6.3) can be integrated numerically in the case of the general motion of the force.

We shall consider as a *particular case* the motion of a force where the contact point follows a quadratic function of time t

$$u(t) = x_0 + ct + \frac{1}{2} at^2,$$

$$\dot{u}(t) = c + at, \quad \ddot{u}(t) = a \quad (6.4)$$

where x_0 — coordinate of the point of application of force F at time $t = 0$,
- c — initial velocity,
- a — constant acceleration of motion.

Equation (6.4) expresses the motion which is either constant acceleration ($a > 0$) or constant deceleration ($a < 0$).

As it has been shown in [69], equation (6.3) together with equation (6.4) may be solved in a closed form for the simply supported beam, for which

$$v_j(x) = \sin j\pi x/l, \quad M_j = \mu l/2 \quad (6.5)$$

where the following notation has been used:
$$\xi_0 = j\pi x_0/l,$$
$$\omega = j\pi c/l, \qquad \Omega^2 = j\pi|a|/(2l),$$
$$\xi_1 = \xi_0 - \omega_j t, \qquad \xi_2 = \xi_0 + \omega_j t,$$
$$r_1 = \frac{1}{2}(\omega + \omega_j), \qquad r_2 = \frac{1}{2}(\omega - \omega_j),$$
$$b_1 = r_1/\Omega, \qquad b_2 = r_2/\Omega. \tag{6.6}$$

In this particular case, the solution of equation (6.3) yelds, according to [69]

$$v(x,t) = \sum_{j=1}^{\infty} \frac{F}{\mu l \omega_j \Omega} \left(\frac{\pi}{2}\right)^{1/2}$$
$$\cdot \sin j\pi x/l \, \{\cos(\pm\xi_1 - b_1^2)[C(\Omega t \pm b_1) - C(\pm b_1)]$$
$$- \sin(\pm\xi_1 - b_1^2)[S(\Omega t \pm b_1) - S(\pm b_1)]$$
$$- \cos(\pm\xi_2 - b_2^2)[C(\Omega t \pm b_2) - C(\pm b_2)]$$
$$+ \sin(\pm\xi_2 - b_2^2)[S(\Omega t \pm b_2) - S(\pm b_2)]\} \tag{6.7}$$

where the upper sign applies to accelerated ($a > 0$), the lower sign to decelerated ($a < 0$) motion, respectively. Equation (6.7) contains Fresnel integrals

$$S(x) = \left(\frac{2}{\pi}\right)^{1/2} \int_0^x \sin t^2 \, dt,$$
$$C(x) = \left(\frac{2}{\pi}\right)^{1/2} \int_0^x \cos t^2 \, dt. \tag{6.8}$$

The dimensionless velocity parameter (5.4) was introduced in numerical calculations, i.e.

$$\alpha = \frac{c}{2 f_1 l} \tag{6.9}$$

and the dimensionless parameter of acceleration or deceleration

$$B = \frac{al}{c^2}. \tag{6.10}$$

Equation (6.3) together with (6.2) were integrated numerically for various values of parameters α and B for three cases:

1. Force F arrives at the left-hand beam end at an initial velocity c and then its motion is regularly accelerated or retarded. A sample calculation is shown in Fig. 6.8.

2. Force F starts to move on the left-hand side of the beam at an initial velocity equal zero, see Fig. 6.9.

3. Force *F* starts at midspan with zero initial velocity, Fig. 6.10 (v_0 being determined by equation (5.3)).

The calculations of 75 different combinations of parameters α and *B* in the above three cases have resulted in the following conclusions [69]:

In many cases the beam vibrates more intensively during retarded motion of the force than during accelerated motion. If the force starts to move from zero initial speed, the beam hardly vibrates at all. The variable speed of motion will be also encountered in Chapter 8.

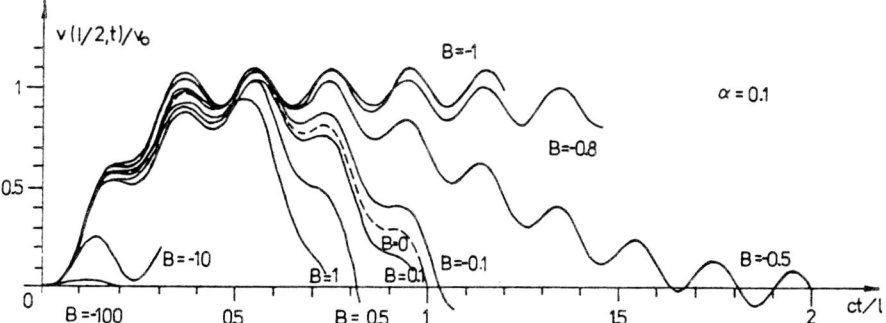

Fig. 6.8. Time-history of deflection at midspan of the beam, $v(l/2, t)/v_0$, loaded by force *F* at variable speed for various *B*, $\alpha = 0.1$.

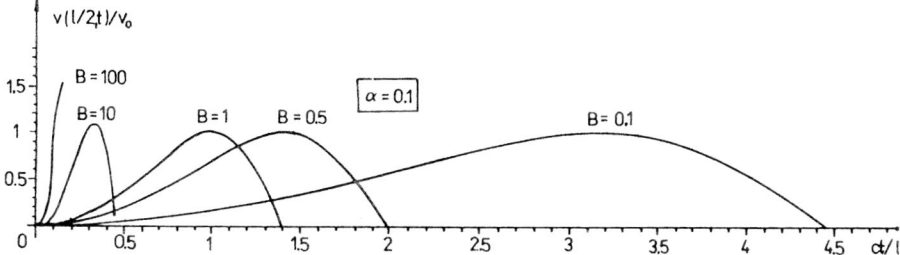

Fig. 6.9. Time-history of deflection at midspan of the beam, $v(l/2, t)/v_0$, when force *F* starts on the left-hand side of the beam at zero initial velocity for various accelerations *B*, $\alpha = 0.1$.

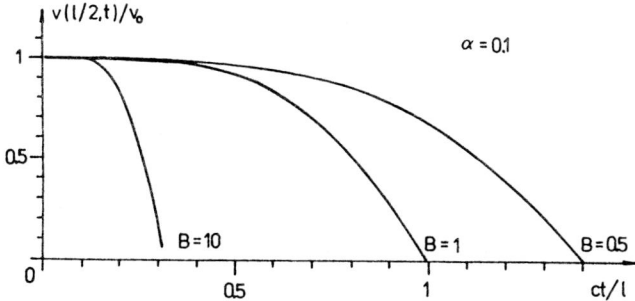

Fig. 6.10. Time-history of deflection at midspan of the beam, $v(l/2, t)/v_0$, when force *F* starts to move at midspan at zero initial velocity for various accelerations *B*, $\alpha = 0.1$.

7. Influence of track irregularities and other parameters

Track irregularities represent an important source of excitation of bridges during the passage of vehicles. The irregularities consist of deviations of the inside edge of the rail from the ideal geometric rail contour and may occur both in an unloaded position and in a loaded position (i.e. deviations from the geometric position even during the passage of the vehicle at a very low speed). The differences between irregularities in unloaded and loaded positions may be considerable at times; they depend chiefly on the clearances between the individual elements of the permanent way and the bridge and their elastic or non-elastic properties.

According to Fig. 7.1 four types of track irregularities can be distinguished:
1. elevation irregularity (vertical profile)

$$\frac{1}{2}(y_1 + y_2) \tag{7.1}$$

which is the mean elevation of two rails,
2. alignment irregularity

$$\frac{1}{2}(z_1 + z_2) \tag{7.2}$$

which is the mean lateral position of two rails,
3. superelevation irregularity (cross level)

$$\frac{1}{2}(y_1 - y_2) \tag{7.3}$$

which is the difference in elevations of two rails,
4. gauge irregularity

$$\frac{1}{2}(z_1 - z_2) \tag{7.4}$$

which is the horizontal distance between the inside edges of two rails measured perpendicularly to them 14 mm below the top of rails.

In equations (7.1) to (7.4) and in Fig. 7.1 the letters y_i, z_i denote the co-ordinates of the left-hand (index 1) and right-hand (index 2) rail at the place x, respectively. Other definitions of irregularities are sometimes used as well.

The elevation and superelevation irregularities influence chiefly the vertical vibrations of vehicles and of the bridge, while the alignment, gauge, and superelevation irregularities initiate horizontal transverse vibrations of vehicles and bridges and the torsion of bridges.

All railway administrations limit these deviations from the ideal state by tolerances, stipulated for the straight track and for curves, and which may also depend on speed.

The irregularity distribution along the length x (Fig. 7.1) may be periodic or entirely irregular (stochastic).

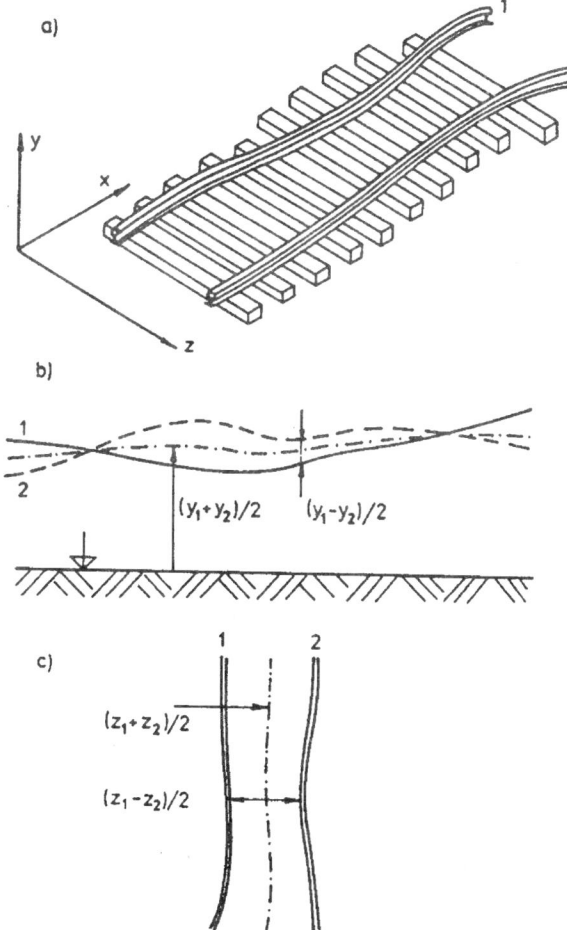

Fig. 7.1. Track irregularities: a) scheme of track with coordinate axes, b) elevation and superelevation irregularities, c) alignment and gauge irregularities,
(*1*) – left-hand rail, (*2*) – right-hand rail.

7.1 Periodic irregularities

Periodic irregularities of the track are analytically described by trigonometric Fourier series

$$r(x) = \frac{1}{2} a_0 + \sum_{n=1}^{\infty} (a_n \cos nx + b_n \sin nx), \tag{7.5}$$

with coefficients

$$a_n = \frac{1}{\pi} \int_{-\pi}^{\pi} r(x) \cos nx \, dx, \qquad n = 0, 1, 2, \ldots,$$

and

$$b_n = \frac{1}{\pi} \int_{-\pi}^{\pi} r(x) \sin nx \, dx, \qquad n = 1, 2, 3, \ldots, \qquad (7.6)$$

obtained from measurements.

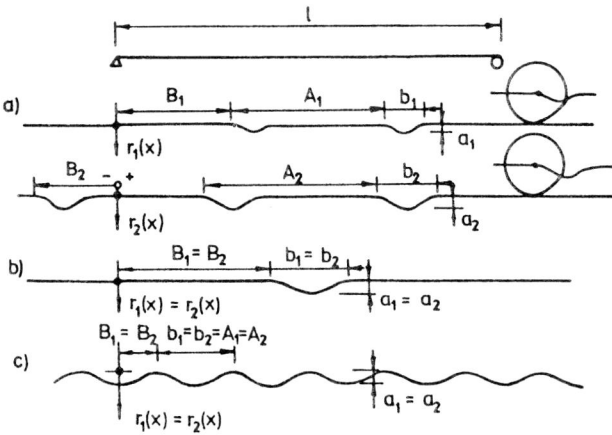

Fig. 7.2. Periodic track irregularities: a) rail joints or flat wheel effects, b) isolated irregularities, c) undulated rail surface (according to [68]).

Fig. 7.2 shows some typical elevation irregularities of the track as given in [68]. For many cases, they can be expressed by:

$$r_i(x) = \begin{cases} \dfrac{1}{2} a_i \left[1 - \cos \dfrac{2\pi}{b_i}(x - kA_i - B_i) \right] \\ 0 \end{cases} \qquad (7.7)$$

$$\text{for} \quad \begin{cases} B_i + kA_i \leq x \leq B_i + kA_i + b_i, \\ B_i + kA_i + b_i < x < B_i + (k+1)A_i \end{cases}$$

where $i = 1, 2$ means the left-hand and the right-hand rail or the first and the second axle, respectively,

$k = 0, 1, 2, \ldots$.

Equation (7.7) can characterize several types of periodic elevation irregularities (Fig. 7.2):

a) The influence of wheel flats of railway vehicles can be considered as a change of distance between the wheel centroid and the neutral axis of the bridge. Then in equation (7.7):
  a_i – depth of the flat spot,
  b_i – length of the flat spot,
  A_i – wheel circumference,
  B_i – the distance of the first impact of the wheel flat from the origin $x = 0$.

If the wheel flat is only on one (first or second) axle, we substitute $a_2 = 0$ or $a_1 = 0$, respectively, in equation (7.7).

The effect of rail joint can be similarly expressed.

b) An isolated irregularity on rail surface can be characterized by equation (7.7) with the following parameters:

$$a_1 = a_2, \quad b_1 = b_2, \quad A_i > l + d, \quad B_1 = B_2$$

where l is the bridge length and d is the vehicle axle base.

c) A corrugated rail surface can be expressed by equation (7.7) with the following parameters:

$$a_1 = a_2, \quad b_1 = b_2, \quad A_i = b_i, \quad B_1 = B_2.$$

A combination of constants a_i, b_i, A_i, B_i, can be used to characterize also other types of periodic irregularities of the track. The ordinates of irregularity $r(x)$ are then substituted in the contact equations of the type (3.32) in Sects 3.4.1 and 3.4.2.

7.1.1 Impact of wheel flats

Periodic track irregularities can generate resonance vibration of bridges; however, this is a rare phenomenon with contemporary mechanized maintenance. More frequent are isolated irregularities represented in practice by wheel flats or rail joints. Railjoints, according to the regulations of most railway administrations, should not occur on bridges because of their large dynamic effects.

Howewer, wheel flats load bridges by their impacts. Their influence on rails and bridges has been given considerable experimental and theoretical attention [63], [66]. Fig. 7.3 analyzes the time history of force $R(\tau)$ originating between the wheel with a flat spot and the bridge, $\tau = ct/l$ being dimensionless time.

The time history of the force for several dimensionless velocities α (5.4) shows three significant phases:

1. For very low $\alpha < 0.0075$, the wheel with a flat spot is in constant contact with the rail and the dynamic effects increase with increasing α.

2. For the velocity about $\alpha = 0.0125$ (in practice about 30 km h^{-1}), the wheel looses contact with the rail and the recovery of contact generates an impact. The highest dynamic effects occur in this case.

3. For higher velocities, $\alpha > 0.02$, an impact is also generated, but the dynamic effects on bridges decrease.

Similar influence on dynamic stresses of railway bridges is also exerted by rail joints.

For the above reason speed reduction measures which intend to reduce dynamic stresses appear highly problematic. The main prerequisite for their reduction must be the elimination of all irregularities on the bridge and on the track in front of it.

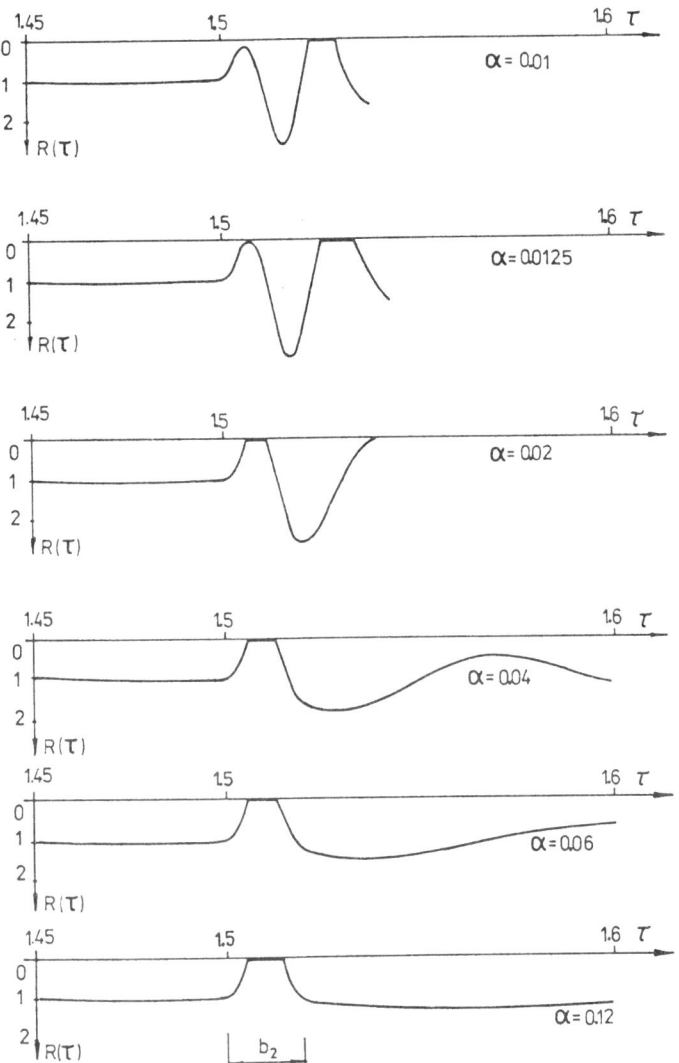

Fig. 7.3. Time-history of dimensionless force $R(\tau)$ between a flat wheel and bridge for various velocities α (according to [68]).

7.1.2 Cross-beam and sleeper effects

The cross-beam and the sleeper effects generate periodic irregularities only in the loaded state. They can generate resonance vibrations in bridges, but this

7.1 Periodic irregularities

TABLE 7.1. Characteristics of geometric position of rails of FRA, USA (according to [81])

Irregularity	Eq.	Parameter		Equation parameters on track of class (1)					
		Notation	Unit	1	2	3	4	5	6
a) Random irregularities (2)									
Elevation	(7.25)	A	10^8 m^3	15.53	8.85	4.92	2.75	1.57	0.98
		Ω_1	10^3 m^{-1}	23.3	23.3	23.3	23.3	23.3	23.3
		Ω_2	10^2 m^{-1}	13.1	13.1	13.1	13.1	13.1	13.1
Alignment	(7.25)	A	10^8 m^3	9.83	5.51	3.15	1.77	0.98	0.59
		Ω_1	10^3 m^{-1}	32.8	32.8	32.8	32.8	32.8	32.8
		Ω_2	10^2 m^{-1}	18.4	18.4	18.4	18.4	18.4	18.4
Superelevation	(7.26)	A	10^8 m^3	4.52	3.15	2.16	1.38	0.98	0.59
		Ω_1	10^3 m^{-1}	23.3	23.3	23.3	23.3	23.3	23.3
		Ω_2	10^2 m^{-1}	13.1	13.1	13.1	13.1	13.1	13.1
Gauge	(7.26)	A	10^8 m^3	9.83	5.51	3.15	1.77	0.98	0.59
		Ω_1	10^3 m^{-1}	29.2	29.2	29.2	29.2	29.2	29.2
		Ω_2	10^2 m^{-1}	23.3	23.3	23.3	23.3	23.3	23.3
b) Isolated irregularities									
Elevation	(7.8)	A	mm	11.4	8.4	6.4	4.8	3.6	2.8
		k	m^{-1}	0.43	0.43	0.46	0.49	0.66	0.82
Alignment	(7.8)	A	mm	8.9	6.9	5.1	3.8	2.8	2.0
		k	m^{-1}	0.39	0.49	0.66	1.1	1.5	1.9

Notes: (1) Track quality rises from Class 1 to Class 6.
(2) The constants of power spectral densities refer to 1 cycle per 1 m.

phenomenon is not frequent because of the different wheel bases of railway vehicles.

The cross-beam and the sleeper effects were described in some detail in Sects 2.5.1 and 2.5.2 and illustrated in Figs 2.10 to 2.14.

7.1.3 Isolated irregularities

Apart from the aforementioned irregularities, there are also other forms of irregularities occurring in rail joints, in waterlogged railway embankments, in rail switches, on bridges, on bridge piers and abutments, on bridge approaches, and so on.

Seven forms of isolated irregularities are described in [81]. They are expressed analytically by the following equations:

$$r(x) = A\,e^{-k|x|}, \tag{7.8}$$

$$r(x) = A\,e^{(-1/2)(kx)^2}, \tag{7.9}$$

$$r(x) = \frac{Akx}{(1 + 4k^2x^2)^{1/2}}, \tag{7.10}$$

$$r(x) = \left(\frac{A^2}{1 + (kx)^8}\right)^{1/2}, \tag{7.11}$$

$$r(x) = Ak\left[(1/k)^2 - x^2\right]^{1/2}, \tag{7.12}$$

$$r(x) = A\,\sin\pi kx, \tag{7.13}$$

$$r(x) = A\,e^{-k|x|}\cos\pi kx. \tag{7.14}$$

The parameters A and k of various irregularity types can be found in [81]. For the most frequent irregularity (7.8), characterizing the irregularity in a rail joint (so-called cusp), the parameters A and k are given in Table 7.1 based on the results of American measurements on tracks of different categories.

7.2 Random irregularities

If the track irregularities are entierly irregular, we consider them as random (stochastic) irregularities which are described by statistical characteristics (Sect. 1.4).

These irregularities are due to wear, clearances, subsidence, insufficient maintenance of the permanent way and the bridge and so on and appear in both unloaded and loaded states.

The irregularities are considered as stationary and ergodic processes in space, i.e. as random functions in the longitudinal coordinate x, and they are charac-

7.2 Random irregularities

terized most frequently by the one-sided power spectral density $G_{rr}(\Omega)$, equation (1.30). The power spectral density depends on the route frequency

$$\Omega = 2\pi/L \tag{7.15}$$

or

$$F = 1/L = \Omega/(2\pi), \tag{7.16}$$

while the length of the irregularity is

$$L = 2\pi/\Omega. \tag{7.17}$$

The circular frequency for the velocity of vehicle motion c is

$$\omega = c\Omega = 2\pi c/L, \tag{7.18}$$

so that the conversion of the power spectral density in route frequency Ω to the power spectral density in circular frequency ω is

$$G_{rr}(\Omega) = c\, G_{rr}(\omega). \tag{7.19}$$

The power spectral density of irregularities is an important quantity in the calculation of vibrations of vehicles and of bridges, because it contains information on track quality. Hence the idealization by random load described in Sect. 3.1.4 and illustrated in Figs 3.5 and 3.6. If the curve characterizing $G_{rr}(\Omega)$ contains sharp peaks, it means that periodic irregularities occur. The functions $G_{rr}(\Omega)$ obtained by measurements are often smoothed by various statistical methods and, as a result, they yield only averaged information on track quality.

The analytical expression of the power spectral density of track irregularities is given in various empirical forms, see Fig. 7.4. A comparison of individual results is difficult, because – apart from the natural differences in the track quality of individual railways – the measuring instruments and evaluation methods are often not described in the literature in sufficient detail.

In comparative computations the formula

$$G_{rr}(\Omega) = A\Omega^{-a}, \quad (\text{m}^3) \tag{7.20}$$

is often used, which is adequate for wave lengths $4 \leq L \leq 40$ m. The constant A is influenced by track quality, while the second constant a usually lies within the limits of $2 \leq a \leq 4$. Both extreme cases for $a = 2$, $a = 4$ and $A = 1 \times 10^{-6}$ are represented by curves 1 and 2 in Fig. 7.4.

According to the Technical University of Transport and Telecommunications, Žilina, the formula

$$G_{rr}(\Omega) = \frac{2.72 \times 10^{-4}}{(1 + 1.185\Omega^2)(1 + 14.388\Omega^2)}, \quad (\text{m}^3), \tag{7.21}$$

was satisfactory for Czechoslovak State Railways (CSD) over a wide frequency range. The formula is represented by curve 3 in Fig. 7.4.

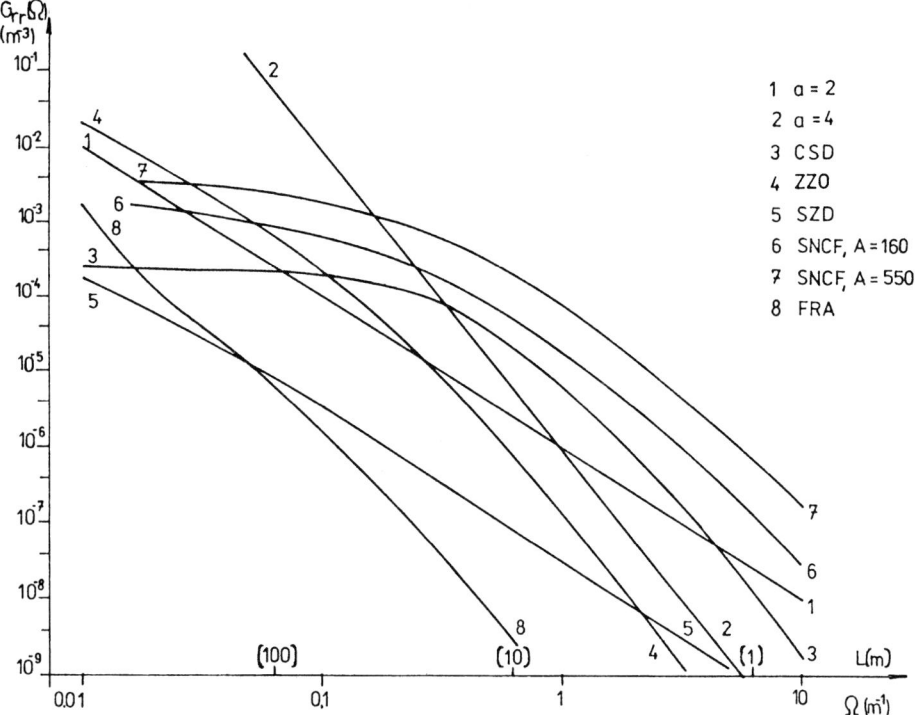

Fig. 7.4. Power spectral densities of elevation track irregularities according to various empirical formulae:
$1 - a = 2, A = 1 \times 10^{-6}$, (7.20), $2 - a = 4, A = 1 \times 10^{-6}$, (7.20), 3 – CSD, (7.21), 4 – ZZO, (7.22), 5 – SZD, (7.23), 6 – SNCF, $A = 160$, (7.24), 7 – SNCF, $A = 550$, (7.24), 8 – FRA, class 4, (7.25).

The measurements carried out by the CKD and by the Railway Research Institute on the Railway Test Circuit in the Czech Republic yielded the following empirical relation:

$$G_{rr}(\Omega) = \frac{0.135}{(1 + 22\Omega^2)(1 + 44\,000\Omega^2)}, \quad (\text{m}^3), \tag{7.22}$$

which is represented by curve 4 in Fig. 7.4.

In the CIS (SZD), the formula

$$G_{rr}(\Omega) = \frac{10^{-6} a}{\pi(\Omega^2 + a^2)}, \quad (\text{m}^3), \tag{7.23}$$

is used, which is adequate for wavelengths of $8 \leqslant L \leqslant 250$ m, while $a = 9.54 \times 10^{-3}\,\text{m}^{-1}$, see curve 5 in Fig. 7.4.

French National Railways (SNCF) have found the relation

$$G_{rr}(\Omega) = \frac{10^{-6} A}{(1 + \Omega/\Omega_0)^3}, \quad (\text{m}^3), \tag{7.24}$$

satisfactory, which is adequate for wave lengths of $2 \leq L \leq 40$ m and the constant $\Omega_0 = 0.307$ m^{-1}. Track quality is expressed by the parameter A; for good track $A = 160$, whilst for poor track $A = 550$; curves 6 and 7 in Fig. 7.4.

On the basis of extensive measurements on railways of the USA (FRA) [81] the empirical formula

$$G_{rr}(\Omega) = \frac{A\Omega_2^2(\Omega^2 + \Omega_1^2)}{\Omega^4(\Omega^2 + \Omega_2^2)}, \quad (\text{m}^3), \tag{7.25}$$

was devised for elevation and alignment irregularities, while for superelevation and gauge irregularities the formula

$$G_{rr}(\Omega) = \frac{A\Omega_2^2}{(\Omega^2 + \Omega_1^2)(\Omega^2 + \Omega_2^2)}, \quad (\text{m}^3), \tag{7.26}$$

has been suggested.

The respective constants in equations (7.25) and (7.26) after conversion to SI units are given in Table 7.1 for various types of irregularities and for the various quality categories into which the FRA railway tracks are classified.

The power spectral density of elevation irregularities (7.25) for FRA class 4 tracks is shown in Fig. 7.4 as curve 8.

Mutual relationships between the individual types of irregularities are described by cross power spectral densities or coherence functions. Most important of them is the mutual relation of alignment irregularities of a single rail and the gauge deviations.

7.3 Further parameters

In addition to the velocity of vehicle motion and track irregularities, the dynamic stresses in railway bridges are influenced by a number of further parameters: the frequency characteristics of the bridge and of vehicles, the damping of vehicles and of the bridge, the mass of the bridge, the mass of the sprung and unsprung vehicle parts, the elastic characteristics of bridge deck, initial conditions of vehicles at the moment of entry on the bridge and further parameters characterizing the complex dynamic bridge/vehicle system.

The study of the influence of the individual parameters has been given great attention, see Chapters 8, 9 and 10 in [68]. The influence of the variations in the magnitude of one or several parameters is difficult to observe experimentally. Therefore, dynamic bridge tests determine only the influence of vehicle speed, while the other parameters are studied theoretically [68] or on laboratory models [159].

In many cases, it is also difficult to ascertain the accurate magnitude of the parameters to be used in the calculations. For this reason, the calculations use a number of input data within the limits of engineering estimates, and the bridge response is then determined within certain limits.

8. Horizontal longitudinal effects on bridges

The movement of railway vehicles along bridges gives rise to horizontal longitudinal forces transferred through friction via the rails and further parts of the permanent way to the bridge superstructure, bearings, piers and abutments.

When the horizontal longitudinal forces are transmitted from the wheels to the bridge by rolling friction at constant vehicle speed, they are relatively small. However, they acquire greater values during the movement at nonuniform speed, which takes place during starting and braking. In this case, significant forces of adhesion act between the vehicle wheels and the rail which are necessary for the starting and braking of vehicles.

8.1 Motion of a disc rolling along a beam taking into account adhesion

Fig. 8.1 shows the simplest model of this problem, i.e. the motion of a disc of mass m, weight $F = mg$, and mass moment of inertia I_0 along a simply supported beam of length l. Apart from its weight F, the disc is also affected by the forces of inertia with the vertical component $m\ddot{v}_0(t)$, horizontal component $m\ddot{u}_0(t)$, inertial moment $I_0\ddot{\varphi}(t)$, the vertical beam reaction $F(t)$ which acts on the arm ρ from the disc centroid (describing the influence of rolling friction), the horizontal force $H(t)$ on the wheel circumference, the traction force $T(t)$, the driving moment $M(t)$ and the resistance to motion $W(t)$.

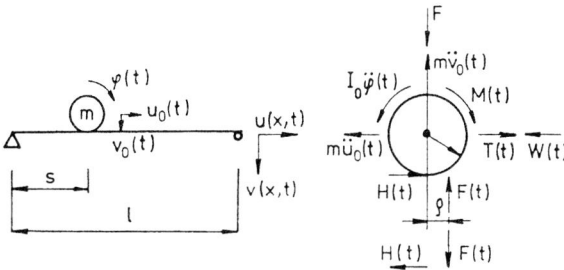

Fig. 8.1. Motion of a rolling disc along a beam: acting forces and moments.

The direction of movement of the disc rolling in Fig. 8.1 is from left to right and the direction of the acting forces and moments is as for accelerated motion.

The beam is affected by the forces $F(t)$ and $H(t)$ in the opposite direction from that of the disc and by other forces assumed in a Bernoulli–Euler beam model with viscous damping. Its deformation is marked $v(x, t)$ in vertical and $u(x, t)$ in horizontal directions.

8.1 Motion of a disc rolling along a beam taking into account adhesion

The motion of the disc is described by the displacement $v_0(t)$ in the vertical and by the displacement $u_0(t)$ in the horizontal directions and $\varphi(t)$ is its rotation. The zero position of the beam is its non-deformed state; for the disc it is when positioned above the left-hand support of the beam.

With these assumptions the following equations hold true for the vibration of the beam and for the disc motion:

$$EI \frac{\partial^4 v(x,t)}{\partial x^4} + \mu \frac{\partial^2 v(x,t)}{\partial t^2} + 2\mu\omega_b \frac{\partial v(x,t)}{\partial t} = \varepsilon \delta(x-s) F(t), \quad (8.1)$$

$$-EA \frac{\partial^2 u(x,t)}{\partial x^2} + \mu \frac{\partial^2 u(x,t)}{\partial t^2} + 2\mu\omega_{bu} \frac{\partial u(x,t)}{\partial t} = -\varepsilon \delta(x-s) H(t), \quad (8.2)$$

$$F - m \frac{d^2 v_0(t)}{dt^2} - F(t) = 0, \quad (8.3)$$

$$-m \frac{d^2 u_0(t)}{dt^2} + T(t) - W(t) + H(t) = 0, \quad (8.4)$$

$$-I_0 \frac{d^2 \varphi(t)}{dt^2} + M(t) - \rho F(t) - r H(t) = 0. \quad (8.5)$$

In addition to the symbols explained earlier and the notation of equation (2.1) the following notation has been used:

$$\varepsilon = \begin{cases} 1 \\ 0 \end{cases} \text{ for } \begin{cases} 0 \leqslant s \leqslant l \\ s < 0,\, s > l \end{cases}$$

where $s = u_0(t)$ – distance of the contact point of the disc with the beam from the left-hand support,

    A – constant area of the beam cross section,

    ω_{bu} – constant of viscous damping during longitudinal vibrations of the beam,

    r – radius of the disc.

Equations (8.1) and (8.2) describe the bending and longitudinal vibrations of the beam which are, after certain simplifications, mutually independent, as has been proved in [70]. The equilibrium of vertical forces affecting the disc is expressed by equation (8.3), the equilibrium of horizontal forces by equation (8.4) and the equilibrium of moments by equation (8.5).

Moreover, it follows from the rolling motion of the disc (without slipping) along the beam that it must hold true that

$$v_0(t) = \varepsilon v(s,t), \quad (8.6)$$

and

$$u_0(t) = r \varphi(t). \quad (8.7)$$

The following condition of disc and beam contact applies for the force $F(t)$

$$F(t) \geqslant 0. \quad (8.8)$$

Should $F(t) < 0$, $F(t) = 0$ must be substituted in equations (8.1) to (8.5) and equation (8.6) ceases to be valid.

Also the force $H(t)$ on the circumference of the wheel is determined by certain conditions concerning the other forces and moments affecting it. Firstly, substitute the relation (8.7) in equation (8.5) and let us find $u_0(t)$ and substitute this value in equation (8.4). We obtain the condition with which the force $H(t)$ must comply, while the disc rotates:

$$H(t) = \frac{m'}{m + m'} \left\{ \frac{m}{m'} \frac{1}{r} [M(t) - \rho F(t)] - T(t) + W(t) \right\} \qquad (8.9)$$

where $m' = I_0/r^2$ is the reduced mass of the disc.

The absolute value of the force $H(t)$, naturally, cannot exceed – according to Coulomb friction law – the value of the product of vertical force $F(t)$ and the coefficient of friction $\mu(t)$, so

$$|R(t)| < \mu(t) \, F(t), \qquad (8.10)$$

where the coefficient of friction is generally a function $f[\ ]$ of the velocity motion

$$\mu(t) = f[\dot{u}_0(t)]. \qquad (8.11)$$

With the knowledge of external forces $T(t)$, $W(t)$ and of the moment $M(t)$ we can find from equation (8.9) the force $H(t)$ on the wheel circumference (except for the influence of rolling friction), as long as it complies with the condition (8.10). If this condition has not been complied with, the disc stops rotating, it begins sliding and the force on its circumference is

$$H(t) = \left[\operatorname{sign} \ddot{u}_0(t) \right] \mu(t) \, F(t). \qquad (8.12)$$

The direction of the force $H(t)$ depends on the sign of the acceleration of the disc m.

It is customary in technical practice for all driving and resistance forces to be referred to the circumference of the railway vehicle wheel, and to neglect the rolling friction ($\rho = 0$) as being very small in comparison with other forces. In this interpretation the force $H(t)$ becomes the given external force. However, the conditions (8.10) and (8.12) remain valid.

In the numerical calculations in [71] the forces $T(t)$ and $W(t)$ in the equation (8.4) were also neglected, which are of smaller importance during the nonuniform motion than the driving or braking force and moment. This means that out of all external forces shown in Fig. 8.1 only the moment $M(t)$ and the force $H(t)$ on the wheel circumference are taken into account.

8.1.1 Solution

The system of equations (8.1) to (8.5) was solved for boundary conditions

$$v(0, t) = v(l, t) = v''(0, t) = v''(l, t) = 0,$$
$$u(0, t) = u'(l, t) = 0 \qquad (8.13)$$

and for the following initial conditions:

$$v(x, 0) = \dot{v}(x, 0) = 0,$$
$$u(x, 0) = \dot{u}(x, 0) = 0,$$
$$v_0(0) = \dot{v}_0(0) = 0,$$
$$u_0(0) = 0, \quad \dot{u}_0(0) = c, \quad (8.14)$$

where c is the initial horizontal velocity of the disc.

A numerical solution was carried out by the Euler and Runge–Kutta method in the dimensionless form of

$$v(x, t)/v_0, \quad u(x, t)/u_0$$

and

$$\tau = \omega_1 t$$

where

$$v_0 = \frac{Fl^3}{48EI}, \quad (8.15)$$

$$u_0 = \frac{Fl}{EA}, \quad (8.16)$$

are the static deformations of the beam due to the force F, and ω_1 is the first circular frequency of the bending vibrations of the beam.

The external force on wheel circumference was assumed to be of the form

$$H(t) = Fh(\tau) \quad (8.17)$$

where

$$h(\tau) = \begin{cases} r_0(r_1 + r_2\tau - r_3 \, e^{-r_4\tau} \cos r_5 \tau) \\ 0 \end{cases} \quad \text{for } \dot{u}_0(\tau) \begin{cases} > 0 \\ \leq 0 \end{cases}. \quad (8.18)$$

The dimensionless force (8.18) permits, by adequate choice of coefficients r_i ($i = 0, 1, 2, 3, 4, 5$), the idealization of the force on wheel circumference and characterizes approximately the braking, $h(\tau) < 0$, or the starting, $h(\tau) > 0$, forces on the wheel circumference, respectively.

The coefficient of adhesion (8.11) was assumed, according to [71], to be of the following empirical form:

$$\mu(\tau) = \begin{cases} \mu_0 \exp\left[-\left(\frac{\pi \dot{u}_0(\tau)}{\alpha_0}\right)^{1/2}\right] \\ 0 \end{cases} \quad \text{for } \dot{u}_0(\tau) \begin{cases} \geq 0 \\ < 0 \end{cases}, \quad (8.19)$$

where the constants μ_0 and α_0 were obtained by experiments [160].

The numerical solution is described in detail in [71] and shown in Figs 8.2 to 8.8, which give the time-histories of some quantities during the braking of a disc on a beam. The input quantities for the calculations represented in Figs 8.2 to 8.8 correspond approximately to the parameters of a steel plate girder railway bridge of span 30 m with a continuous gravel bed (see Table 8.1).

TABLE 8.1. Input data of the basic case No. 5

α	$\varkappa$	γ	ϑ	ϑ_u	α_0	μ_0	B
0.02	2	5	0.15	0.3	0.03	0.3	2×10^{-4}
r_0	r_1	r_2	r_3	r_4	r_5	h	j_{max}
−0.3	1	0	0	0	0	0.1	3

Note: h is the integration step length,
j_{max} is the number of terms in series development.

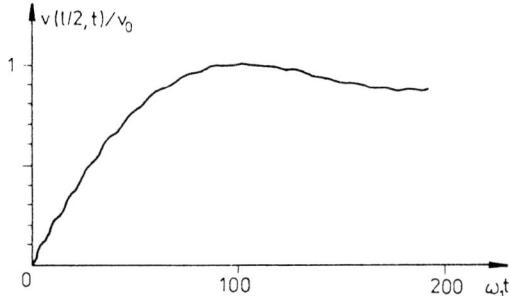
Fig. 8.2. Time-history of vertical beam deflection at midspan.

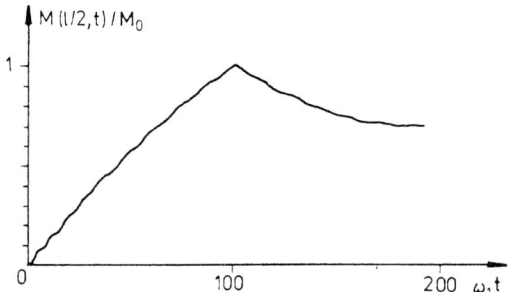
Fig. 8.3. Time-history of beam bending moment at midspan.

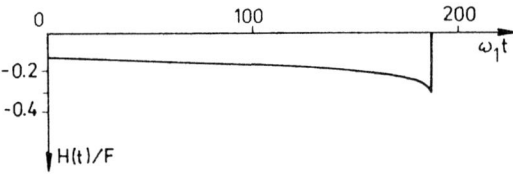
Fig. 8.4. Time-history of horizontal force on disc circumference.

According to Fig. 8.7 the braked disc stops just after the beam midspan, the point $s/l = 0.655$. The dynamic effect of the rolling disc disappears shortly after the arrival at the beginning of the beam and when the disc passes above the beam midspan its velocity is low. For these reasons the deflection $v(x, t)$ and the bending moment $M(x, t)$ of the beam have a quasistatic rather than dynamic character (see Figs 8.2, 8.3, 8.5). This once again confirms the conclusions from Sect. 6.2 about the influence of the force moving along the beam with variable speed.

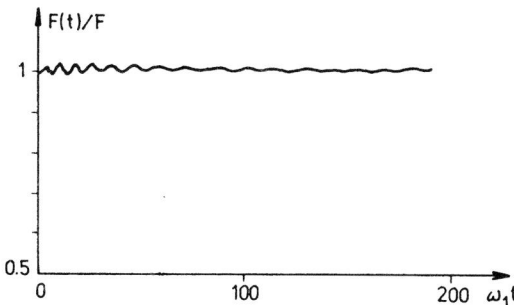

Fig. 8.5. Time-history of vertical force between disc and beam.

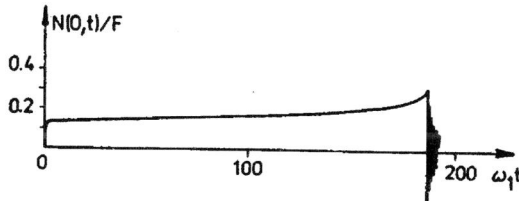

Fig. 8.6. Time-history of horizontal longitudinal force in fixed beam support.

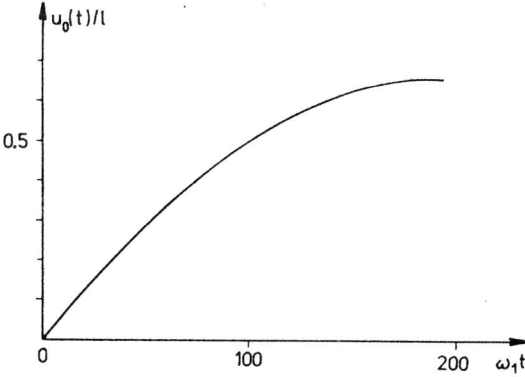

Fig. 8.7. Time-history of disc motion along a beam.

The horizontal longitudinal force $N(0, t) = EAu'(0, t)$ in the left-hand beam support shows a slight increase with time and attains its maximum shortly before the stopping of the disc (Fig. 8.6). When the disc stops, the studied phenomenon disappears, as is suggested by the damped natural vibrations of the beam in a horizontal direction (Fig. 8.8 or Fig. 8.6).

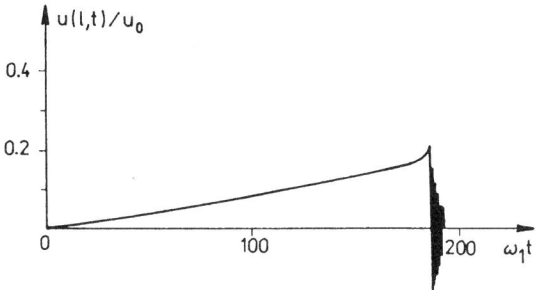

Fig. 8.8. Time-history of motion of movable beam end.

8.1.2 Influence of some parameters

In the study of the influence of some parameters those from Table 8.1 were considered as basic data and only one of them is varied in what follows. In this process the dependence of the maximum horizontal longitudinal force on the fixed beam support $N(0, t)/F$ on one parameter was studied.

The influence of the dimensionless velocity parameter α (5.4) is shown in Fig. 8.9. An increase in α results in a drop in the horizontal force on the fixed beam support. It should be noted, however, that at higher velocities the disc will not stop on the beam and that due to (8.19) the braking force will not attain its highest value while the disc is on the beam.

The dependence of $N(0, t)/F$ on the weight parameter

$$\varkappa = F/G \tag{8.20}$$

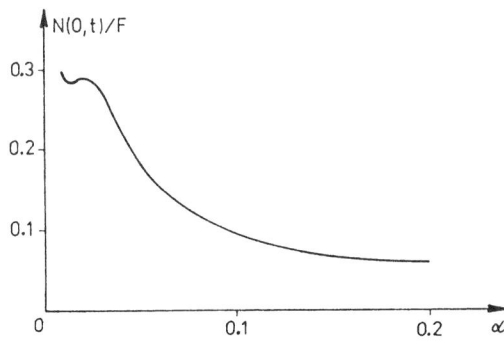

Fig. 8.9. Influence of velocity parameter α (5.4) on the horizontal force $N(0, t)$ at the fixed bearing.

8.1 Motion of a disc rolling along a beam taking into account adhesion

where $G = \mu l g$ is the beam weight, is constant (for the parameters from Table 8.1 and $0.1 \leq \varkappa \leq 5$).

Consequently, it holds approximately true that

$$\max N(0, t) \approx \mu_0 F ,\tag{8.21}$$

which means that the horizontal longitudinal force in the fixed beam support equals approximately the weight of the disc multiplied by the coefficient of adhesion μ_0 and that the studied problem has a quasistatic rather than dynamic character.

The frequency parameter

$$\gamma = f_{1u}/f_1 \tag{8.22}$$

where f_{1u} and f_1 are first natural frequencies of longitudinal and bending beam vibrations, respectively, has no substantial influence on the horizontal longitudinal force on the fixed beam support within the range of $1 \leq \gamma \leq 5$. The same also applies to the bridge parameter

$$B = g/(\omega_1^2 l) \tag{8.23}$$

within the range of $2 \times 10^{-4} \leq B \leq 5 \times 10^{-4}$. For large $B \geq 1 \times 10^{-2}$ (impossible for railway bridges), however, the vibrations of the system are significant.

The logarithmic decrement of damping for bending vibrations, ϑ, and for longitudinal vibrations, ϑ_u, of the beam had very little influence within the studied range $(0.1 \leq \vartheta \leq 0.2; 0.2 \leq \vartheta_u \leq 0.4)$.

A significant role is played by the coefficient of adhesion μ_0 from equation (8.19), as follows from Fig. 8.10. Generally speaking, the growth of μ_0 results also in the growing horizontal longitudinal force in the fixed beam support. On the other hand the parameter α_0 which also appears in equation (8.19), has no significant influence within the range of $0.01 \leq \alpha_0 \leq 0.1$.

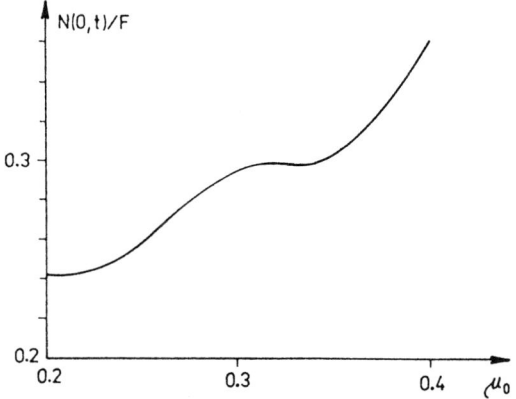

Fig. 8.10. Influence of coefficient of adhesion μ_0 on the horizontal force $N(0, t)$ at the fixed bearing.

The horizontal force on wheel circumference (8.17) and (8.18) was calculated for some braking conditions. Fig. 8.11 shows the horizontal longitudinal force on the fixed beam support as a function of the parameter r_0 with the braking force constant for two values of μ_0. The horizontal force grows linearly until $|r_0| = \mu_0$ and then remains constant.

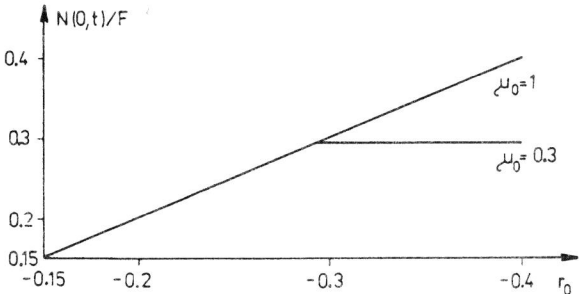

Fig. 8.11. Influence of the parameter of constant braking force r_0 for $\mu_0 = 0.3$ and $\mu_0 = 1$ on the horizontal force $N(0, t)$ at the fixed bearing.

Fig. 8.12 represents the case of the braking force growing linearly with time, i.e. with changing r_2 while $r_1 = r_3 = r_4 = r_5 = 0$.

The simplest railway bridge model characterizing the starting and braking of vehicles results in the following significant conclusions:

As the vehicle velocities at the time of starting and braking (should the vehicle stops on the bridge) are low, the forces acting on the beam are of static rather than dynamic character.

Hence the largest longitudinal beam deformations can be calculated approximately from static horizontal forces equal to the vehicle weight multiplied by the coefficient of adhesion.

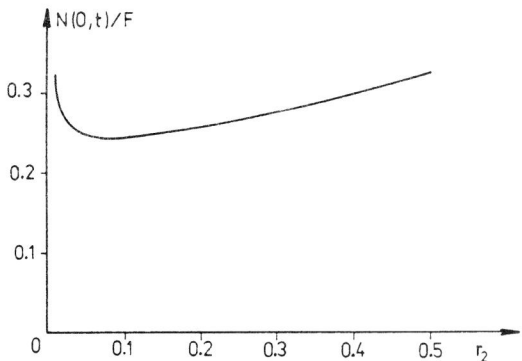

Fig. 8.12. Influence of the parameter of braking force r_2 growing linearly with time ($r_1 = 0$, other parameters according to Table 8.1) on the horizontal force $N(0, t)$ in the fixed bearing.

8.2 Quasistatic model

On the basis of the conclusions from Sect. 8.1 a quasistatic model has been constructed, enabling the calculation of the distribution of starting and braking forces in the rails and in the bridge. The model according to Fig. 8.13 is based on the following assumptions:

1. The model characterizes the static state at the moment when the horizontal longitudinal forces are at their maximum. This is shortly before stopping in the case of braking, or at the moment when the starting forces are highest, i.e. at small velocities (Fig. 8.14).

2. Longitudinal stresses in the bridge and in the rails can be considered separately from bending stresses (for proof see [70]).

3. The bridge is idealized as a beam. In the simplest case, shown in Fig. 8.13, the beam is fixed on the left-hand side and can move freely horizontally on a perfect roller bearing on the right-hand side.

4. The rails are idealized as bars. Both the left-hand and the right-hand ends of the rails are considered horizontally free (perfect joints are assumed).

5. The complex interaction of the rails and bridge in a horizontal direction is approximated by an elastic layer (i.e. by a system of horizontal springs situ-

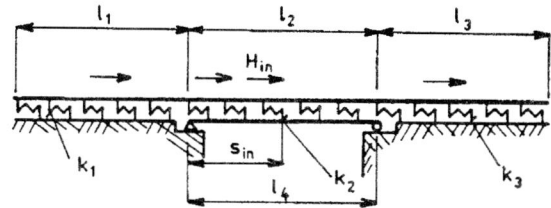

Fog. 8.13. Quasistatic model of railway bridge for loading by horizontal longitudinal forces.

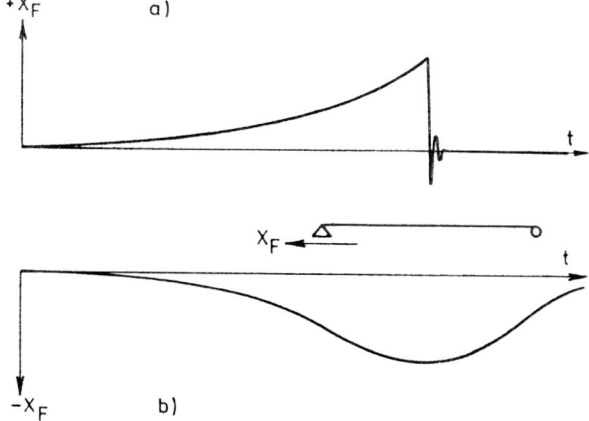

Fig. 8.14. Time-history of horizontal longitudinal force X_F under fixed bridge bearing a) during braking, b) during starting; direction of motion from left to right.

ated infinitely closely together, which can provide different characteristics in front of, on and behind the bridge – see Fig. 8.13). The purpose of this idealization is to characterize the distribution of horizontal forces along the bridge and the track fields in front of and behind the bridge. The elastic layer idealizes the complex characteristics of the permanent way in the horizontal direction which has not yet been sufficiently investigated. These characteristics are influenced by vertical and horizontal loads, friction, structural details of the bridge and of the permanent way, the gravel bed, climatic conditions, maintenance and numerous other factors, the majority of which have non-linear effects.

Figure 8.15 shows, according to [2], the dependence of the coefficient of horizontal elastic foundation k on horizontal load q for several values of the vertical load f. The tests were made on a Russian railways (SZD) track with type R 65 rails, prestressed concrete sleepers with the number of 1840 per 1 km of track with 1524 mm gauge. Fig 8.15 shows that the coefficient k of horizontal elastic foundation increases with increasing vertical load per unit track length, but decreases with increasing horizontal load. Fig. 8.15 can be used to estimate the magnitude of the coefficient k. Similar results were also found on Czech and Slovak railway tracks, where k varies from about 0.1×10^4 to 1×10^4 kNm^{-2}.

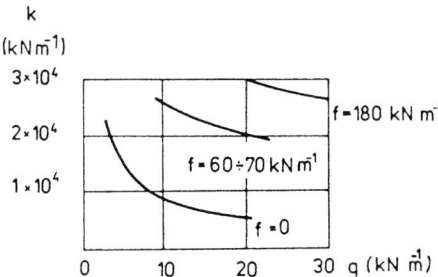

Fig. 8.15. Coefficient of elastic foundation in horizontal direction k as a function of horizontal load q and vertical load f (according to [2]).

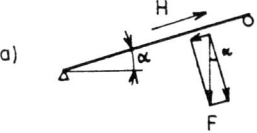

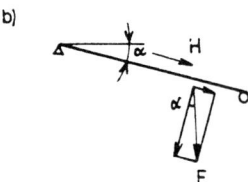

Fig. 8.16. Bridge a) in ascending track, b) in descending track.

8.2 Quasistatic model

6. The external forces include the horizontal longitudinal forces H obtained according to Coulomb's law from vertical forces F by the multiplication with the coefficient of adhesion for the quasistatic solution:

$$H = \mu_0 F . \tag{8.24}$$

The ascending or descending track on the bridge influences the magnitude of starting or braking forces according to Fig. 8.16. In an ascending track these forces decrease during braking (the upper sign in equation (8.25)), while in descending track they increase (lower sign) by the component of vehicle force parallel with the centre line of the bridge.

$$H = \mu_0 F \cos\alpha \mp F \sin\alpha . \tag{8.25}$$

H is now the component of the force parallel with the centre line of the beam and α is the track gradient. As track gradient is usually small

$$\cos\alpha \approx 1 \quad \text{and} \quad \sin\alpha \approx h/1000 \tag{8.26}$$

where h is the track gradient per mille.

After the substitution of (8.26) in (8.25) we obtain

$$H = F(\mu_0 \mp h/1000) . \tag{8.27}$$

During starting from the left to right the force H increases (the lower sign in equations (8.25) or (8.27) on an ascending track and decreases (upper sign) on a descending track by the component $F \sin\alpha$.

8.2.1 Solution

According to Fig. 8.13, N_i horizontal longitudinal forces H_{in} ($n = 1,2, ..., N_i$) act in every span i ($i = 1, 2, 3$) at a distance s_{in} from the left-hand end of the ith bar.

The bars are of lengths l_i and the coefficient of elastic foundation are k_i (Winkler's elastic foundation in horizontal direction). The bar of length l_4 represents the bridge ($l_4 = l_2$, $k_4 = k_2$).

The horizontal longitudinal deformations $u_i(x)$ of the system of beams in Figs. 8.13 are found from the equations of equilibrium:

$$-E_i A_i \frac{d^2 u_i(x)}{dx^2} + k_i u_i(x) = q_i(x), \quad i = 1, 3 ,$$

$$-E_2 A_2 \frac{d^2 u_2(x)}{dx^2} + k_2 [u_2(x) - u_4(x)] = q_2(x) ,$$

$$-E_4 A_4 \frac{d^2 u_4(x)}{dx^2} + k_2 [u_4(x) - u_2(x)] = 0 \tag{8.28}$$

where E_i and A_i are the modulus of elasticity and the cross section area, respectively, in the ith bar, and

$$q_i(x) = \sum_{n=1}^{N_i} \delta(x - s_{in}) H_{in}$$

is the external load in the ith span.

Further

$$\lambda_i^2 = \frac{k_i}{E_i A_i}, \quad \text{where } i = 1, 2, 3, 4,$$

and

$$\lambda^2 = \lambda_2^2 + \lambda_4^2. \tag{8.29}$$

The solution of the system of equations (8.28) carried out in [72] by the method of Laplace–Carson integral transformations according to [68] has the form

$$u_i(x) = -\frac{1}{E_i A_i \lambda_i} \sum_{n=1}^{N_i} H_{in} U(x - s_{in}) \sinh \lambda_i (x - s_{in}) + u_i(0) \cosh \lambda_i x$$

$$+ \frac{u_i'(0)}{\lambda_i} \sinh \lambda_i x, \quad i = 1, 3,$$

$$u_2(x) = -\frac{1}{E_2 A_2 \lambda^3} \sum_{n=1}^{N_2} H_{2n} U(x - s_{2n})[\lambda_2^2 \sinh \lambda(x - s_{2n}) + \lambda_4^2 \lambda(x - s_{2n})]$$

$$+ \frac{1}{\lambda^2}\left[u_2(0)(\lambda_4^2 + \lambda_2^2 \cosh \lambda x) + \frac{u_2'(0)}{\lambda}(\lambda_4^2 \lambda x + \lambda_2^2 \sinh \lambda x)\right]$$

$$- \frac{\lambda_2^2}{\lambda^2}\left[u_4(0)(\cosh \lambda x - 1) + \frac{u_4'(0)}{\lambda}(\sinh \lambda x - \lambda x)\right],$$

$$u_4(x) = \frac{\lambda_4^2}{E_2 A_2 \lambda^3} \sum_{n=1}^{N_2} H_{2n} U(x - s_{2n})[\sinh \lambda(x - s_{2n}) - \lambda(x - s_{2n})]$$

$$- \frac{\lambda_4^2}{\lambda^2}\left[u_2(0)(\cosh \lambda x - 1) + \frac{u_2'(0)}{\lambda}(\sinh \lambda x - \lambda x)\right]$$

$$+ \frac{1}{\lambda^2}\left[u_4(0)(\lambda_2^2 + \lambda_4^2 \cosh \lambda x) + \frac{u_4'(0)}{\lambda}(\lambda_4^2 \sinh \lambda x + \lambda_2^2 \lambda x)\right],$$

$$\tag{8.30}$$

where

$$U(x) = \begin{cases} 1 \\ 0 \end{cases} \text{ for } x \begin{cases} > 0 \\ < 0 \end{cases} \tag{8.31}$$

is the Heaviside unit function.

The system of equations (8.28) must satisfy the following boundary conditions (formulated in points 3 and 4 of Sect. 8.2):

$$N_1(0) = 0, \qquad u_1(l_1) = u_2(0),$$
$$u_2(l_2) = u_3(0), \qquad N_3(l_3) = 0,$$
$$N_1(l_1) - H_{l_1} = N_2(0), \qquad N_2(l_2) - H_{l_2} = N_3(0),$$
$$u_4(0) = 0, \qquad N_4(l_4) = 0, \tag{8.32}$$

8.2 Quasistatic model

where the normal axial force in the ith bar is

$$N_i(x) = E_i A_i \frac{\mathrm{d}u_i(x)}{\mathrm{d}x} \tag{8.33}$$

and H_{l_i} is the external force acting at the beginning of the ith bar.

There are unknown quantities $u_i(0)$ and $u'_i(0)$ in equation (8.30), which can be found from the initial conditions (8.32). This produces eight equations with eight unknown quantities which may be modified into the following form:

$$u'_1(0) = 0, \quad u_4(0) = 0,$$

$$u_1(0) \cosh \lambda_1 l_1 - u_2(0) = \frac{1}{E_1 A_1 \lambda_1} \sum_{n=1}^{N_1} H_{1n} U(l_1 - s_{1n}) \sinh \lambda_1 (l_1 - s_{1n}),$$

$$u_2(0) \frac{1}{\lambda^2}(\lambda_4^2 + \lambda_2^2 \cosh \lambda l_2) + u'_2(0) \frac{1}{\lambda^3}(\lambda_4^2 \lambda l_2 + \lambda_2^2 \sinh \lambda l_2) - u_3(0)$$

$$- u'_4(0) \frac{\lambda_4^2}{\lambda^3}(\sinh \lambda l_2 - \lambda l_2)$$

$$= \frac{1}{E_2 A_2 \lambda^3} \sum_{n=1}^{N_2} H_{2n} U(l_2 - s_{2n})[\lambda_2^2 \sinh \lambda(l_2 - s_{2n}) + \lambda_4^2 \lambda(l_2 - s_{2n})],$$

$$u_3(0)\lambda_3 \sinh \lambda_3 l_3 + u'_3(0) \cosh \lambda_3 l_3 = \frac{1}{E_3 A_3} \sum_{n=1}^{N_3} H_{3n} U(l_3 - s_{3n}) \cosh \lambda_3 (l_3 - s_{3n}),$$

$$- u_2(0) \frac{\lambda_4^2}{\lambda} \sinh \lambda l_4 - u'_2(0) \frac{\lambda_4^2}{\lambda^2}(\cosh \lambda l_4 - 1) + u'_4(0) \frac{1}{\lambda^2}(\lambda_4^2 \cosh \lambda l_4 + \lambda_2^2)$$

$$= \frac{-\lambda_4^2}{E_2 A_2 \lambda^2} \sum_{n=1}^{N_2} H_{2n} U(l_2 - s_{2n})[\cosh \lambda(l_2 - s_{2n}) - 1],$$

$$u_1(0) E_1 A_1 \lambda_1 \sinh \lambda_1 l_1 - E_2 A_2 u'_2(0) = \sum_{n=1}^{N_1} H_{1n} U(l_1 - s_{1n}) \cosh \lambda_1 (l_1 - s_{2n}),$$

$$u_2(0) \frac{E_2 A_2 \lambda_2^2}{\lambda} \sinh \lambda l_2 + u'_2(0) \frac{E_2 A_2}{\lambda^2}(\lambda_4^2 + \lambda_2^2 \cosh \lambda l_2) - E_3 A_3 u'_3(0)$$

$$- u'_4(0) \frac{E_2 A_2 \lambda_2^2}{\lambda^2}(\cosh \lambda l_2 - 1)$$

$$= \frac{1}{\lambda^2} \sum_{n=1}^{N_2} H_{2n} U(l_2 - s_{2n})[\lambda_2^2 \cosh \lambda(l_2 - s_{2n}) + \lambda_4^2]. \tag{8.34}$$

Equations (8.34) yield the unknown $u_i(0)$ and equations (8.30) the bar deformations $u_i(x)$; equation (8.33) can be used for the calculation of axial forces $N_i(x)$.

Particular cases may also be derived from the model in Fig. 8.13:
- in a bridge with perfect rail joints or expansions on both ends: $l_1 = l_3 = 0$,
- if long-welded rails are on the bridge: $l_1 = l_3 \to \infty$,
- if the displacement of rails along the bridge is perfectly free: $k_2 = 0$,
- if the rails are fixed to the bridge: $k_2 \to \infty$.

All cases can be obtained from equations (8.28) to (8.34) by including the respective limit.

Numerical evaluation of equation (8.30) was carried out on a computer. According to Fig. 8.17 the following quantities were calculated:

$$X_F = N_4(0), \qquad X_M = N_4(l_4),$$
$$X = |X_F| + |X_M|, \qquad S = |S_F| + |S_M|,$$
$$S_F = N_2(0), \qquad S_M = N_3(0),$$
$$Z_F = \frac{1}{\mu_0 l_4} \sum_{n=1}^{N_2} |H_{2n}|(l_4 - s_{2n}), \qquad Z_M = \frac{1}{\mu_0 l_4} \sum_{n=1}^{N_2} |H_{2n}| s_{2n},$$
$$Z = Z_F + Z_M,$$
$$\mu_F = \frac{|X_F|}{Z_F + G_F}, \qquad \mu_M = \frac{|X_M|}{Z_M + G_M},$$
$$\mu_B = \frac{X}{F}, \qquad \mu_T = \frac{X + S}{F}, \qquad (8.35)$$

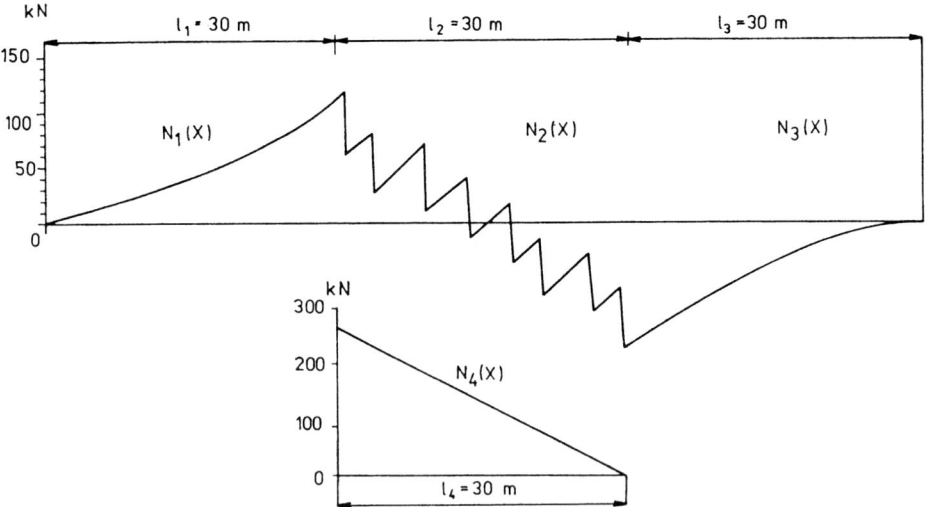

Fig. 8.17. Distribution of force H into rails S, bridge bearings X, gravel bed D. Vertical bridge reactions are Z. Indices: F – fixed bearing, M – movable bearing.

Fig. 8.18. Distribution of horizontal forces in rails $N_i(x)$, $i = 1, 2, 3$, and in the bridge $N_4(x)$ for Case No. 48 in Table 8.2.

8.2 Quasistatic model

where F – total vehicle weight on the bridge,
G – bridge weight,
X_i – horizontal forces under bridge bearings ($X_M = 0$ according to assumptions),
S_i – horizontal forces in rails,
Z_i – vertical reactions in bridge bearings,
μ_i – ratio of horizontal to vertical forces,
i – index: F – fixed bearing,
M – movable bearing.

TABLE 8.2. Input data computation

Quantity	Unit	Case No. 48	Bridge $l = 30$ m	Bridge $l = 16.2$ m
l_1	m	30	30	30
$l_2 = l_4$	m	30	30	16.2
l_3	m	30	30	30
$E_1 A_1$	kN	2.5×10^6	2.35×10^6	1.86×10^6
$E_2 A_2$	kN	2.5×10^6	2.35×10^6	1.86×10^6
$E_3 A_3$	kN	2.5×10^6	2.35×10^6	1.86×10^6
$E_4 A_4$	kN	5×10^7	5.06×10^7	1.85×10^7
k_1	kN m^{-2}	10^4	10^4	0.5×10^4
k_2	kN m^{-2}	10^4	10^4	0.5×10^4
k_3	kN m^{-2}	10^4	10^4	0.5×10^4
G	kN	1400	1360	270
F	kN	1600	1760	720
μ_0	1	0.3	0.352	0.285
H_{1n}	kN	0	0	0
S_{1n}	m	0	0	0
H_{21}	kN	60	77.4	51.25
H_{22}	kN	60	77.4	51.25
H_{23}	kN	60	77.4	51.25
H_{24}	kN	60	77.4	51.25
H_{25}	kN	60	77.4	0
H_{26}	kN	60	77.4	0
H_{27}	kN	60	77.4	0
H_{28}	kN	60	77.4	0
S_{21}	m	28	28.82	13.8
S_{22}	m	25	25.49	11.4
S_{23}	m	20	20.65	4.8
S_{24}	m	17	17.32	2.4
S_{25}	m	13	12.68	0
S_{26}	m	10	9.35	0
S_{27}	m	5	4.51	0
S_{28}	m	2	1.18	0
H_{3n}	kN	0	0	0
S_{3n}	m	0	0	0

An example of the calculation of horizontal longitudinal forces and their distribution in the individual bars is illustrated in Fig. 8.18. The input data are given in Table 8.2, case No. 48.

8.2.2 Influence of some parameters

The problem solved in the preceding section depends on a number of parameters. Therefore, their influence was studied for case No. 48 from Table 8.2 which, after rounding-off, represents a bridge of span 30 m loaded by two four-axle locomotives which braked in the middle of the bridge. During the calculations only one parameter is varied, the others remaining the same as in Table 8.2. In Figs 8.19 through 8.37, the notation (8.35) is used according to Fig. 8.17.

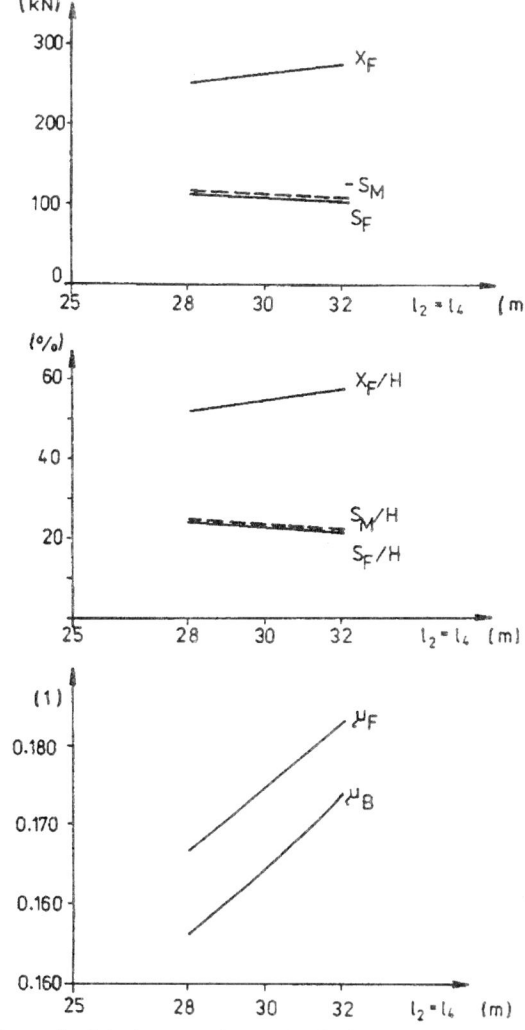

Fig. 8.19. Influence of span l_4 of the beam on forces X_F, S_F, S_M.

8.2 Quasistatic model

The influence of the span l_4 is investigated in Fig. 8.19, which shows that the force X_F under the fixed bearing grows slightly while the forces in the rails S_F and S_M decrease slightly with increasing span l_4. On the other hand, it is possible to observe a step rise of μ_F and μ_B, because the denominator in the respective equations (8.35) decreases with increasing span. Within the observed range of 4×10^7 to 6×10^7 kN the parameter $E_4 A_4$ exerts little influence on the quantities (8.35).

The influence of rail lengths l_1 and l_3 in front of and behind the bridge was investigated over the range 2 m to 90 m for three values $k = k_1 = k_2 = k_3$ ($k = 0.2 \times 10^4, 0.5 \times 10^4, 1 \times 10^4$ kN m^{-2}). The results are given in Fig. 8.20 which shows that with increasing rail lengths the force X_F in the bearing decreases, while the forces in the rails S_F and S_M increase. However, this applies only to the

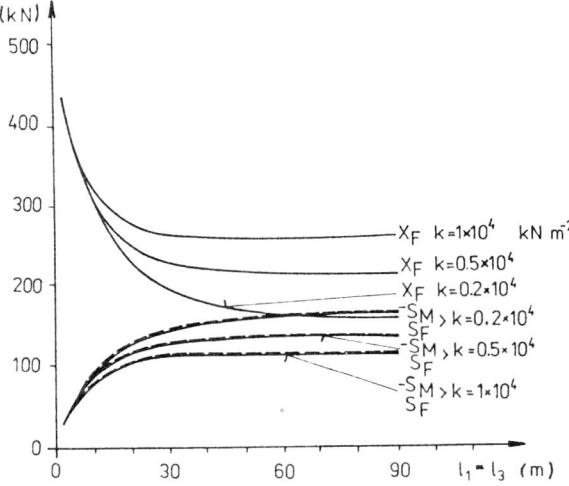

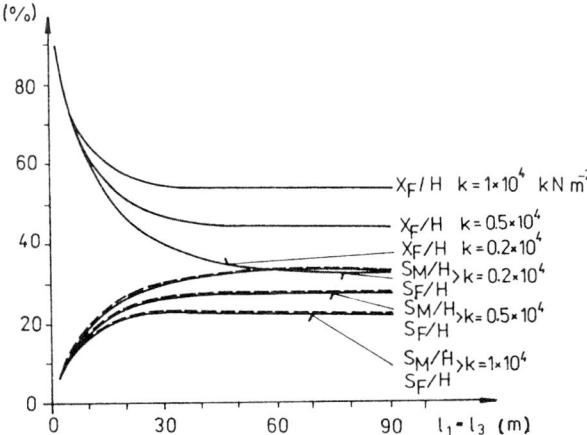

Fig. 8.20 (continuation on page 148).

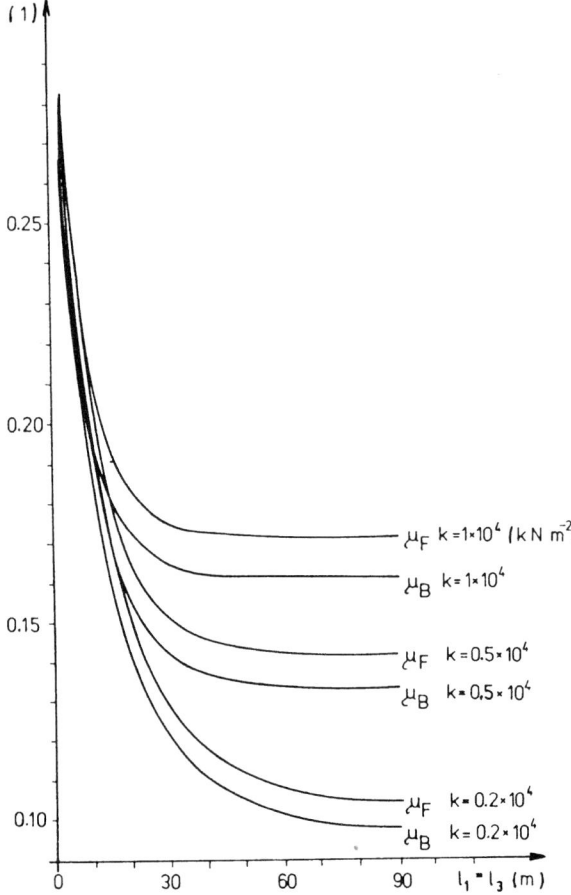

Fig. 8.20. Influence of rail length in front of, on and behind the bridge $l_1 = l_3$ for various values of k on forces X_F, S_F, S_M.

length of 30 m; the change of these forces is small beyond that limit. The rail length, which absorbs almost the whole horizontal force, depends also on the coefficient k. When the horizontal rigidity of the track is lower, the active rail length increases. The coefficients μ_F and μ_B decrease with the increasing lengths l_1 and l_3 but the asymptote is attained once again when the length of the rails is about 30 m.

The influence of the product of $E_1A_1 = E_2A_2 = E_3A_3$ is investigated in Fig. 8.21, where the coefficients μ_F and μ_B drop steeply with increasing E_iA_i.

The rigidity of the horizontal elastic foundation in front of, on and behind the bridge was considered similarly to the first study ($k = k_1 = k_2 = k_3$, Fig. 8.22). Its increase resulted in an increase of X_F but a drop of S_F and S_M. With high rigidity, however, all quantities vary very little.

8.2 Quasistatic model

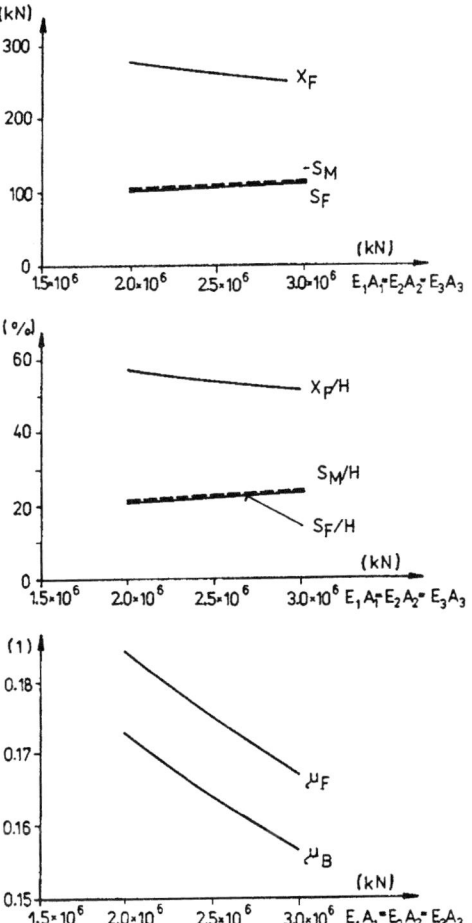

Fig. 8.21. Influence of modulus of elasticity E_i and cross section area of the rails A_i, $i = 1, 2, 3$, on forces X_F, S_F, S_M.

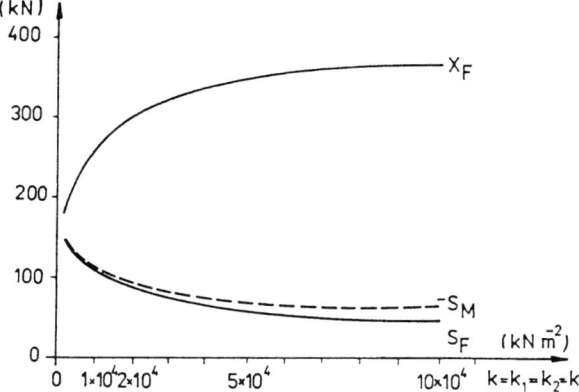

Fig. 8.22 (continuation on page 150).

150   8. Horizontal longitudinal effect on bridges

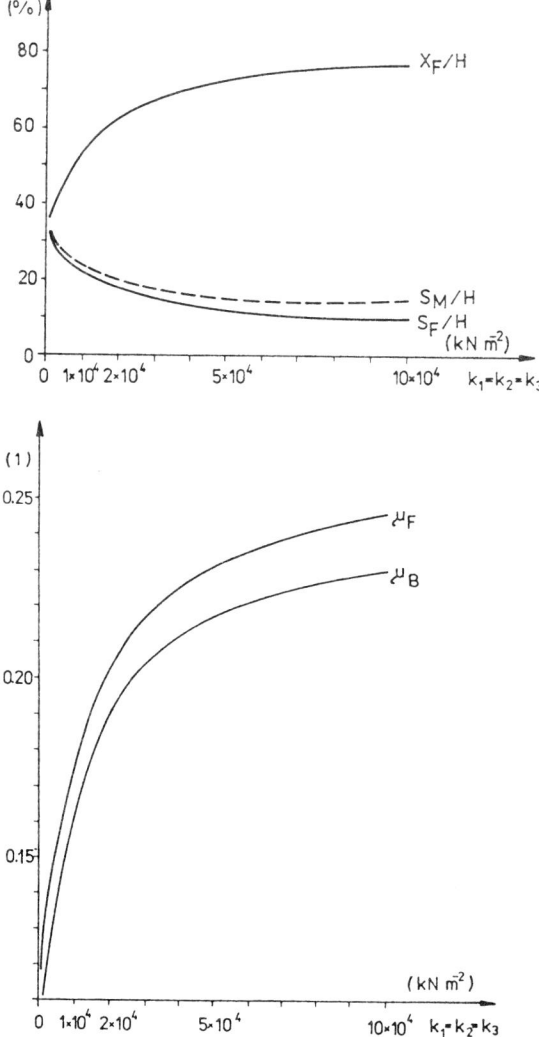

Fig. 8.22. Influence of horizontal elastic foundation $k = k_1 = k_2 = k_3$ on forces X_F, S_F, S_M.

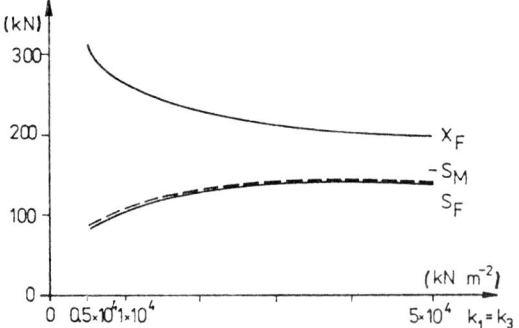

Fig. 8.23 (continuation on page 151).

8.2 Quasistatic model

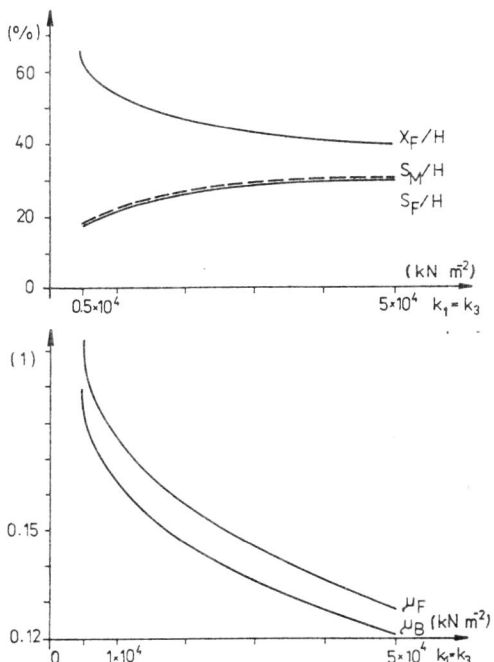

Fig. 8.23. Influence of rigidity of elastic foundation in front of and behind the bridge $k_1 = k_3$, $k_2 = 1 \times 10^4$ kN m^{-2} on forces X_F, S_F, S_M.

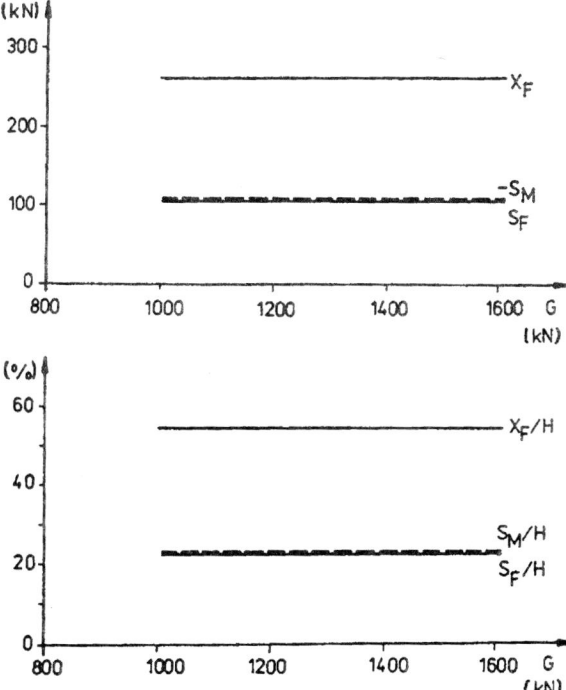

Fig. 8.24 (continuation on page 152)

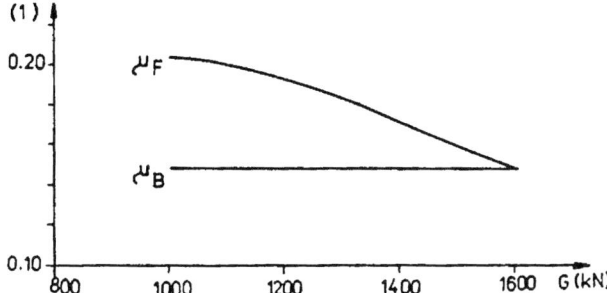

Fig. 8.24. Influence of bridge weight G on forces X_F, S_F, S_M.

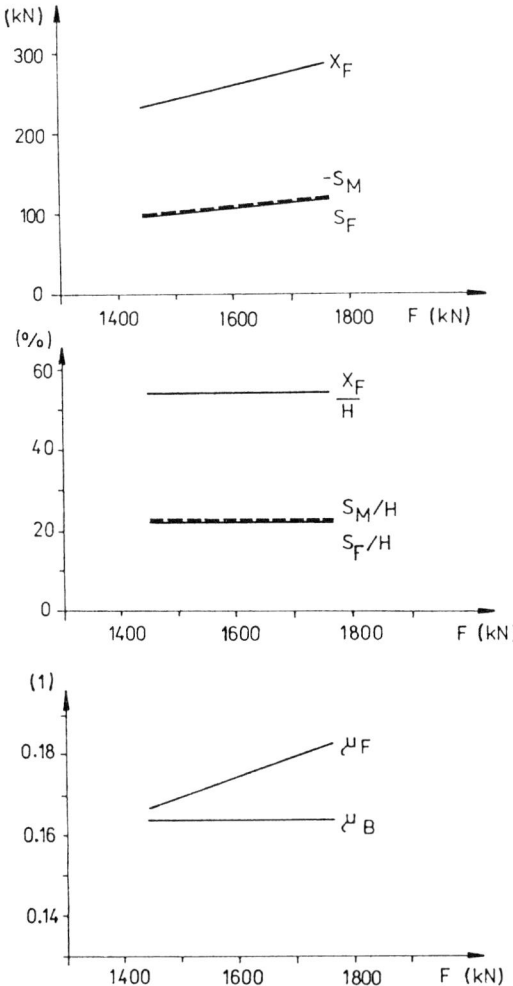

Fig. 8.25. Influence of vehicle weight F on forces X_F, S_F, S_M.

8.2 Quasistatic model

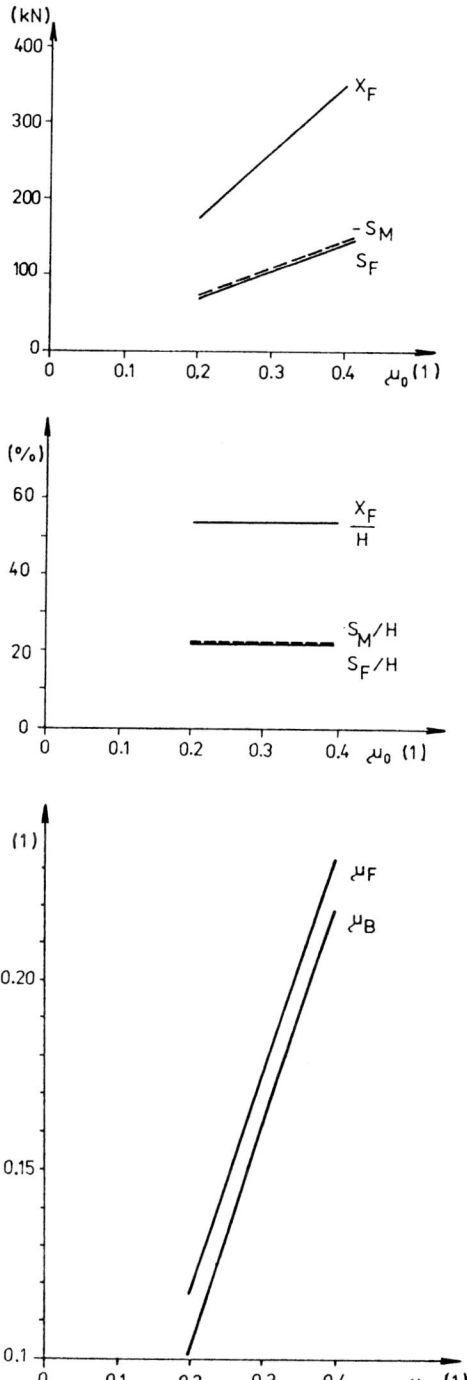

Fig. 8.26. Influence of the coefficient of adhesion μ_0 on forces X_F, S_F, S_M.

Fig. 8.23 investigates the case, when the rigidity $k_2 = 1 \times 10^{-4}$ kN m^{-2} is constant on the bridge, but the rigidities $k_1 = k_3$ in front and behind the bridge vary. When $k_1 = k_3$ is less than k_2, the force X_F increases and the forces S_F and S_M decrease. When $k_1 = k_3$ exceeds k_2 the opposite situation arises to that when $k_1 = k_1 = k_3$. The parameters μ_F and μ_B decrease with increasing $k_1 = k_3$.

The bridge weight G influences only the coefficient μ_F, Fig. 8.24. The weight of the vehicle F in Fig. 8.25 was studied with the assumption that the number of braked axles is constant. The quantities X_F, S_F, S_M, and μ_F increase with the increasing F. On the other hand, the forces referred to the total horizontal force H and the coefficient μ_B are not influenced by the variation of F.

The influence of the coefficient of adhesion μ_0 is shown in Fig. 8.26. The forces X_F, S_F, and S_M and the coefficient μ_F and μ_B increase linearly with increasing μ_0 but the relative values of forces with reference to the force H are independent of μ_0.

The results of the basic case No. 48 from Table 8.2 are given in Fig. 8.27 and this case was further studied in detail with reference to the direction of drive (Fig. 8.28), position of vehicle on the bridge (Fig. 8.29 and Fig. 8.30), number of axles in front of (Fig. 8.31), on (Fig. 8.32) and behind (Fig. 8.33) the bridge, the braking in front of (Fig. 8.34) and behind (Fig. 8.35) the bridge with reference to the number of braking axles on the bridge (Fig. 8.36) with two six-axle locomotives of the same total weight and adhesion coefficient as the basic case.

The horizontal longitudinal force under the fixed bearing X_F acquires its highest values when the braked axles are distributed in front of, on and behind the bridge (Fig. 8.32). In this case the rails transmit the smallest part of horizontal forces. This has also been proved by experiments.

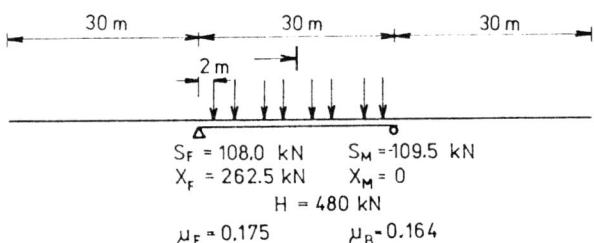

Fig. 8.27. Basic case No. 48 from Table 8.2.

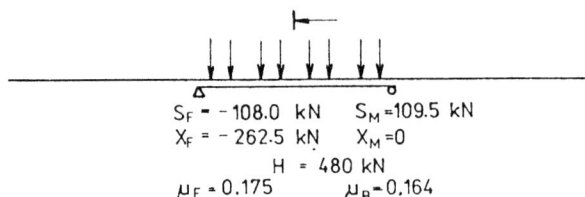

Fig. 8.28. Direction of motion from right to left.

8.2 Quasistatic model

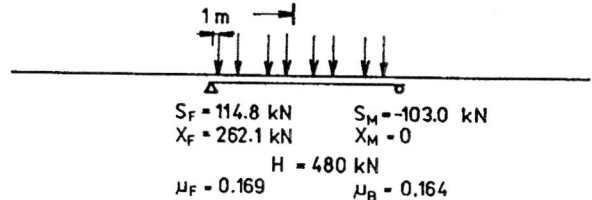

Fig. 8.29. The first force 1 m from the left-hand abutment.

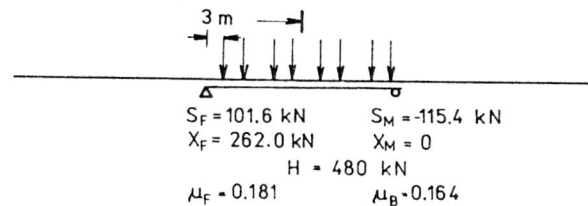

Fig. 8.30. The first force 3 m from the left-hand abutment.

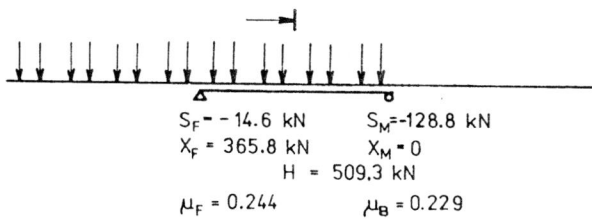

Fig. 8.31. The first and the second spans loaded by eight forces.

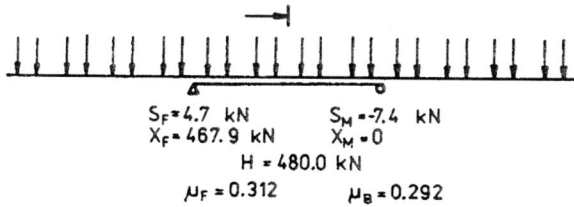

Fig. 8.32. The first, second and third spans loaded by eight forces.

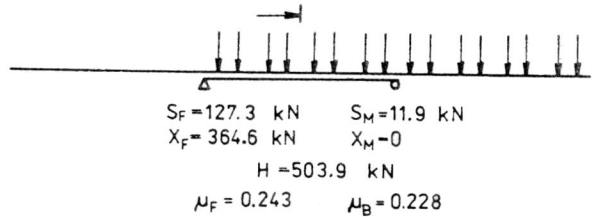

Fig. 8.33. The second and third spans loaded by eight forces.

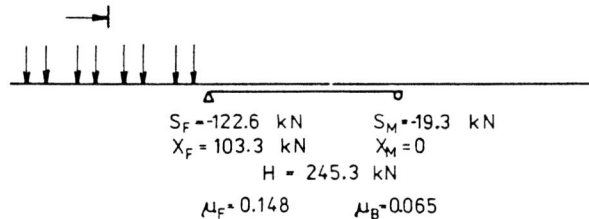

Fig. 8.34. The first span loaded by eight forces.

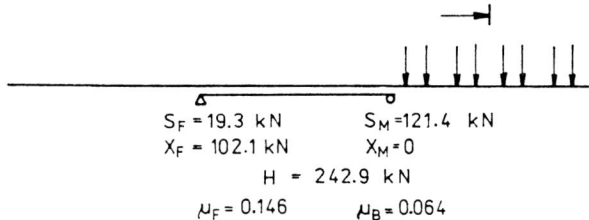

Fig. 8.35. The third span loaded by eight forces.

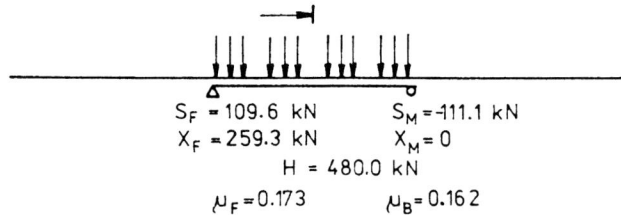

Fig. 8.36. The second span loaded by twelve forces.

8.2.3 Experiments on bridges

Considerable experimental research in the international program of the ORE D 101 [160] was carried out to ascertain the effect of starting and braking forces on bridges. Seven European railway administrations participated (CSD, DB, NS, ÖBB, RATP, SBB and SNCB). Their experiments included the tests on seven bridges of various types and spans [75], [160].

The experiments involved the measurements of the forces X_F, X_M, Z_F, Z_M, S_F, S_M, D (horizontal force transmitted by the gravel bed or by the closing wall, Fig. 8.17), acceleration a on the engine body and other data.

During evaluation it was also possible to calculate

$$\mu_0 = a/g ,$$

and

$$H = am , \qquad (8.36)$$

where $g = 9.81$ m s^{-2} is the acceleration due to gravity, and m is the mass of the set braked on the bridge.

8.2 Quasistatic model

TABLE 8.3. Comparison of theoretical and experimental results

Bridge span	T – Theory E – Experiment	X_F	X_M	S_F	S_M	D	μ_F	μ_B	μ
		kN					1		
$l = 30$ m	T	338.6	0	139.3	141.3		0.217	0.192	0.352
	E	303.0	21.5	139.2	148.0	4.6	0.193	0.186	0.352
$l = 16.2$ m	T	68.9	0	68.0	68.1		0.139	0.096	0.285
	E	56.8	11.8	42.2	102.7		0.134	0.146	0.285

Table 8.3 gives a comparison of theoretical calculations according to Sect. 8.2.1 with the results of measurements on two bridges of CSD. The first bridge of span 30 m was a steel plate girder bridge with continuous gravel bed and orthotropic bridge deck, the other, of span 16.2 m, was a steel plate girder bridge with an open bridge deck. The computation input data are summarized in Table 8.2.

Table 8.3 reveals that the theoretical results are in very good agreement with the measurements of the 30 m bridge. Major differences can be observed only in some quantities ascertained by measurement on the 16.2 m bridge. This is due, probably, to a poorer state of track in the latter case which caused great scatter of experimental results.

The theory was in good agreement with measurements particularly in test series, when the bridge was under load. In the cases, when the locomotive stopped in front of the bridge, the differences were greater.

The calculations were carried out for several values of the coefficient of horizontal elastic foundation k_i, however, it has come to light that constant $k_1 = k_2 = k_3$ has proved most satisfactory, i.e. for the 30 m bridge $k = 1 \times 10^{-4}$ kN m^{-2} (good track state) and $k = 0.5 \times 10^{-4}$ kN m^{-2} for the 16.2 m bridge (poor track state).

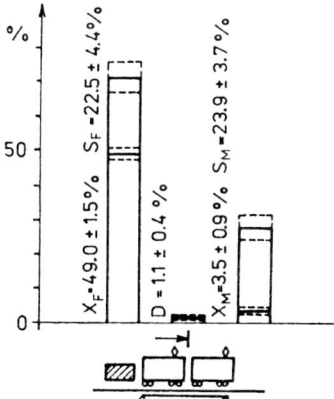

Fig. 8.37. Percentage distribution of braking forces on a CSD bridge with span 30 m.

It has also been discovered by experiments that under moving bearings a certain small part of the braking and starting force is applied (Fig. 8.37) which is transferred by friction in the case of imperfect bearings. This had not been considered in theoretical computations, where $X_M = 0$.

The horizontal longitudinal force is distributed in about 30 m of rails in a well maintained track. In a poorly maintained track, however, this distribution length drops, on the bridge of span 16.2 m, for instance, to as little as 7 m.

8.3 Starting and braking forces on bridges

Theoretical and experimental research on the actions of horizontal longitudinal forces on bridges has yielded the following conclusions:

The characteristic feature of horizontal longitudinal forces during braking is their slow increase and their sudden drop at the moment of stopping. During the starting process these forces slowly increase in accordance with the increasing traction force of the engine. Maximum values coincide with small velocities in both cases. The starting force is also limited by the capacity of the coupling.

The percentage distribution of horizontal forces is approximately the same for starting and braking. According to static laws the horizontal longitudinal forces are mostly carried by those members which are most rigid: these include fixed steel bearings and rails. The soft and yielding members (neoprene bearings, gravel bed, etc.) carry a substantially smaller part.

The adhesion coefficient μ_0 is the most important factor in (8.35) and (8.36). Due to the interaction of the permanent way and the bearings only a part of the external forces is transmitted to the bridge structure itself, which is characterized by the factor μ_B (8.35).

The structural system and bridge material exercise no substantial influence. The bridge span does not influence the relative magnitude of braking forces, as the bridge may be loaded by the braking train along its whole length. On the other hand, the greatest starting forces may be caused by two driving locomotives, which may attain the permissible force on the coupling. The length of these two locomotives is about 25 to 30 m; for this reason, the starting forces drop relatively in those bridges with spans over 30 m.

The rails transfer relatively high forces in the case of smaller span bridges, but their participation in interaction drops with increasing span. The track gradient increases or decreases the adhesion coefficient, but this change is fairly insignificant.

Horizontal longitudinal forces influence the design of bridge piers and abutments. The bridge ends transfer the forces to the backfill soil, which are higher if the soil has been compacted than in the case of fresh, insufficiently compacted soil. Earth pressure below the piers and abutments were found to be very low.

The principal influence on the distribution of horizontal longitudinal forces is exerted by the type of bearing. The major part of these forces is sustained by

8.3 Starting and braking forces on bridges

hinged steel bearings, while movable (roller) bearings can sustain only about 3–5% of vertical forces applied to them. Pot bearings are less rigid members than steel bearings, their action eliminating the difference between fixed and movable bearings. This difference does not arise with neoprene bearings, because they are sufficiently yielding in a horizontal direction and, therefore, transmit relatively small horizontal forces. The influence of the location of steel bearings on piers and abutments is important and must be taken into account in design.

The rails may transfer a considerable part of the horizontal longitudinal forces to a distance of some 30 m; therefore, their interaction must be taken into account. On the other hand, the gravel bed may transfer only a very low longitudinal force. The overall quality of the permanent way is described by the coefficient of longitudinal elastic rail fastening k, which depends considerably on the state of the track. The weaker the permanent way and the poorer its state, the more the rails are stressed by braking and starting forces. On the other hand, the increasing rigidity of fastening increases the horizontal forces in bridge bearings and decreases these forces in rails.

The growing axle forces of railway vehicles and the growing adhesion coefficient linearly influence the horizontal forces in bridge bearings and rails. The greatest horizontal forces in bridge bearings originate when the braked or driven sets cover the rail field in front of, on and behind the bridge. In such a case, the rails take over the smallest part of horizontal force as compared with the case when only the bridge is loaded.

The coefficient of adhesion possesses higher values (0.3 to 0.4) during starting than during braking (0.15 to 0.3), and μ_0 of electric locomotive is higher than that of diesel-electric engines. The magnitude is influenced by a number of factors, such as the weather, moisture, state and fouling of rails, wheel tyres, type of brakes, etc.

The magnitude of horizontal longitudinal forces in bridges and rails is also influenced by the mode of driving, place of stopping or starting, the braking regime, the method of wagon coupling and other factors.

On the basis of all these conclusions it is possible to consider the starting and braking force at the rail top level as

$$H = \gamma_f \mu_0 F , \qquad (8.37)$$

where γ_f – load factor; according to experiments $\gamma_f = 1.3$ with 97.7% reliability,

$$\mu_0 = \begin{cases} 0.2 & \text{for braking} \\ 0.35 & \text{for starting} \end{cases} \text{ is the coefficient of adhesion },$$

F – sum of vertical axle forces which can be placed on the loading length of the bridge L (equal to the span in the case of simply supported girders or the sum of spans of the individual fields in the case of continuous girders).

TABLE 8.4. Coefficient μ_B calculated according to Eq. (8.39)

Drive type	μ_0	Bearings	Rail joints (*)	Loading bridge length L in metres						
				≤ 25	30	40	50	60	80	100
				Coefficient μ_B						
Braking	0.2	Steel	− / //				0.10 / 0.12 / 0.14			
		Pot	− / //				0.09 / 0.11 / 0.13			
		Neoprene	− / //				0.08 / 0.10 / 0.11			
Starting	0.35	Steel	− / //	0.18 / 0.21 / 0.25	0.15 / 0.18 / 0.20	0.11 / 0.13 / 0.15	0.09 / 0.11 / 0.12	0.07 / 0.09 / 0.10	0.05 / 0.07 / 0.08	0.04 / 0.05 / 0.06
		Pot	− / //	0.16 / 0.19 / 0.22	0.13 / 0.16 / 0.18	0.10 / 0.12 / 0.14	0.08 / 0.09 / 0.11	0.07 / 0.08 / 0.09	0.05 / 0.06 / 0.07	0.04 / 0.05 / 0.06
		Neoprene	− / //	0.14 / 0.17 / 0.20	0.12 / 0.14 / 0.16	0.09 / 0.11 / 0.12	0.07 / 0.08 / 0.10	0.06 / 0.07 / 0.08	0.04 / 0.05 / 0.06	0.04 / 0.04 / 0.05

(*) Rail joints: − joined or welded rail,
   / rail with expansion or free joint on one bridge end,
   // rail with expansions or free joints on both bridge ends.

8.3 Starting and braking forces on bridges

With regard to interaction with rails, the force

$$X = \gamma_f \mu_B F, \qquad (8.38)$$

is applied in the bridge bearing plan, where the coefficient μ_B can be given in the form

$$\mu_B = b_1 b_2 b_3 \mu_0. \qquad (8.39)$$

The coefficients b_i characterize:
– the type of bearings:

$$b_1 = \begin{cases} 0.5 & \text{for steel bearings} \\ 0.45 & \text{for pot bearings} \\ 0.4 & \text{for neoprene bearings} \end{cases},$$

– the interaction with rails:

$$b_2 = \begin{cases} 1 & \text{for jointed or welded rail} \\ 1.2 & \text{for rail with expansion or entirely free joint on one bridge end} \\ 1.4 & \text{for rail with expansions or entirely free joints on both bridge ends,} \end{cases}$$

– the loading length of bridge:

$$b_3 = \begin{cases} 1 & \\ 1 \\ 25/L \end{cases} \text{during braking} \quad \text{during starting for loading length} \quad L \begin{cases} \leq 25 \text{ m} \\ > 25 \text{ m}. \end{cases}$$

The coefficient μ_B has been found by this method and tabulated in Table 8.4. This table reveals that in single-track bridges up to the loading length of 43.75 m the starting forces prevail, while for major spans the braking forces yield higher values.

The force X (8.38) is resolved according to bearing types as follows:
– fixed steel bearing take up the whole force X,
– movable steel bearings (roller bearings) transmit an horizontal force equal 5% of vertical load applied to these bearings,
– pot and neoprene bearings take up the force X uniformly.

The difference between the forces H (8.37) and X (8.38) is transferred by the rails, rail bed, etc.

9. Horizontal transverse effects on bridges

Horizontal transverse forces are generated by lateral movements of railway vehicles from two sources in a straight track: horizontal track irregularities and the sinusoidal motion of conical wheels along cylindrical rail heads. The load can be characterized by a system of horizontal transverse random forces which are variable with time, moving on the rail head level. Their number corresponds with the number of wheels of the vehicle or train.

Apart from these two sources, which are called lateral impacts, centrifugal forces also originate in a curved track which act on the bridge in the outward direction.

The problem of the movement of vertical random forces along a bridge was dealt with by J. Sláma [191] and B. Sniady [195]. In the sections which follow we shall derive the solution for horizontal random forces applied to railway bridges idealized by two models.

9.1 Beam

The most usual model of railway bridges is the Bernoulli–Euler beam loaded by a system of N forces which are random functions of time and which arrive at the bridge with deterministic spacings d_n according to Fig. 9.1. Vertical, horizontal and torsional vibrations can be investigated separately in this beam.

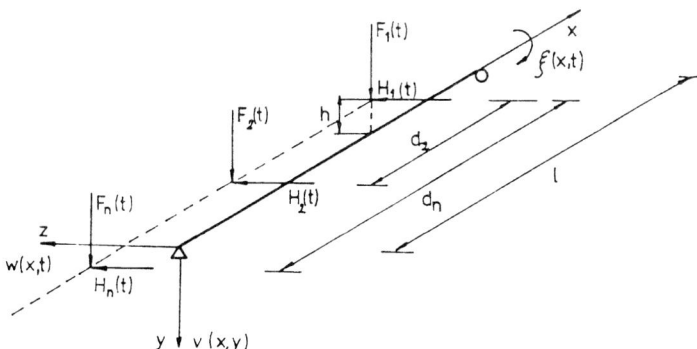

Fig. 9.1. Movement of vertical, $F_n(t)$, and horizontal, $H_n(t)$, random forces spaced at d_n along a beam with span l.

9.1.1 Vertical vibration

The beam vibrating in a vertical direction is affected by axle forces $F_n(t)$, $n = 1, 2, ..., N$, according to Fig. 9.1. The deformation of the beam is described by the Bernoulli-Euler differential equation

$$EI\ v^{IV}(x, t) + \mu\ \ddot{v}(x, t) = \sum_{n=1}^{N} \varepsilon_n \delta(x + d_n - ct)\ F_n(t),\qquad (9.1)$$

where $v(x, t)$ is the vertical deflection of the beam at point x and time t,
$\quad EI$ – constant flexural rigidity of the beam,
$\quad \mu$ – constant mass of the beam per unit length,
$\quad \varepsilon_n = \begin{cases} 1 \\ 0 \end{cases}$ for $\begin{cases} d_n/c \leq t \leq l/c + d_n/c \\ t < d_n/c;\ t > (l + d_n)/c \end{cases}$ function expressing the presence of the force $F_n(t)$ on the beam,

$\quad l$ – beam span,
$\quad c$ – constant velocity of motion,
$\quad d_n$ – distance of the nth force from the first, $d_1 = 0$.

The axle force $F_n(t)$ is resolved into its mean value F_n (static axle force – deterministic) and its centred random component $\mathring{F}_n(t)$

$$F_n(t) = F_n + \mathring{F}_n(t),\qquad (9.2)$$

where $F_n = E[F_n(t)]$; E is of the mean value.

a) Approximate solution

An approximate solution to equation (9.1) may be derived for low velocities of motion and small damping. Both of these prerequisites are complied with by railway bridges, because the velocities of railway vehicles at present are relatively low in comparison with the critical velocity (compare with equation (5.4) and Sect. 6.1). The approximate solution characterizes very well the action of the system of forces travelling along the bridge.

Neglecting the second term in the left-hand side of equation (9.1) we obtain its solution in the form

$$v(x, t) = \sum_{j=1}^{\infty} \sum_{n=1}^{N} \varepsilon_n v_{0j}\ \overline{F}_n(t) \sin \frac{ct - d_n}{L_j} \sin \frac{x}{L_j},\qquad (9.3)$$

which can be considered, under the above mentioned assumptions, as an approximate solution (quasistatic) of equation (9.1) with boundary conditions for a simply supported beam with zero initial conditions – compare with equation (1.30) in [68].

In equation (9.3) the following symbols were used:

$v_{0j} = \dfrac{2Fl^3}{j^4 \pi^4 EI}$ – midspan deflection due to the mean axle force F applied at midspan,

$\overline{F}_n(t) = \dfrac{F_n(t)}{F}$ – dimensionless axle force,

$$L_j = \frac{l}{j\pi}, \quad L = \frac{l}{\pi}. \qquad (9.4)$$

Analogously with equations (9.3) and (9.4), the centred random component is obtain as:

$$\mathring{v}(x, t) = \sum_{j=1}^{\infty} \sum_{n=1}^{N} v_{0j} \varepsilon_n \bar{F}_n(t) \sin \frac{ct - d_n}{L_j} \sin \frac{x}{L_j}. \qquad (9.5)$$

The covariance of deflection is then found in accordance with its definition (1.27)

$$C_{vv}(x_1, x_2, t_1, t_2) = E[\mathring{v}(x_1, t_1) \mathring{v}(x_2, t_2)] = \sum_{j=1}^{\infty} \sum_{k=1}^{\infty} \sum_{n=1}^{N} \sum_{m=1}^{N} v_{0j} v_{0k} \varepsilon_n \varepsilon_m$$

$$\cdot \sin \frac{ct_1 - d_n}{L_j} \sin \frac{ct_2 - d_m}{L_k} \sin \frac{x_1}{L_j} \sin \frac{x_2}{L_k} C_{F_n F_m}(t_1, t_2), \qquad (9.6)$$

where $C_{F_n F_m}(t_1, t_2)$ are cross covariances of the forces $F_n(t)$.

The variance of the deflection is found from equation (9.6)

$$\sigma_v^2(x, t) = C_{vv}(x, x, t, t) = \sum_{j=1}^{\infty} \sum_{k=1}^{\infty} \sum_{n=1}^{N} \sum_{m=1}^{N} v_{0j} v_{0k} \varepsilon_n \varepsilon_m$$

$$\cdot \sin \frac{ct - d_n}{L_j} \sin \frac{ct - d_m}{L_k} \sin \frac{x}{L_j} \sin \frac{x}{L_k} C_{F_n F_m}(t, t). \qquad (9.7)$$

In equation (9.7), only the first term of the series is considered in the approximate solution, $j = k = 1$, $L = L_{01}$, $v_0 = v_{01}$ because the convergence of the series for deflection is very rapid:

$$\sigma_v^2(x, t) = \sum_{n=1}^{N} \sum_{m=1}^{N} v_0^2 \varepsilon_n \varepsilon_m \sin \frac{ct - d_n}{L} \sin \frac{ct - d_m}{L} \sin^2 \frac{x}{L} C_{F_n F_m}(t, t). \qquad (9.8)$$

With the assumption of equality of cross covariances of two adjacent axle forces $C_{F_n F_{n+1}} = C_{F_{n+1} F_n}$ and very low covariances of more distant axle forces the dual series in equation (9.8) can be simplified further to a single series

$$\sigma_v^2(x, t) \approx \sum_{n=1}^{N} v_0^2 \varepsilon_n \sin^2 \frac{ct - d_n}{L} \sin^2 \frac{x}{L} C_{F_n F_n}(t, t)$$

$$+ \sum_{n=1}^{N-1} 2 v_0^2 \varepsilon_n \sin \frac{ct - d_n}{L} \sin \frac{ct - d_{n+1}}{L} \sin^2 \frac{x}{L} C_{F_n F_{n+1}}(t, t). \qquad (9.9)$$

All the most important statistical characteristics of the first and the second order have been obtained in this way for the deflection of the bridge loaded by a series of moving random forces.

b) Dynamic solution

The solution of equation (9.1) without making simplifying assumptions can be derived from equation (4.5) in [68] in the following form:

$$v(x, t) = \sum_{j=1}^{\infty} \sum_{n=1}^{N} \frac{2\varepsilon_n}{\mu l \omega_j} \sin \frac{x}{L_j} \int_0^t F_n(\tau) \sin \frac{c\tau - d_n}{L_j} \sin \omega_j(t - \tau) \, d\tau, \qquad (9.10)$$

9.1 Beam

where the natural circular frequency of the simply supported beam is

$$\omega_j^2 = \frac{j^4 \pi^4}{l^4} \frac{EI}{\mu}, \quad f_j = \frac{\omega_j}{2\pi}. \tag{9.11}$$

The natural frequencies used in Chapter 9 are found for the two investigated bridges in Table 9.1.

The covariance of deflection is calculated by the same procedure as that used in point a), i.e.

$$C_{vv}(x_1, x_2, t_1, t_2) =$$

$$= \sum_{j=1}^{\infty} \sum_{k=1}^{\infty} \sum_{n=1}^{N} \sum_{m=1}^{N} \frac{4\varepsilon_n \varepsilon_m}{\mu^2 l^2 \omega_j \omega_k} \sin \frac{x_1}{L_j} \sin \frac{x_2}{L_k} \int_0^{t_1} \int_0^{t_2} \sin \frac{c\tau_1 - d_n}{L_j} \sin \frac{c\tau_2 - d_m}{L_k}$$

$$\cdot \sin \omega_j(t_1 - \tau_1) \sin \omega_k(t_2 - \tau_2) C_{F_n F_m}(\tau_1, \tau_2) \, d\tau_1 \, d\tau_2. \tag{9.12}$$

A comparison of equation (9.12) with equation (9.6) reveals that the approximate solution (9.6) depends directly on the covariances of forces, while in the dynamic solution (9.12) these covariances are below the integration symbol. The double integral in (9.12) can be found in a closed form for the following covariance (white noise)

$$C_{F_n F_m}(t_1, t_2) = S_{F_{nm}} \delta(t_2 - t_1). \tag{9.13}$$

The variance of deflection is found from equation (9.12) for $j = k = 1$ after the substitution of equation (9.13)

$$\sigma_v^2(x, t) = \sum_{n=1}^{N} \sum_{m=1}^{N} \varepsilon_n \varepsilon_m \alpha^{-1} v_0^2 \sin^2 \frac{x}{L} Z_{nm}^2 V_{F_{nm}}^2, \tag{9.14}$$

where $\alpha = c/(2f_1 l)$ is a dimensionless parameter of velocity (5.4),

$$Z_{nm}^2 = \int_0^z \sin^2 a(z - x) \sin(x - D_n) \sin(x - D_m) \, dx$$

$$= \frac{1}{4}\left[\left(z - \frac{1}{2a} \sin 2az\right) \cos(D_n - D_m) - \frac{a^2}{2(a^2 - 1)} \sin(2z - D_n - D_m) \right.$$

$$- \frac{1}{2} \sin(D_n + D_m) + \frac{a}{2(a^2 - 1)} \sin 2az \cos(D_n + D_m) - \frac{1}{2(a^2 - 1)}$$

$$\left. \cdot \cos 2az \sin(D_n + D_m)\right], \tag{9.15}$$

$$z = ct/L,$$
$$x = c\tau/L,$$
$$a = 1/\alpha,$$
$$D_n = d_n/L,$$
$$V_{F_{nm}} = \frac{(S_{F_{nm}} \omega_1)^{1/2}}{F} \text{ is the coefficient of variation for forces } F_n(t).$$

The coefficient of variation for deflection is then

$$V_{v_{nm}}(x, t) = \frac{\sigma_v(x, t)}{v_0} = \left[\sum_{n=1}^{N}\sum_{m=1}^{N}\varepsilon_n\varepsilon_m\alpha^{-1}\sin^2\frac{x}{L}Z_{nm}^2 V_{F_{nm}}^2\right]^{1/2}. \quad (9.16)$$

9.1.2 Horizontal vibration

Horizontal vibrations of the beam in a transverse direction $w(x, t)$ according to Fig. 9.1 are generated by lateral random forces $H_n(t)$ due to random irregularities and sinusoidal motion. The differential equation for the deformation of the beam is analogous with equation (9.1)

$$EI_y\, w^{IV}(x, t) + \mu\, \ddot{w}(x, t) = \sum_{n=1}^{N}\varepsilon_n\delta(x + d_n - ct)H_n(t), \quad (9.17)$$

where I_y is the moment of inertia of the beam cross section with regard to the vertical axis y. The circular natural frequency of horizontal vibrations of the simply supported beam is

$$\omega_{hj}^2 = \frac{j^4\pi^4}{l^4}\frac{EI_y}{\mu}, \quad f_{hj} = \frac{\omega_{hj}}{2\pi}, \quad (9.18)$$

see Table 9.1.

a) Approximate solution

The approximate solution is obtained from equation (9.17) where the second term on the left-hand side is neglected. Thus, a quasistatic solution is obtained, analogous with equation (9.3):

$$w(x, t) = \sum_{j=1}^{\infty}\sum_{n=1}^{N}\varepsilon_n\frac{2}{\mu l\omega_{hj}^2}\sin\frac{x}{L_j}H_n(t)\sin\frac{ct - d_n}{L_j}. \quad (9.19)$$

The horizontal forces $H_n(t)$ have zero mean values, hence the covariance of horizontal deformations of the beam is

$$C_{ww}(x_1, x_2, t_1, t_2) = \sum_{j=1}^{\infty}\sum_{k=1}^{\infty}\sum_{n=1}^{N}\sum_{m=1}^{N}\varepsilon_n\varepsilon_m\frac{4}{\mu^2 l^2\omega_{hj}^2\omega_{hk}^2}$$

$$\cdot \sin\frac{x_1}{L_j}\sin\frac{x_2}{L_k}\sin\frac{ct_1 - d_n}{L_j}\sin\frac{ct_2 - d_m}{L_k}C_{H_nH_m}(t_1, t_2), \quad (9.20)$$

where $C_{H_nH_m}(t_1, t_2)$ are cross covariances of horizontal forces $H_n(t)$.

The variance is then

$$\sigma_w^2(x, t) = \sum_{j=1}^{\infty}\sum_{k=1}^{\infty}\sum_{n=1}^{N}\sum_{m=1}^{N}\varepsilon_n\varepsilon_m\frac{4}{\mu^2 l^2\omega_{hj}^2\omega_{hk}^2}$$

$$\cdot \sin\frac{x}{L_j}\sin\frac{x}{L_k}\sin\frac{ct - d_n}{L_j}\sin\frac{ct - d_m}{L_k}C_{H_nH_m}(t, t), \quad (9.21)$$

9.1 Beam

and for $j = k = 1$ may be simplified to

$$\sigma_w^2(x, t) = \sum_{n=1}^{N}\sum_{m=1}^{N} \varepsilon_n \varepsilon_m v_0^2 \varphi_h^4 \sin^2 \frac{x}{L} \sin \frac{ct - d_n}{L} \sin \frac{ct - d_m}{L} V_{H_n H_m}^2(t, t),$$

(9.22)

where

$$\varphi_h = \frac{\omega_1}{\omega_{h_1}},$$

$$V_{H_n H_m}^2(t, t) = C_{H_n H_m}(t, t) / F^2.$$

(9.23)

The coefficient of variation yields

$$V_h(x, t) = \frac{\sigma_w(x, t)}{v_0}$$

$$= \left[\sum_{n=1}^{N}\sum_{m=1}^{N} \varepsilon_n \varepsilon_m \varphi_h^4 \sin^2 \frac{x}{L} \sin \frac{ct - d_n}{L} \sin \frac{ct - d_m}{L} V_{H_n H_m}^2(t, t) \right]^{1/2},$$

(9.24)

which can be simplified as equation (9.9).

b) Dynamic solution

The solution of equation (9.17) may be found as (see equation (9.10)):

$$w(x, t) = \sum_{j=1}^{\infty}\sum_{n=1}^{N} \varepsilon_n \frac{2}{\mu l \omega_{hj}} \sin \frac{x}{L_j} \int_0^t H_n(\tau) \sin \frac{c\tau - d_n}{L_j} \sin \omega_{hj}(t - \tau) d\tau,$$

(9.25)

and the covariance of horizontal transverse deformations is

$$C_{ww}(x_1, x_2, t_1, t_2) = \sum_{j=1}^{\infty}\sum_{k=1}^{\infty}\sum_{n=1}^{N}\sum_{m=1}^{N} \varepsilon_n \varepsilon_m \frac{4}{\mu^2 l^2 \omega_{hj} \omega_{hk}}$$

$$\cdot \sin \frac{x_1}{L_j} \sin \frac{x_2}{L_k} \int_0^{t_1}\int_0^{t_2} \sin \frac{c\tau_1 - d_n}{L_j} \sin \frac{c\tau_2 - d_m}{L_k} \sin \omega_{hj}(t_1 - \tau_1)$$

$$\cdot \sin \omega_{hk}(t_2 - \tau_2) C_{H_n H_m}(\tau_1, \tau_2) d\tau_1 d\tau_2.$$

(9.26)

The variance for $j = k = 1$ and $C_{H_n H_m}(t_1, t_2) = S_{H_n H_m}\delta(t_2 - t_1)$ is calculated from equation (9.26)

$$\sigma_w^2(x, t) = \sum_{n=1}^{N}\sum_{m=1}^{N} \varepsilon_n \varepsilon_m v_0^2 \varphi_h^2 \alpha^{-1} \sin^2 \frac{x}{L} Z_{nm}^2 V_{H_n H_m}^2,$$

(9.27)

where Z_{nm}^2 is determined by equation (9.15) with $a = 1/(\alpha\varphi_h)$ and

$$V_{H_n H_m} = (S_{H_{nm}} \omega_1)^{1/2} / F.$$

(9.28)

is the coefficient of variation for forces $H_n(t)$.

The coefficient of variation of horizontal deflection gives:

$$V_{w_{nm}}(x, t) = \frac{\sigma_w(x, t)}{v_0} = \left[\sum_{n=1}^{N} \sum_{m=1}^{N} \varepsilon_n \varepsilon_m \varphi_h^2 \alpha^{-1} \sin^2 \frac{x}{L} Z_{nm}^2 V_{H_{nm}}^2 \right]^{1/2}. \quad (9.29)$$

9.1.3 Torsional vibration

The horizontal lateral forces of railway vehicles act on the rail top level, i.e. outside the cross section centroid of the bridge in the majority of cases. Let the difference of elevations according to Fig. 9.1 be h. Consequently, they affect the bridge by twisting moments $h\,H_n(t)$. The differential equation of a beam due to simple torsion is

$$-GI_\xi \xi''(x, t) + \mu_\xi \ddot{\xi}(x, t) = \sum_{n=1}^{N} \varepsilon_n \delta(x + d_n - ct) h\, H_n(t), \quad (9.30)$$

where $\xi(x, t)$ is the rotation about the longitudinal beam axis x,
$\quad G$ – modulus of elasticity in shear,
$\quad GI_\xi$ – moment of torsional rigidity per unit length,
$\quad \mu_\xi$ – mass polar moment of inertia with regard to axis x per unit length.

The natural circular frequency of the simply supported beam in torsion is

$$\omega_{\xi j}^2 = \frac{j^2 \pi^2}{l^2} \frac{GI_\xi}{\mu_\xi}, \qquad f_{\xi j} = \frac{\omega_{\xi j}}{2\pi}, \quad (9.31)$$

see Table 9.1.

a) Approximate solution

The approximate quasistatic solution of equation (9.30) is when neglecting the second term in the left hand side,

$$\xi(x, t) = \sum_{j=1}^{\infty} \sum_{n=1}^{N} \varepsilon_n \frac{2h}{\mu_\xi l \omega_{\xi j}^2} \sin \frac{x}{L_j} H_n(t) \sin \frac{ct - d_n}{L_j}. \quad (9.32)$$

The same procedure yields the covariance of torsion

$$C_{\xi\xi}(x_1, x_2, t_1, t_2) = \sum_{j=1}^{\infty} \sum_{k=1}^{\infty} \sum_{n=1}^{N} \sum_{m=1}^{N} \varepsilon_n \varepsilon_m \frac{4h^2}{\mu_\xi^2 l^2 \omega_{\xi j}^2 \omega_{\xi k}^2}$$

$$\cdot \sin \frac{x_1}{L_j} \sin \frac{x_2}{L_k} \sin \frac{ct_1 - d_n}{L_j} \sin \frac{ct_2 - d_m}{L_k} C_{H_n H_m}(t_1, t_2), \quad (9.33)$$

and the variance

$$\sigma_\xi^2(x, t) = \sum_{j=1}^{\infty} \sum_{k=1}^{\infty} \sum_{n=1}^{N} \sum_{m=1}^{N} \varepsilon_n \varepsilon_m \frac{4h^2}{\mu_\xi^2 l^2 \omega_{\xi j}^2 \omega_{\xi k}^2}$$

$$\cdot \sin \frac{x}{L_j} \sin \frac{x}{L_k} \sin \frac{ct - d_n}{L_j} \sin \frac{ct - d_m}{L_k} C_{H_n H_m}(t, t). \quad (9.34)$$

9.1 Beam

For $j = k = 1$ the equation (9.34) will be simplified to

$$\sigma_\xi^2(x, t) = \sum_{n=1}^{N} \sum_{m=1}^{N} \varepsilon_n \varepsilon_m \frac{v_0^2}{h^2} \gamma^2 \varphi_\xi^4 \sin^2 \frac{x}{L} \sin \frac{ct - d_n}{L} \sin \frac{ct - d_m}{L} V_{H_n H_m}^2(t, t),$$

(9.35)

where $V_{H_n H_m}^2(t, t) = C_{H_n H_m}(t, t)/F^2$ coefficient of variation of forces $H_n(t)$,

$$\varphi_\xi = \frac{\omega_1}{\omega_{\xi 1}},$$

$$\gamma = \frac{\mu h^2}{\mu_\xi}.$$

(9.36)

With these symbols the coefficient of variation for beam torsion yields

$$V_\xi(x, t) = \frac{\sigma_\xi(x, t)}{v_0/h}$$

$$= \left[\sum_{n=1}^{N} \sum_{m=1}^{N} \varepsilon_n \varepsilon_m \gamma^2 \varphi_\xi^4 \sin^2 \frac{x}{L} \sin \frac{ct - d_n}{L} \sin \frac{ct - d_m}{L} V_{H_n H_m}^2(t, t) \right]^{1/2}.$$ (9.37)

The double series in equation (9.37) can be simplified once again to a single series by the same approach as that used for equation (9.9).

b) Dynamic solution

The solution of equation (9.30) representing torsional vibration of a beam loaded by a series of moving random forces is

$$\xi(x, t) = \sum_{j=1}^{\infty} \sum_{n=1}^{N} \varepsilon_n \frac{2h}{\mu_\xi l \omega_{\xi j}} \sin \frac{x}{L_j} \int_0^t \sin \omega_{\xi j}(t - \tau) \sin \frac{c\tau - d_n}{L_j} H_n(\tau) \, d\tau.$$

(9.38)

The covariance of beam torsion is calculated from its definition and equation (9.38):

$$C_{\xi\xi}(x_1, x_2, t_1, t_2) = \sum_{j=1}^{\infty} \sum_{k=1}^{\infty} \sum_{n=1}^{N} \sum_{m=1}^{N} \varepsilon_n \varepsilon_m \frac{4h^2}{\mu_\xi^2 l^2 \omega_{\xi j} \omega_{\xi k}} \sin \frac{x_1}{L_j}$$

$$\cdot \sin \frac{x_2}{L_k} \int_0^{t_1} \int_0^{t_2} \sin \omega_{\xi j}(t_1 - \tau_1) \sin \omega_{\xi k}(t_2 - \tau_2) \sin \frac{c\tau_1 - d_n}{L_j} \sin \frac{c\tau_2 - d_m}{L_k}$$

$$\cdot C_{H_n H_m}(\tau_1, \tau_2) \, d\tau_1 \, d\tau_2.$$ (9.39)

Equation (9.39) can be calculated in closed form for white noise $C_{H_n H_m}(t_1, t_2) = S_{H_{nm}} \delta(t_2 - t_1)$. For $j = k = 1$, the variance of torsional vibration gives

$$\sigma_\xi^2(x,\,t) = \sum_{n=1}^{N}\sum_{m=1}^{N} \varepsilon_n \varepsilon_m \frac{4h^2 S_{H_{nm}}}{\mu_\xi^2 l^2 \omega_{\xi 1}^2} \sin^2 \frac{x}{L} \int_0^t \sin^2 \omega_\xi(t-\tau)\sin\frac{c\tau - d_n}{L}$$

$$\cdot \sin\frac{c\tau - d_m}{L}\,d\tau = \sum_{n=1}^{N}\sum_{m=1}^{N} \varepsilon_n \varepsilon_m \frac{v_0^2}{h^2}\gamma^2\varphi_\xi^2\alpha^{-1}\sin^2\frac{x}{L} Z_{nm}^2 V_{H_{nm}}^2, \quad (9.40)$$

where the symbols of equation (9.36) and equation (9.15) with $a = 1/(\alpha\varphi_\xi)$ and $V_{H_n H_m} = (S_{H_{nm}}\omega_1)^{1/2}/F$ as the coefficient of variation of horizontal forces have been used.

The coefficient of variation for the beam torsion is then

$$V_\xi(x,\,t) = \frac{\sigma_\xi(x,\,t)}{v_0/h} = \left[\sum_{n=1}^{N}\sum_{m=1}^{N}\varepsilon_n\varepsilon_m\gamma^2\varphi_\xi^2\alpha^{-1}\sin^2\frac{x}{L} Z_{nm}^2 V_{H_{nm}}^2\right]^{1/2}. \quad (9.41)$$

9.2 Thin-walled bar with vertical axis of symmetry

The idealization of the bridge by a mass beam in Sect. 9.1 made it possible to investigate separately the vertical, horizontal and torsional vibrations. Actually, however, these motions are mutually coupled. The theoretical model characterizing this circumstance is the Vlasov and Umanskiĭ thin-walled bar; the assumptions for its derivation are described in greater detail in Chapter 22 of [68]. Most bridges approximately satisfy these assumptions.

The majority of bridges possess a vertical axis of symmetry, which brings considerable simplifications. In the first place, vertical vibrations become independent of other motions (for their solution see Sect. 9.1.1) and horizontal longitudinal vibrations can be analyzed separately. In this chapter, only the lateral actions of vehicles are of interest and for this case horizontal forces and their twisting moments influence the coupled bending and torsional vibrations in the direction of axis z and around the axis x according to Fig. 9.2.

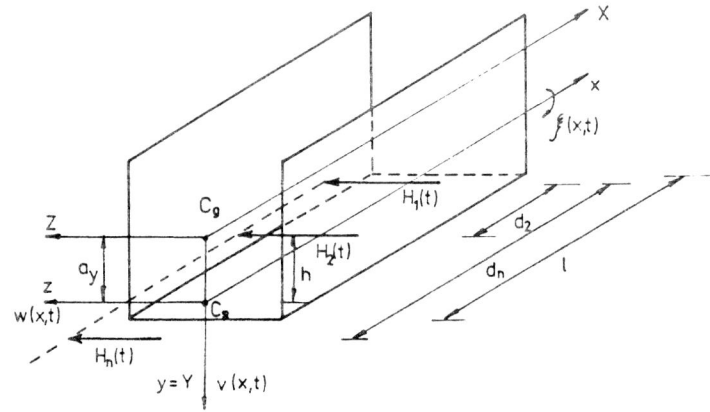

Fig. 9.2. Movement of horizontal random forces $H_n(t)$ spaced at d_n along a thin-walled bar of the length l.

9.2. Thin-walled bar with vertical axis of symmetry

The differential equations for both of these motions are (following Vlasov [216])

$$EI_y w^{IV}(x, t) + \mu r_y^2 \ddot{w}''(x, t) + \mu \ddot{w}(x, t) - \mu a_y \ddot{\xi}(x, t)$$

$$= \sum_{n=1}^{N} \varepsilon_n \delta(x + d_n - ct) H_n(t),$$

$$EI_\varphi \xi^{IV}(x, t) - GI_\xi \xi''(x, t) - \mu r_\varphi^4 \ddot{\xi}''(x, t) + \mu r^2 \ddot{\xi}(x, t) - \mu a_y \ddot{w}(x, t)$$

$$= \sum_{n=1}^{N} \varepsilon_n \delta(x + d_n - ct) h H_n(t), \qquad (9.42)$$

where, in addition to the symbols explained in the preceding text:
I_ξ – polar moment of inertia of the cross section with regard to axis x,
$I = \int_A \varphi^2 \, dA$ – section moment of inertia,
φ – sector coordinate equal to double the sector area the pole of which coincides with the centre of bending C_s whose zero position can be found from the condition $\int_A \varphi \, dA = 0$ where A is the cross section area; the centre of bending C_s in thin-walled bars differs from the cross-section centroid C_g – see Fig. 9.2,
a_y – coordinate of the centre of bending C_s,

$$r_y^2 = \frac{I_y}{A}, \quad r_z^2 = \frac{I_z}{A}, \quad r_\varphi^4 = \frac{I_\varphi}{A}, \quad r^2 = a_y^2 + r_y^2 + r_z^2.$$

For the solution of the problem of movement of a series of horizontal random forces $H_n(t)$ along a thin-walled bar with vertical axis of symmetry, described by equations (9.42), the following symbols will be used (see Table 9.1):

$$\mu_y = \mu \left(1 + \frac{j^2 \pi^2 r_y^2}{l^2}\right), \quad \mu_\varphi = \mu \left(r^2 + \frac{j^2 \pi^2 r_\varphi^4}{l^2}\right),$$

$$b_y = \mu a_y / \mu_y, \quad b_\varphi = \mu a_y / \mu_\varphi,$$

$$\omega_{yj}^2 = \frac{j^4 \pi^4}{l^4} \frac{EI_y}{\mu_y}, \quad \omega_{\varphi j}^2 = \frac{j^4 \pi^4}{l^4} \frac{EI_\varphi}{\mu_\varphi} + \frac{j^2 \pi^2}{l^2} \frac{GI_\xi}{\mu_\varphi},$$

$$f_{yj} = \omega_{yj} / (2\pi), \quad f_{\varphi j} = \omega_{\varphi j} / (2\pi). \qquad (9.43)$$

The boundary conditions of a simply supported beam and zero initial conditions are assumed. In a thin-walled beam, the natural vibration frequencies in the direction of axis z is ω_{yj}, and in torsion about axis x, $\omega_{\varphi j}$.

9.2.1 Approximate solution

Once again, the approximate solution of the system of equations (9.42) will be sought by neglecting (with regard to the low velocity of motion) all terms with derivatives in respect of time. Thus, the quasistatic solution is obtained:

$$w(x, t) = \sum_{j=1}^{\infty} \sum_{n=1}^{N} \varepsilon_n \frac{2}{\mu_y l \omega_{yj}^2} \sin \frac{x}{L_j} H_n(t) \sin \frac{ct - d_n}{L_j},$$

$$\xi(x, t) = \sum_{j=1}^{\infty} \sum_{n=1}^{N} \varepsilon_n \frac{2h}{\mu_\varphi l \omega_{\varphi j}^2} \sin \frac{x}{L_j} H_n(t) \sin \frac{ct - d_n}{L_j}. \quad (9.44)$$

It is assumed that the horizontal forces $H_n(t)$ have zero mean values, so that the mean values of the functions $w(x, t)$ and $\xi(x, t)$ are zero as well. Statistical characteristics of the second order are obtained from equation (9.44) and from the definition of covariance

$$C_{ww}(x_1, x_2, t_1, t_2) = \sum_{j=1}^{\infty} \sum_{k=1}^{\infty} \sum_{n=1}^{N} \sum_{m=1}^{N} \varepsilon_n \varepsilon_m \frac{4}{\mu_y^2 l^2 \omega_{yj}^2 \omega_{yk}^2}$$

$$\cdot \sin \frac{x_1}{L_j} \sin \frac{x_2}{L_k} \sin \frac{ct_1 - d_n}{L_j} \sin \frac{ct_2 - d_m}{L_k} C_{H_n H_m}(t_1, t_2),$$

$$C_{\xi\xi}(x_1, x_2, t_1, t_2) = \sum_{j=1}^{\infty} \sum_{k=1}^{\infty} \sum_{n=1}^{N} \sum_{m=1}^{N} \varepsilon_n \varepsilon_m \frac{4h^2}{\mu_\varphi^2 l^2 \omega_{\varphi j}^2 \omega_{\varphi k}^2}$$

$$\cdot \sin \frac{x_1}{L_j} \sin \frac{x_2}{L_k} \sin \frac{ct_1 - d_n}{L_j} \sin \frac{ct_2 - d_m}{L_k} C_{H_n H_m}(t_1, t_2). \quad (9.45)$$

The variance is found from equation (9.45)

$$\sigma_w^2(x, t) = \sum_{j=1}^{\infty} \sum_{k=1}^{\infty} \sum_{n=1}^{N} \sum_{m=1}^{N} \varepsilon_n \varepsilon_m \frac{4}{\mu_y^2 l^2 \omega_{yj}^2 \omega_{yk}^2}$$

$$\cdot \sin \frac{x}{L_j} \sin \frac{x}{L_k} \sin \frac{ct - d_n}{L_j} \sin \frac{ct - d_m}{L_k} C_{H_n H_m}(t, t),$$

$$\sigma_\xi^2(x, t) = \sum_{j=1}^{\infty} \sum_{k=1}^{\infty} \sum_{n=1}^{N} \sum_{m=1}^{N} \varepsilon_n \varepsilon_m \frac{4h^2}{\mu_\varphi^2 l^2 \omega_{\varphi j}^2 \omega_{\varphi k}^2}$$

$$\cdot \sin \frac{x}{L_j} \sin \frac{x}{L_k} \sin \frac{ct - d_n}{L_j} \sin \frac{ct - d_m}{L_k} C_{H_n H_m}(t, t). \quad (9.46)$$

The coefficient of variation is obtained from equation (9.46) for $j = k = 1$

$$V_w(x, t) = \frac{\sigma_w(x, t)}{v_0}$$

$$= \left[\sum_{n=1}^{N} \sum_{m=1}^{N} \varepsilon_n \varepsilon_m \gamma_y^2 \varphi_y^4 \sin^2 \frac{x}{L} \sin \frac{ct - d_n}{L} \sin \frac{ct - d_m}{L} V_{H_n H_m}^2(t, t) \right]^{1/2},$$

$$V_\xi(x, t) = \frac{\sigma_\xi(x, t)}{v_0 / h}$$

$$= \left[\sum_{n=1}^{N} \sum_{m=1}^{N} \varepsilon_n \varepsilon_m \gamma_\varphi^2 \varphi_\varphi^4 \sin^2 \frac{x}{L} \sin \frac{ct - d_n}{L} \sin \frac{ct - d_m}{L} V_{H_n H_m}^2(t, t) \right]^{1/2}. \quad (9.47)$$

where
$$v_0 = \frac{2F}{\mu l \omega_1^2}, \qquad V_{H_n H_m}^2(t, t) = C_{H_n H_m}(t, t)/F^2,$$

$$\gamma_y = \frac{\mu}{\mu_y}, \qquad \gamma_\varphi = \frac{\mu h^2}{\mu_\varphi},$$

$$\varphi_y = \frac{\omega_1}{\omega_{y_1}}, \qquad \varphi_\varphi = \frac{\omega_1}{\omega_{\varphi_1}}. \tag{9.48}$$

9.2.2 Dynamic solution

The solution of equations (9.42) for a general time history of the forces $H_n(t)$ may be derived by the integral transformation method by means of equations (27.5), (27.32), (27.33), (27.67) and (27.74) from [68]. The method results in the equations for the horizontal deflection and torsion of a thin-walled bar under lateral forces in the form

$$w(x, t) = \sum_{j=1}^{\infty} \sum_{n=1}^{N} \frac{2 E_{jn}}{l} \sin \frac{x}{L_j} \int_0^t H_n(\tau) \sin \frac{c\tau - d_n}{L_j}$$
$$\cdot \left[A_j \sin \omega'_{yj}(t - \tau) + B_j \sin \omega'_{\varphi j}(t - \tau) \right] d\tau,$$

$$\xi(x, t) = \sum_{j=1}^{\infty} \sum_{n=1}^{N} \frac{2 E_{jn}}{l} \sin \frac{x}{L_j} \int_0^t H_n(\tau) \sin \frac{c\tau - d_n}{L_j}$$
$$\cdot \left[C_j \sin \omega'_{yj}(t - \tau) + D_j \sin \omega'_{\varphi j}(t - \tau) \right] d\tau, \tag{9.49}$$

where

$$\left. \begin{array}{c} \omega'^2_{yj} \\ \omega'^2_{\varphi j} \end{array} \right\} = \frac{\omega^2_{yj} + \omega^2_{\varphi j}}{2(1 - b_y b_\varphi)} \mp \left[\frac{1}{4} \left(\frac{\omega^2_{yj} + \omega^2_{\varphi j}}{1 - b_y b_\varphi} \right)^2 - \frac{\omega^2_{yj} \omega^2_{\varphi j}}{1 - b_y b_\varphi} \right]^{1/2},$$

$$f'_{yj} = \omega'_{yj}/(2\pi); \qquad f'_{\varphi j} = \omega'_{\varphi j}/(2\pi), \qquad \text{see Table 9.1,}$$

$$A_j = \frac{\omega'_{yj}}{\mu_y} - \frac{\omega^2_{\varphi j}}{\mu_y \omega'_{yj}} + \frac{hb_y \omega'_{yj}}{\mu_\varphi},$$

$$B_j = -\frac{\omega'_{\varphi j}}{\mu_y} + \frac{\omega^2_{\varphi j}}{\mu_y \omega'_{\varphi j}} - \frac{hb_y \omega'_{\varphi j}}{\mu_\varphi},$$

$$C_j = \frac{h \omega'_{yj}}{\mu_\varphi} - \frac{h \omega^2_{yj}}{\mu_\varphi \omega'_{yj}} + \frac{b_\varphi \omega'_{yj}}{\mu_y},$$

$$D_j = -\frac{h \omega'_{\varphi j}}{\mu_\varphi} + \frac{h \omega^2_{yj}}{\mu_\varphi \omega'_{\varphi j}} - \frac{b_\varphi \omega'_{\varphi j}}{\mu_y},$$

$$E_{jn} = \frac{\varepsilon_n}{(1 - b_y b_\varphi)(\omega'^2_{yj} - \omega'^2_{\varphi j})}. \tag{9.50}$$

Equations (9.49) can be used for the derivation of the equations for covariances

$$C_{ww}(x_1, x_2, t_1, t_2) = \sum_{j=1}^{\infty}\sum_{k=1}^{\infty}\sum_{n=1}^{N}\sum_{m=1}^{N} \frac{4E_{jn}E_{km}}{l^2} \sin\frac{x_1}{L_j} \sin\frac{x_2}{L_k} \int_0^{t_1}\int_0^{t_2} C_{H_n H_m}(\tau_1, \tau_2)$$

$$\cdot \sin\frac{c\tau_1 - d_n}{L_j} \sin\frac{c\tau_2 - d_m}{L_k} \left[A_j \sin\omega'_{yj}(t_1 - \tau_1) + B_j \sin\omega'_{\varphi j}(t_1 - \tau_1)\right]$$

$$\cdot \left[A_k \sin\omega'_{yk}(t_2 - \tau_2) + B_k \sin\omega'_{\varphi k}(t_2 - \tau_2)\right] d\tau_1 d\tau_2 ,$$

$$C_{\xi\xi}(x_1, x_2, t_1, t_2) = \sum_{j=1}^{\infty}\sum_{k=1}^{\infty}\sum_{n=1}^{N}\sum_{m=1}^{N} \frac{4E_{jn}E_{km}}{l^2} \sin\frac{x_1}{L_j} \sin\frac{x_2}{L_k} \int_0^{t_1}\int_0^{t_2} C_{H_n H_m}(\tau_1, \tau_2)$$

$$\cdot \sin\frac{c\tau_1 - d_n}{L_j} \sin\frac{c\tau_2 - d_m}{L_k} \left[C_j \sin\omega'_{yj}(t_1 - \tau_1) + D_j \sin\omega'_{\varphi j}(t_1 - \tau_1)\right]$$

$$\cdot \left[C_k \sin\omega'_{yk}(t_2 - \tau_2) + D_k \sin\omega'_{\varphi k}(t_2 - \tau_2)\right] d\tau_1 d\tau_2 . \qquad (9.51)$$

The variance is found from equation (9.51)

$$\sigma_w^2(x, t) = \sum_{j=1}^{\infty}\sum_{k=1}^{\infty}\sum_{n=1}^{N}\sum_{m=1}^{N} \frac{4E_{jn}E_{km}}{l^2} \sin\frac{x}{L_j} \sin\frac{x}{L_k} \int_0^{t}\int_0^{t} C_{H_n H_m}(\tau_1, \tau_2)$$

$$\cdot \sin\frac{c\tau_1 - d_n}{L_j} \sin\frac{c\tau_2 - d_m}{L_k} \left[A_j \sin\omega'_{yj}(t - \tau_1) + B_j \sin\omega'_{\varphi j}(t - \tau_1)\right]$$

$$\cdot \left[A_k \sin\omega'_{yk}(t - \tau_2) + B_k \sin\omega'_{\varphi k}(t - \tau_2)\right] d\tau_1 d\tau_2 ,$$

$$\sigma_\xi^2(x, t) = \sum_{j=1}^{\infty}\sum_{k=1}^{\infty}\sum_{n=1}^{N}\sum_{m=1}^{N} \frac{4E_{jn}E_{km}}{l^2} \sin\frac{x}{L_j} \sin\frac{x}{L_k} \int_0^{t}\int_0^{t} C_{H_n H_m}(\tau_1, \tau_2)$$

$$\cdot \sin\frac{c\tau_1 - d_n}{L_j} \sin\frac{c\tau_2 - d_m}{L_k} \left[C_j \sin\omega'_{yj}(t - \tau_1) + D_j \sin\omega'_{\varphi j}(t - \tau_1)\right]$$

$$\cdot \left[C_k \sin\omega'_{yk}(t - \tau_2) + D_k \sin\omega'_{\varphi k}(t - \tau_2)\right] d\tau_1 d\tau_2 . \qquad (9.52)$$

For white noise $C_{H_n H_m}(t_1, t_2) = S_{H_n H_m}\delta(t_2 - t_1)$ and for $j = k = 1$ the variance will simplify to

$$\sigma_w^2(x, t) = \sum_{n=1}^{N}\sum_{m=1}^{N} \frac{4S_{H_{nm}}E_{1n}E_{1m}}{l^2} \sin^2\frac{x}{L} \int_0^{t} \sin\frac{c\tau - d_n}{L} \sin\frac{c\tau - d_m}{L}$$

$$\cdot \left[A_1 \sin\omega'_{y1}(t - \tau) + B_1 \sin\omega'_{\varphi 1}(t - \tau)\right]^2 d\tau ,$$

$$\sigma_\xi^2(x, t) = \sum_{n=1}^{N}\sum_{m=1}^{N} \frac{4S_{H_{nm}}E_{1n}E_{1m}}{l^2} \sin^2\frac{x}{L} \int_0^{t} \sin\frac{c\tau - d_n}{L} \sin\frac{c\tau - d_m}{L}$$

$$\cdot \left[C_1 \sin\omega'_{y1}(t - \tau) + D_1 \sin\omega'_{\varphi 1}(t - \tau)\right]^2 d\tau . \qquad (9.53)$$

9.2.3 Thin-walled bar with two axes of symmetry

Should the thin-walled bar possesses two axes of symmetry, which is a rare case in bridges, the centre of bending C_s coincides with the cross section centroid C_g, $a_y = 0$, see Fig. 9.2. In that case both the vertical, horizontal and torsion vibrations are mutually independent.

Equations (9.42) differ from equations (9.17) and (9.30) only in that the former include the influence of the warping of the thin-walled bar cross section, which is not considered for the beam.

However, the solutions of equations (9.42) for $a_y = 0$ are identical with the expressions in Sects 9.1.2 and 9.1.3. In the case of horizontal vibrations, however, it is necessary to substitute μ_y and ω_{yj} for μ and ω_{hj}, respecively (in Sect. 9.1.2) and μ_φ and $\omega_{\varphi j}$ for μ_ξ and $\omega_{\xi j}$, respectively (in Sect. 9.1.3), in accordance with equation (9.43).

9.3 Experiments on bridges

The lateral action of vehicles due to horizontal rail irregularities and sinusoidal motion are of a typically random character and can be replaced by horizontal transverse forces with zero mean values.

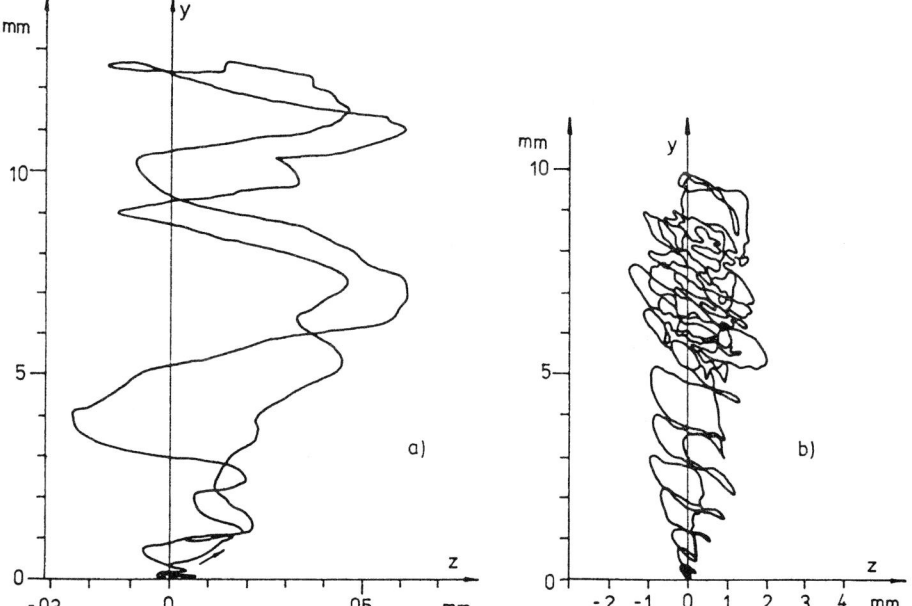

Fig. 9.3. Vertical (y) and horizontal (z) deflection of a steel lattice bridge with span $l = 48.4$ m at the point $x = 0.4l$ (according to the measurements by J. Sláma):
a) during the drive of the T 678.0009 locomotive weighing 1100 kN at the velocity of 31.3 km h^{-1} from the fixed to the movable bearing, b) during the drive of a mixed freight train, consisting of the T 435 locomotive (weight 610 kN) and 12 cars, at a velocity of 46.4 km h^{-1} from the fixed to the movable bearing.

The response to such a load is spatial stochastic vibrations of railway bridges. This is illustrated in Fig. 9.3, which shows the spatial motion of one point of a bridge with a span $l = 48.4$ m in the vertical plane y, z perpendicular to the longitudinal bridge axis x (according to the orientation of axes in Fig. 9.2). On this bridge, the vertical displacements $v(x, t)$ in the direction of axis y and the horizontal displacements $w(x, t)$ in the direction of axis z were measured separately and independently on each other; in Fig. 9.3 both of these displacements have been summed as vectors. Fig. 9.3 shows that every point of the bridge is undergoing a complex spatial motion of random character.

TABLE 9.1. Spatial vibrations of steel truss bridges with open bridge deck

Vibration type	Symbol	Equation	j	Bridge with span	
				$l = 25.85$ m	$l = 48.4$ m
Vertical	f_j	(9.11)	1	8.7	5.4
			2	34.7	21.5
			3	78.1	48.3
Horizontal	f_{hj}	(9.18)	1	15.3	4.7
			2	61.1	19.0
			3	137.5	42.7
	f_{vj}	(9.43)	1	14.8	4.7
			2	53.9	18.1
			3	107.4	38.8
	f'_{vj}	(9.50)	1	14.7	4.6
			2	52.5	17.5
			3	101.7	36.4
Torsional	$f_{\bar{z}j}$	(9.31)	1	35.7	19.2
			2	71.3	38.4
			3	107.0	57.7
	$f_{\varphi j}$	(9.43)	1	34.9	16.7
			2	73.5	35.4
			3	117.4	56.9
	$f'_{\varphi j}$	(9.50)	1	36.5	19.7
			2	77.6	41.9
			3	126.4	67.5
Constants	γ_v	(9.48)		0.934	0.977
	γ_φ	(9.48)		0.118	0.087
	φ_y	(9.48)		0.588	1.147
	φ_φ	(9.48)		0.244	0.279

In addition, another bridge of span $l = 25.85$ m was investigated experimentally. Both bridges are steel lattice structures with open bridge deck. The horizontal transverse wheel forces on the second bridge have a mean value equal to about one third of the vertical wheel forces and are analyzed in some detail in Sect. 10.1.3.

Table 9.1 presents the first three natural frequencies of vertical, horizontal and torsional vibration of a mass beam and of the thin-walled bar, idealizing the two investigated bridges analysed according to the various formulas from Sects 9.1 and 9.2.

The natural frequencies of vertical and horizontal vibrations are relatively close to each other and in some cases are difficult to separate.

On the other hand, the natural frequences of torsional vibrations are usually higher than the corresponding frequencies of vertical and horizontal vibrations.

9.4 Coefficients of variation for horizontal and torsional vibrations

The approximate expressions for the coefficient of variation of horizontal and torsional vibrations, equations (9.47), at midspan of the beam $x = l/2$ can be modified in the following form:

$$V_w\left(\frac{l}{2}, t\right) = \frac{\sigma_w\left(\frac{l}{2}, t\right)}{v_0} = \gamma_y \varphi_y^2 \text{ var}(\tau),$$

$$V_\xi\left(\frac{l}{2}, t\right) = \frac{\sigma_\xi\left(\frac{l}{2}, t\right)}{v_0/h} = \gamma_\varphi \varphi_\varphi^2 \text{ var}(\tau), \tag{9.54}$$

where the function var (τ) characterizing the variance is

$$\text{var}(\tau) = \left[\sum_{n=1}^{N}\sum_{m=1}^{N} \varepsilon_n \varepsilon_m \sin \pi(\tau - D_n) \sin \pi(\tau - D_m) V_{H_n H_m}^2(\tau)\right]^{1/2} \tag{9.55}$$

and the coefficients γ_y, γ_φ, φ_y, φ_φ for two investigated cases have been found numerically according to equations (9.48); they are given in Table 9.1.

In this process, the dimensionless time

$$\tau = ct/l \tag{9.56}$$

and the dimensionless axle distances (Fig. 9.2)

$$D_n = \frac{d_n}{l}, \quad D_1 = 0 \tag{9.57}$$

were introduced, so that the function ε_n appears in the form:

$$\varepsilon_n = \begin{cases} 1 \\ 0 \end{cases} \text{ for } \begin{cases} D_n \leq \tau \leq 1 + D_n \\ \tau < D_n, \tau > 1 + D_n \end{cases} \tag{9.58}$$

For the covariance of horizontal transverse forces, equation (9.48), the following expression was used provisionally

$$V^2_{H_n H_m}(t, t) = V_{nm} H \, e^{-T\tau}, \qquad (9.59)$$

until more experimental data are available (see also Sect. 10.1.3).

In equation (9.59):

V_{nm} is the matrix of mutual dependence of the nth and the mth horizontal forces (for example see equation (9.63)),

T – dimensionless coefficient characterizing the reduction of the coefficient of variation with time,

$H = (\bar{H}/\bar{F})^2$ is the square of the ratio of mean horizontal and vertical forces.

The form of the function var (τ) is studied in some details in Figs 9.4 through 9.7.

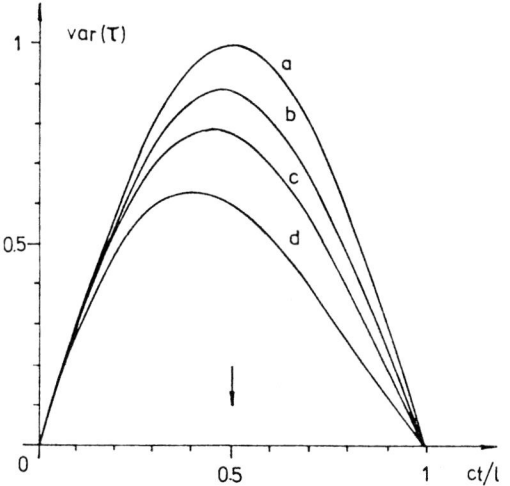

Fig. 9.4. Time-history of the function var (τ) under one force: influence of the damping coefficient T in equation (9.59):
a) $T = 0$; b) $T = 0.5$; c) $T = 1$; d) $T = 2$.

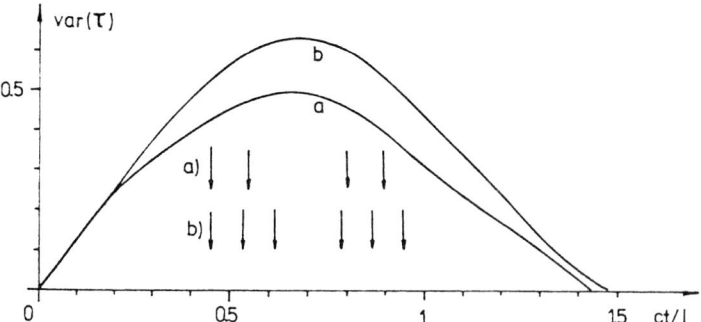

Fig. 9.5. Time-history of the function var (τ) on the 25.85 m span bridge during the drive of:
a) four-axle locomotive type T 478.3, b) six-axle locomotive type T 679.1.

9.4. Coefficients of variation for horizontal and torsional vibrations

The infuence of the damping coefficient T was investigated for the case of the movement of a single horizontal random force; the following coefficients were used in equations (9.55) and (9.59):

$$N = 1; \quad D_n = 0; \quad V_{nn} = 1; \quad V_{nm} = 0; \quad H = 1;$$

for a) $T = 0$; b) $T = 0.5$; c) $T = 1$ and d) $T = 2$. (9.60)

Fig. 9.4 shows that increasing the coefficient T reduces the coefficient of variation.

Fig. 9.5 compares the function var (τ) for a four-axle and a six-axle locomotive in the case of the 25.85 m span bridge (see Table 9.1). In this case, the exponential decay of the coefficients V_{nm} was used (e^{-n+1}, n being the number of the axle):

a) $N = 4; \quad D_n = 0, 0.093, 0.348, 0.441;$

$\quad V_{nm} = 1, 0.368, 0.135, 0.050; \quad H = 0.1; \quad T = 1,$ (9.61)

b) $N = 6; \quad D_n = 0, 0.081, 0.162, 0.333, 0.414, 0.495;$

$\quad V_{nm} = 1, 0.368, 0.135, 0.050, 0.018, 0.007;$

$\quad H = 0.1; \quad T = 1,$ (9.62)

Fig 9.5 shows that on the same bridge and with the same parameters the six-axle locomotive produces greater horizontal and torsional actions than the four axle locomotive.

Fig. 9.6 investigates the influence of the matrix V_{nm} of the mutual relations of the nth and the mth forces on the 48.8 m span bridge using a six-axle diesel-electric locomotive of the type T 678.0. In the case of the exponential decay the matrix V_{nm} presents, for instance, the following values:

$$V_{nm} = \begin{bmatrix} 1 & 0.368 & 0.135 & 0.050 & 0.018 & 0.007 \\ 0.368 & 1 & 0.368 & 0.135 & 0.050 & 0.018 \\ 0.135 & 0.368 & 1 & 0.368 & 0.135 & 0.050 \\ 0.050 & 0.135 & 0.368 & 1 & 0.368 & 0.135 \\ 0.018 & 0.050 & 0.135 & 0.368 & 1 & 0.368 \\ 0.007 & 0.018 & 0.050 & 0.135 & 0.368 & 1 \end{bmatrix}.$$ (9.63)

Similarly in other cases, the matrix V_{nm} being defined by the first line only. Further parameters are:

$N = 6; \quad D_n = 0, 0.041, 0.083, 0.196, 0.238, 0.279;$

a) $V_{nm} = 1, 1, 1, 1, 1, 1$;

b) $V_{nm} = 1, 0.8, 0.6, 0.4, 0.2, 0$;

c) $V_{nm} = 1, 0.368, 0.135, 0.050, 0.018, 0.007$;

d) $V_{nm} = 1, 0, 0, 0, 0, 0$;

$H = 0.1; \quad T = 1,$ (9.64)

Fig. 9.6 shows that the lower the coefficients V_{nm} in more distant axles, the lower the magnitude of the function var (τ).

Finally, the time-history of the function var (τ) was calculated for a number of cases of bridge spans within the limits of $l/d_n = 0.5$ to 10, d_n being the distance between the first and the last axles of a four-axle locomotive. The following coefficients were used in equations (9.55) and (9.59);

$$N = 4; \quad D_n = 2 \text{ to } 0,1; \quad V_{nm} = 1, 0.368, 0.135, 0.050;$$
$$H = 0.1; \quad T = 1. \tag{9.65}$$

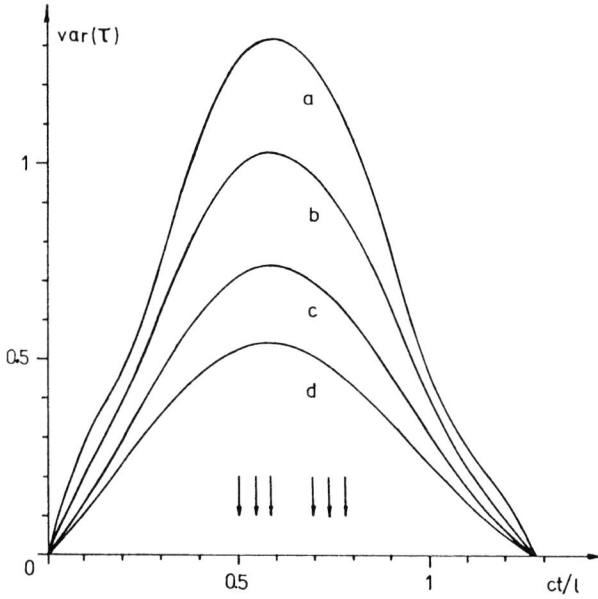

Fig. 9.6. Time-history of the function var (τ) on the 48.4 m span bridge; influence of coefficients V_{nm} in equation (9.59):
a) constant, $V_{nm} = 1$, b) linearly decreasing, $V_{nm} = 1; 0.8; 0.6; 0.4; 0.2; 0.2; 0$, c) exponentially decreasing, $V_{nm} = 1; 0.368; 0.135; 0.050; 0.018; 0.007$, d) zero except for V_{nn}, i.e. $V_{nm} = 1; 0; 0; 0; 0; 0$.

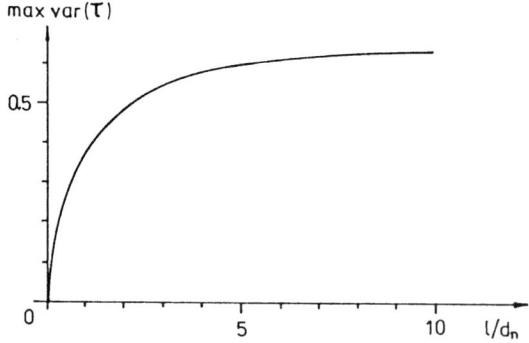

Fig. 9.7. Maximum values of var (τ) plotted against span l; d_n – distance between the first and the last

9.5 Centrifugal forces

If the track on the bridge is curved or in a transition curve, the movement of vehicles produces centrifugal forces.

Movement of a mass $m = F/g$ along a circular curve of radius r at a velocity c generates a centrifugal force

$$F_{cfg} = \frac{mc^2}{r}. \qquad (9.66)$$

After the substitution of the respective SI units, equation (9.66) is used for the formula given in the national standards

$$F_{cfg} = \frac{Fv^2}{127 r}. \qquad (9.67)$$

where F_{cfg} – centrifugal force (kN),
F – vertical moving load (kN),
v – velocity (km h^{-1}),
r – radius of curve (m).

Centrifugal force is of a deterministic character and acts horizontally at the centroid of vehicles, in a direction outwards from the curve.

If the centrifugal force F_{cfg} is to be calculated for the standard load scheme according to [212], the right-hand side of equation (9.67) is multiplied by the reduction coefficient

$$\psi = 1 - \frac{v - 120}{1000}\left(\frac{814}{v} + 1.75\right)\left[1 - \left(\frac{2.88}{l}\right)^{1/2}\right], \qquad (9.68)$$

where l – span in (m),
v – design velocity in (km h^{-1}).

The reduction coefficient ψ characterizes the fact that the passenger trains, which achieve high speeds, have considerably lower axle forces than the standard load and that lower centrifugal forces originate on bridges in reality than those calculated by equation (9.67). For $v \leqslant 120$ km h^{-1} and $l \leqslant 2.28$ m the reduction coefficient $\psi = 1.00$.

The centrifugal force acts always simultaneously with the vertical moving load. The simultaneity of action of centrifugal force and lateral impacts (i.e. forces due to horizontal rail irregularities and sinusoidal motion) has not yet been investigated sufficiently. Some authors believe that a high centrifugal force presses the wheel flange to the rail head so that the sinusoidal motion can no longer occur. The situation depends obviously on the velocity, superelevation and on the radius of curvature.

According to the majority of national standards, the action of centrifugal force and lateral impacts is assessed as a combination of two mutually independent loads with the application of the factor of load combination because the simultaneous occurrence is a random phenomenon. The centrifugal force is considered without any dynamic coefficient because the influence of velocity is already incorporated in equation (9.67).

10. Traffic loads on railway bridges

From Newton's laws and d'Alembert's principles railway bridges are loaded by vehicles according to:
– whether the applied forces are static or dynamic loads,
– the direction of the applied forces: vertical, horizontal longitudinal and horizontal transverse loads,
– the magnitude of the applied forces caused by traffic, standard and extreme loads.

In this and the subsequent three Chapters we shall consider traffic loads, i.e., the loads currently occurring in everyday operation which are important chiefly for the assessment of bridges to fatigue.

European railway administrations have accepted the mass of the useful load and vehicles including engines traversing the given railway bridge in a year as unit of traffic load. It is expressed in millions of tonnes per year. This unit has two advantages:

1. It is generally recorded in railway statistics (or can be computed from the gross tonnes-km by dividing them by the length of the respective railway section in km).

2. The fatigue damage of bridges is approximately proportional to the traffic load.

According to [162] the traffic loads of bridges are usually classified as follows:

very heavy	over	60×10^6 t/year,
heavy	over	30×10^6 t/year,
medium	over	10×10^6 or 20×10^6 t/year,
light	over	2×10^6 t/year,
very light	up to	2×10^6 t/year.

The former Czechoslovak railways ranked among the most highly loaded in the world and in several sections the traffic load considerably exceeded 60×10^6 t/year. Fig. 10.1 shows the development of traffic load on one track of one of the main railway lines in late Czechoslovakia over the past 100 years. The graph shows the slow increase of traffic load until the post-war period and its rapid growth since 1950.

The development of the mean traffic load in the whole Czech and Slovak railway networks converted to one track is similar. This is illustrated in Fig. 10.2 based on statistics [121].

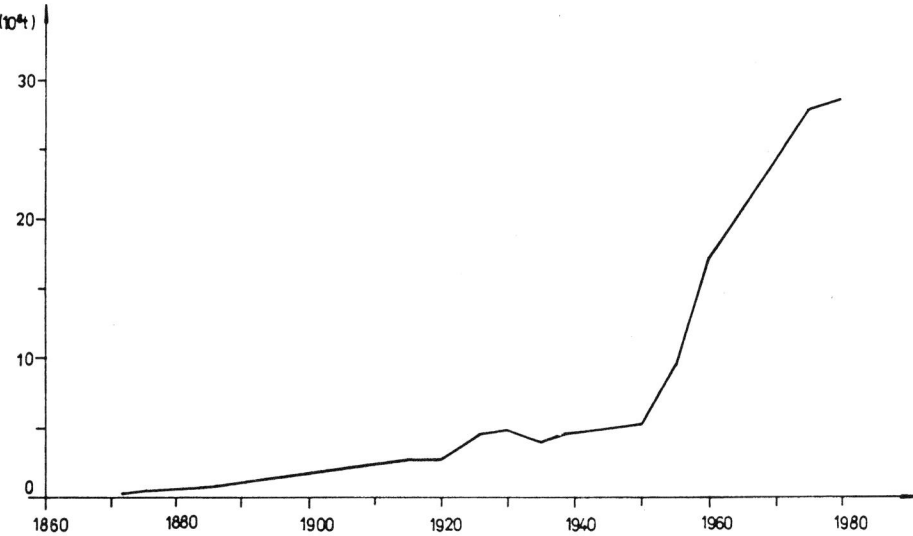

Fig. 10.1. Development of traffic load in million tonnes per year on one of the main CSD lines over the past 100 years.

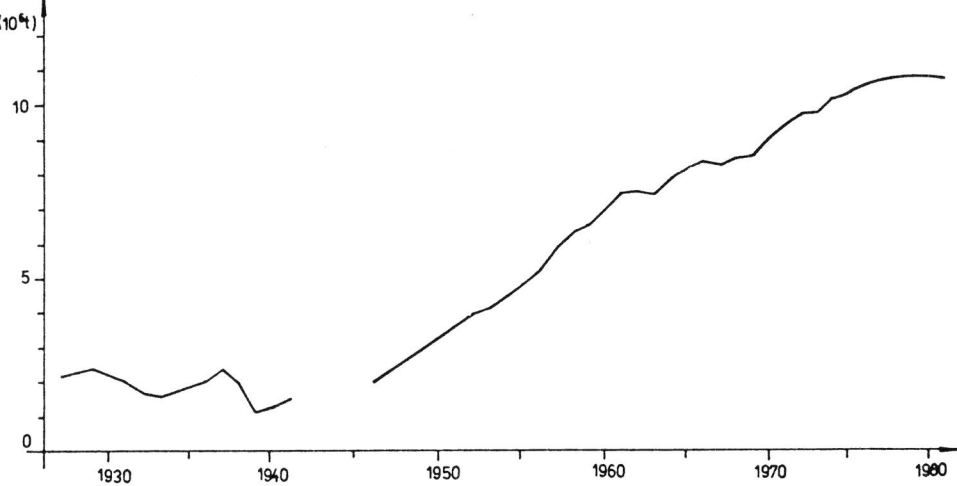

Fig. 10.2. Mean traffic load in million tonnes per year on CSD lines and its development over the past 50 years (converted to one track).

The increase of traffic on railway bridges, illustrated in Figs. 10.1 and 10.2, is due to the rise of the technical and economic standard of every state. The bridges themselves, naturally, are unfavourably influenced by this tendency, particularly in the form of reduction of the fatigue life of their structures and the reduction of the possibilities of maintenance of both the permanent way and the bridges.

The traffic load is characterized by axle forces, axle distances and speed.

10.1 Axle forces

The past period is characterized by the increase of not only traffic loads, but also axle or wheel forces of railway vehicles. Until 1870, bridges in the former Austria–Hungary were designed for loads agreed between the customer (i.e., the railway company) and the bridge builder.

It was not until 1870 that an ordinance of the Austro-Hungarian Ministry of Trade was issued, according to which railway bridges were to be designed for a uniformly distributed load the magnitude of which depended on the bridge span. This load amounted to some 180 kN m^{-1} for the spans of from 1 m upwards and decreased to 36 kN m^{-1} for the spans exceeding 30 m. Apart from that, the structure had also to withstand axle forces of 116 kN.

Since that time several bridge codes and standards for the loads of bridges have been issued in which the axle forces continuously increased to reach the present day value of 250 kN and the uniformly distributed load increased to the present value of 80 kN m^{-1} according to the international recommendations of the UIC [211], prescribed for the design of new bridges. Some national standards contain even higher loads.

10.1.1 Measurements of axle forces

The actual axle or wheel forces may be measured in three ways:

1. Weighing on a weighbridge by which the static vertical forces are obtained.
2. Strain measurements on wheel discs or axles or bearing housings [221], [155]; in this way, the time history of wheel forces is obtained which, however, only applies to the particular vehicle selected for experiments.
3. Strain measurements on rails, which yield the instantenous dynamic axle or wheel forces at the moment of the passage of any vehicle over the measured place.

Every experimental method obviously has its advantages and disadvantages, depending on the purpose of measurements. With respect to traffic loads, the third method is most advantageous. Therefore, we shall describe it in greater detail for three components of force:

a) *Vertical axle forces* are measured as shear forces in the section between two supports (sleepers) according to Fig. 10.3.

Sixteen active strain gauges are fastened on the neutral axis of the rail at an angle of 45° so as not to be influenced by supports (Fig. 10.3a). This means that a line is drawn at 45° from the edges of a sole plate and at its crossing point with the neutral axis two strain gauges are mounted (preferably a rectangular rosette on each side of the rail web).

The strain gauges are numbered *1–16* in Fig. 10.3. They are connected to a full Wheatstone bridge according to Fig. 10.3b. Calibration is carried out by the very slow passage of a vehicle of known weight.

The result of the mesurements, shown in Fig. 10.3c, is the time history of the shear force: in the section between the strain gauges the shear force equals the axle force, outside them it equals zero.

Similarly, vertical wheel forces are obtained by connecting only eight strain gauges, Nos *1–8* or *9–16*, on one rail to a single Wheatstone bridge.

The connection of strain gauges according to Fig. 10.3 is time consuming, but the results are usually very good, because the described method eliminates the influence of the bending of rail webs and further disturbing factors (the influence of temperature is small in dynamic tests of short duration).

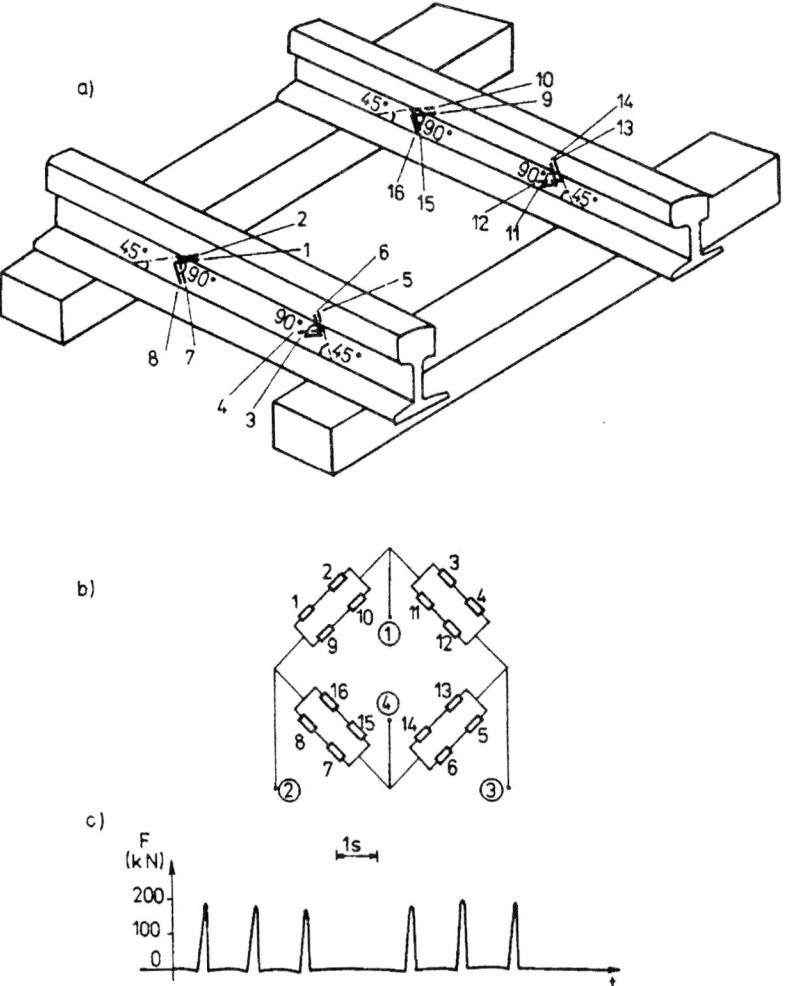

Fig. 10.3. Measurements of vertical axle forces
a) fastening of 16 strain gauges to the rail, b) their connection to a full Wheatstone bridge, c) instantaneous dynamic axle forces F during the passage of a 6-axle engine (oscilogram of the shear force).

10.1 Axle forces

b) *Horizontal longitudinal forces* are measured as the axial force in the rail according to Fig. 10.4. Two strain gauges are fastened horizontally on the neutral axis of the rail and are connected to opposite branches of a Wheatstone bridge. In this way the stress σ in the neutral axis of the rail of a cross section area A is obtained with double sensitivity. The horizontal longitudinal force is then

$$H = \sigma A. \tag{10.1}$$

An example of the history of horizontal longitudinal force in a rail during braking and starting is shown in Fig. 8.14.

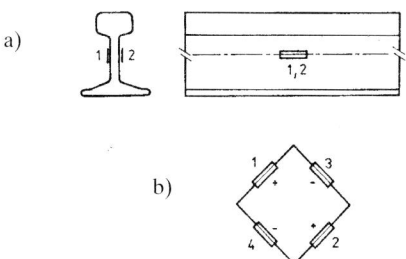

Fig. 10.4. Measurements of horizontal longitudinal forces:
a) fastening of two longitudinal strain gauges in the neutral axis of the rail, b) connection to a Wheatstone bridge. Active strain gauges *1, 2*, compensation strain gauges *3, 4*.

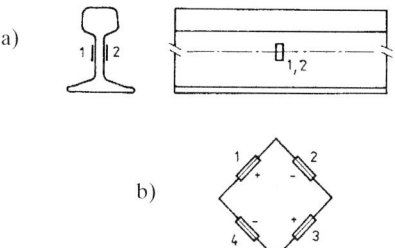

Fig. 10.5. Measurements of horizontal transverse forces:
a) fastening of two vertical strain gauges in the neutral axis of the rail, b) connection to a Wheatstone bridge. Active strain gauges *1* and *2*, strain ganges *3* and *4* are superfluous.

The method of measurements of horizontal longitudinal forces is very simple, but is influenced by local wheel action and by the influence of several axles. Temperature is of little importance in short-term dynamic tests; otherwise further arrangements must be made [78].

c) *Horizontal transverse forces* are assessed on the basis of bending stresses in the rail web according to Fig. 10.5. Two strain gauges are fastened vertically on the neutral axis of the rail according to Fig. 10.5 and are connected to adjoining branches of a Wheatstone bridge. These strain gauges then show the difference of strains $\varepsilon_1 - \varepsilon_1$; the bending stress is found from the equation

$$\sigma = E(\varepsilon_1 - \varepsilon_2), \tag{10.2}$$

where E is the modulus of elasticity of the rail.

The bending moment M producing the stress σ is

$$M = W\sigma, \tag{10.3}$$

where W is the resistance modulus of the interacting rail cross section. This cross section is a rectangle defined by the web thickness t and the width b, so that

$$W = \frac{1}{6} bt^2. \tag{10.4}$$

The width b can be obtained from oscillographic records according to Fig. 10.6. Actually, only about one half this width operates (with regard to rail length) so that

$$b_{red} = b/2 \tag{10.5}$$

is substituted for b in equation (10.4).

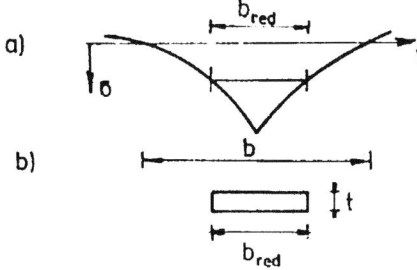

Fig. 10.6. a) Oscillogram of bending stress σ in rail web, b) reduced interacting width b_{red}.

The determination of the horizontal transverse force H_y from the vertical force F follows from the moment condition of equilibrium at the measured place which, according to Fig. 10.7, is given by

$$M = Fa + H_y h. \tag{10.6}$$

From this equation

$$H_y = \frac{1}{h}(M - Fa). \tag{10.7}$$

or, using σ and W,

$$H_y = \frac{1}{h}(W\sigma - Fa). \tag{10.8}$$

10.1 Axle forces

The distances a and h (fig. 10.7) are variable, but our purpose will be satisfied approximately, if they are considered constant:

$$a = b_r/2 - r,$$
$$h = v - e - r,$$

where b_r is the rail head width,
- r – radius of rail head,
- v – rail height,
- e – distance of neutral axis from rail base.

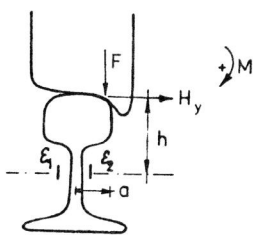

Fig. 10.7. Forces acting between the wheel and the rail.

Figure 10.7 and equation (10.8) represent the force in the left-hand rail (in driving direction). The condition of equilibrium in the right-hand rail (10.6) is

$$M = -Fa + H_y h, \qquad (10.9)$$

so that the horizontal transverse force is

$$H_y = \frac{1}{h}(W\sigma + Fa). \qquad (10.10)$$

The positive direction of vertical force is assumed downwards, horizontal force to the right, and positive moment clockwise.

The described method for measurements and assessment is simple, but it has some disadvantages: with regard to its fastening the interacting rail width is rather variable, the necessity of simultaneous measurements of the vertical force, etc. Therefore, more accurate, but also more complicated methods for measurements of horizontal transverse forces were devised [47].

10.1.2 Vertical axle forces

Instantaneous dynamic vertical axle forces were measured at midspan of many bridges, so that the data of about 30 000 axles were obtained.

An example of the measurements on a bridge with freight transport is shown in Fig. 10.8, while an analogous histogram of vertical axle forces on a bridge with major passenger transport is given in Fig. 10.9. For the numbers of tested bridges see Table 12.6. The traffic load characteristics pertaining to these two bridges are tabulated in Table 10.1. The measurements on both bridges lasted 24 hours; annual data were obtained by extrapolation.

Extensive measurements of traffic loads and vertical axle forces were carried out on one of the most heavily trafficked bridges of the CSD, where – according to Table 10.2 – the annual traffic load amounts to 39.6×10^6 tonnes per year. The measurements concerned 180 trains with 13 733 axles, a survey of which is given in Table 10.3.

In this particular case the axle forces were measured 53.13 m in front of the bridge midspan (*F1*) and at midspan (*F2*). Detailed results are shown in Tables 10.3 and 10.4.

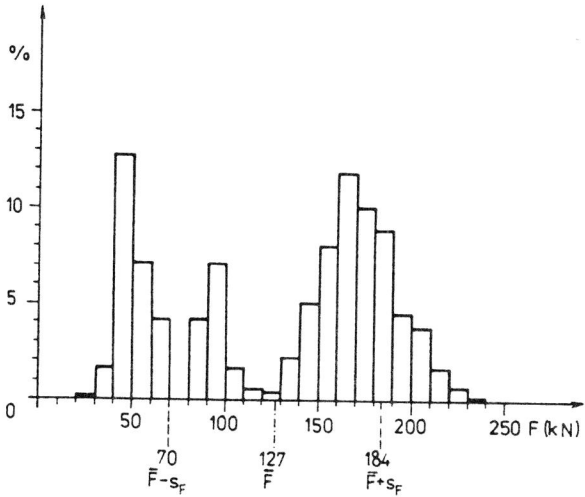

Fig. 10.8. Frequency histogram of vertical axle forces F on a bridge with prevailing freight transport. 100% = 5242 axles, traffic load 23.7×10^6 tonnes per year, $\bar{F}$ – mean value, s_F – standard deviation (steel plate girder bridge, No 2, $l = 2 \times 16.2$ m).

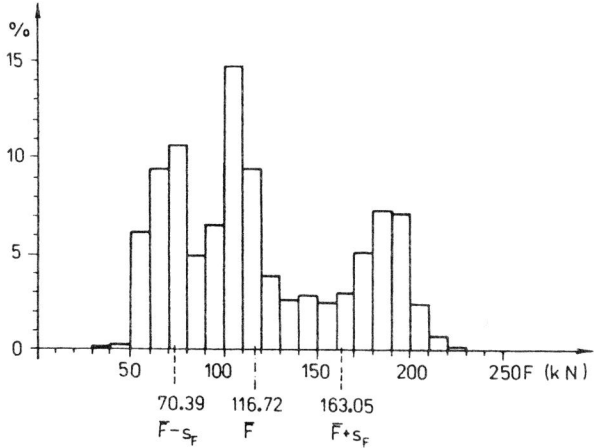

Fig. 10.9. Frequency histogram of vertical axle forces F on a bridge with prevailing passenger transport, 100% = 2998 axles, traffic load 12.75×10^6 tonnes per year, $\bar{F}$ – mean value, s_F – standard deviation (steel truss bridge, No 3, $l = 30.32$ m).

10.1 Axle forces

TABLE 10.1. Traffic load characteristics

Prevailing traffic	Trains	Number of trains			Number of axles			Sum of axle forces			Mean number of axles per train	Mean weight of one train (kN)
		Per day	%	Per year	Per day	%	Per year	Per day (kN)	%	Mass per year (10^6 t)		
Freight (bridge No 2, $l = 2 \times 16.2$ m)	Passenger	53	36.81	19 345	1016	19.38	370 840	92 025	14.17	3.3	19	1736
	Freight	91	63.19	33 215	4226	80.62	1 542 490	557 680	85.83	20.4	46	6128
	Together	144	100	52 560	5242	100	1 913 330	649 705	100	23.7	36.4	4512
Passenger (bridge No 3, $l = 30.32$ m)	Passenger	23	45.10	8 395	732	24.42	267 180	81 525	23.34	2.98	32	3545
	Freight	28	54.90	10 220	2266	75.58	827 090	267 760	76.66	9.77	81	9563
	Together	51	100	18 615	2998	100	1 094 270	349 285	100	12.75	59	6849

TABLE 10.2. Traffic statistics of a very heavy trafficked CSD bridge (No. 1, l = 32 m)

	Trains		Together
	Passenger	Freight	
Number of trains per year	11 480	25 378	37 218
Mass of trains per year (t)	5.3×10^6	34.3×10^6	39.6×10^6

TABLE 10.3. List of investigated trains on a very heavily trafficked CSD bridge (No. 1, l = 32 m)

Trains	Number of trains	Number of axles	Weight of trains		From traffic statistics
			Measured		
			In front of the bridge	On the bridge	
			$F1$	$F2$	
			kN	kN	kN
Passenger	57	2 272	259 770	282 510	301 330
Freight	95	11 306	1 643 640	1 736 700	1 631 870
Work	10	55	10 905	11 265	10 515
Calibration	18	100	18 270	19 170	17 310
Together	180	13 733	1 932 585	2 049 645	1 961 025

Extensive experiments have yielded the following most important conclusions concerning vertical axle forces:
– instantaneous dynamic axle forces acquire random values depending on the vibration of both the vehicles and the bridge;
– histograms of axle forces, usually possess three peaks, the ocurrence and magnitude of which depends on the type of traffic on the given railway line. The first peak of about 180–200 kN characterizes the locomotives and fully loaded freight cars. The second peak of about 100 kN suggests the presence of passenger cars and not fully loaded freight cars, while the third peak of about 50 kN characterizes empty freight cars;
– dynamic axle forces on the bridge and in front of it do not differ much, although in many cases the forces on the bridge were slightly higher. These differences, due to the dynamic interaction of vehicles with the bridge, were covered by a large variation in experimental results;
– the load data obtained from traffic statistics are in relatively good agreement with the sum of the measured axle forces.

10.1 Axle forces

TABLE 10.4. Statistics of axle forces on a very heavily trafficked CSD bridge (No. 1, l = 32 m)

Trains	Quantity	Unit	Axle force	
			In front of the bridge $F1$	On the bridge $F2$
Passenger	n	1	2 272	2 272
	max F	kN	255	255
	$\bar{F}$	kN	114.3	124.3
	S_F	kN	35.97	36.49
	V	1	0.315	0.293
	$S_{\bar{F}}$	kN	0.755	0.766
Freight	n	1	11 306	11 306
	max F	kN	255	285
	$\bar{F}$	kN	145.4	153.6
	S_F	kN	49.27	53.20
	V	1	0.339	0.346
	$S_{\bar{F}}$	kN	0.463	0.500
Work	n	1	55	55
	max F	kN	225	255
	$\bar{F}$	kN	198.3	204.8
	S_F	kN	36.8	41.66
	V	1	0.186	0.203
	$S_{\bar{F}}$	kN	5.00	5.62
Calibration	n	1	100	100
	max F	kN	195	225
	$\bar{F}$	kN	182.7	191.7
	S_F	kN	19.1	14.07
	V	1	0.105	0.073
	$S_{\bar{F}}$	kN	1.91	1.41
Together	n	1	13 733	13 733
	max F	kN	225	285
	$\bar{F}$	kN	140.7	149.3
	S_F	kN	48.78	51.99
	V	1	0.347	0.348
	$S_{\bar{F}}$	kN	0.416	0.444

Symbols:
- n — number measurements
- max F — maximum axle force
- $\bar{F}$ — mean axle force
- S_F — standard deviation
- $V = S_F/\bar{F}$ — coefficient of variation
- $S_{\bar{F}}$ — standard deviation of mean axle force

TABLE 10.5. Statistical evaluation of vertical and horizontal wheel forces (steel truss bridge, $l = 25.85$ m)

Characteristic	Symbol	Unit	Locomotive type				Usual traffic	
			T 478.3		T 679.1			
			\multicolumn{4}{c}{Rail}			
			Left	Right	Left	Right	Left	Right
Number of measurements	n		158	156	143	149	143	160
Mean value of horizontal forces	$\bar{H}_v$	kN	−24.83	24.62	−30.30	28.32	−21.91	14.53
Mean value of vertical forces	$\bar{F}$	kN	93.68	94.78	94.30	92.84	69.05	64.43
Standard deviation of horizontal forces	S_{H_v}	kN	2.16	3.75	3.01	3.96	6.83	7.62
Standard deviation of vertical forces	S_F	kN	5.94	6.79	5.78	6.15	16.99	19.13
Variation coefficient of horizontal forces	V_{H_v}	%	8.70	15.24	9.95	13.97	31.19	52.42
Variation coefficient of vertical forces	V_F	%	6.35	7.17	6.13	6.63	24.61	29.69
Covariance	S_{HF}	kN²	7.79	16.47	11.57	9.26	90.35	124.30
Correlation coefficient	χ_{HF}	1	0.61	0.65	0.66	0.38	0.78	0.85
Ratio of forces	$\bar{H}_v/\bar{F}$	1	0.265	0.260	0.321	0.305	0.317	0.226
Average ratio	$\bar{H}_v/\bar{F}$	1	\multicolumn{6}{c}{0.282}					

10.1.3 Horizontal transverse wheel forces

Horizontal transverse wheel forces measured on the bridge in a straight line according to the method described in Sect. 10.1.1 have yielded values within the limits of about 10 kN and 30 kN with a mean value of 24 kN. They act mostly outward from the longitudinal bridge axis, but their absolute values on the right-hand and the left-hand rails differ very little (see Table 10.5).

The dependence of horizontal transverse forces H_y on vertical forces F was analyzed statistically. Table 10.5 shows that this dependence is statistically fairly significant, which follows from the small standard deviations and the relatively high correlation coefficient.

For the same reason the ratio of horizontal and vertical forces was found for Table 10.5, which amounts, on average, to

$$\overline{H}_y / \overline{F} = 0.28 . \tag{10.11}$$

Also linear regression of horizontal transverse forces H_y as a function of velocity V and vertical wheel forces F was investigated according to the equation

$$H_y = a + bV + cF \tag{10.12}$$

where a, b, c are regression coefficients.

The results of the calculations are given in Table 10.6 which shows that in the investigated range of low velocities between 0 and 60 km h^{-1} the horizontal transverse forces do not depend significantly on the velocity. However, the regression analysis has also confirmed the conclusion that the horizontal transverse forces amount to about one third of the magnitude of vertical wheel forces.

The horizontal longitudinal forces are dealt with in detail in Chapter 8.

TABLE 10.6. Linear regression of horizontal wheel forces H_y (kN) as a function of velocity V (km h^{-1}) and vertical wheel forces F (kN): $H_y = a + bV + cF$ (steel truss bridge, $l = 25.85$ m)

Characteristic	Symbol	Unit	Locomotive type		Usual traffic
			T 478.3	T 679.1	
Number of measurements	n	1	314	292	303
Coefficient of determination	R^2	1	0.383	0.254	0.611
Regression coefficients	a	kN	−2.801	0.699	−8.686
	b	kN h km^{-1}	0.004	−0.004	0.076
	c	1	0.291	0.307	0.345

10.2 Axle spacing

The axle spacing of railway vehicles is a deterministic quantity in the case when known vehicles travel along the bridge. This particularly applies to passenger trains. In the case of current traffic – especially if it involves mixed freight trains – the vehicle composition is unknown; therefore, the axle spacing is also considered as a stochastic quantity.

This phenomenon is shown in Figs 10.10 and 10.11 which give three-dimensional presentations of the frequency of the occurrence of axle forces F and the respective axle spacings a. The first axle has been omitted to make the number of axles correspond to the respective number of axle distances.

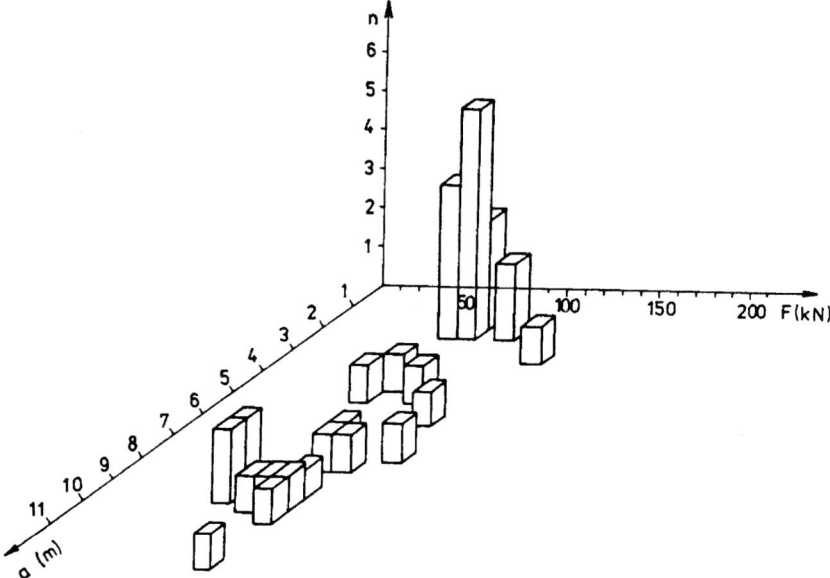

Fig. 10.10. Frequency histogram of axle spacing a and axle forces F, passenger trains (steel plate girder bridge, No 6, $l = 35$ m).

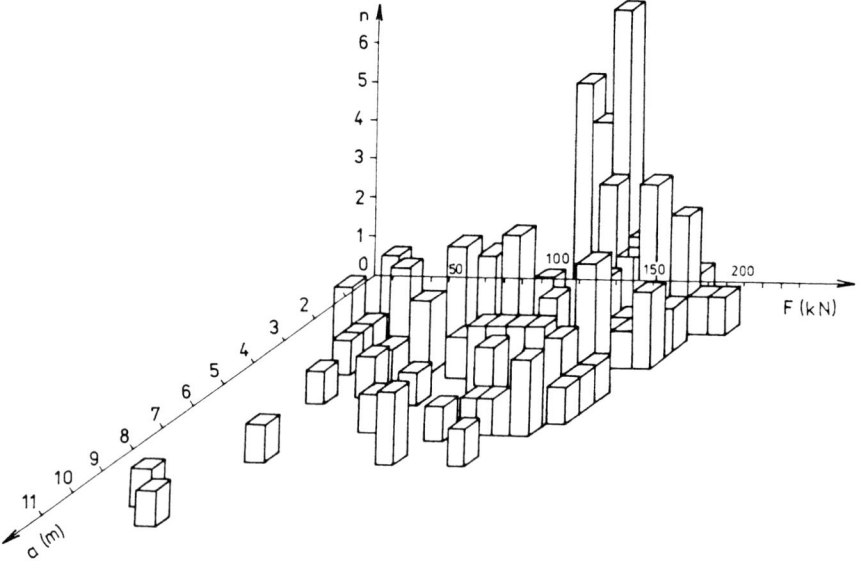

Fig. 10.11. Frequency histogram of axle spacing a and axle forces F, freight trains (steel plate girder, No 6, $l = 35$ m).

10.2 Axle spacing

Figure 10.10 applies to passanger trains (light-weight locomotive units), while Fig. 10.11 applies to freight trains. Both figures illustrate the above mentioned statement, i.e., the deterministic spacing of axles in the case of passenger trains and the mostly stochastic spacing in the case of freight trains.

The basic statistical data on axle spacing for the given case are tabulated in Table 10.7. The spacings were determined from the oscillographs of axle forces with the assumption of constant velocity of motion which was also measured by an independent method.

TABLE 10.7. Statistical characteristics of axle spacing (steel plate girder bridge, No. 6, $l = 35$ m)

Characteristic	Symbol	Unit	Trains		
			Passenger	Freight	Together
Number of measurements	n	1	35	124	159
Mean spacing	$\bar{a}$	m	5.54	3.83	4.21
Standard deviation	s_a	m	3.67	1.95	2.52
Variation coefficient	$V = s_a/\bar{a}$	%	66.32	50.87	59.96

10.3 Velocities

The speed of train on bridges depends on the maximum permitted track velocity in the given section and/or local conditions, such as gradient, curve, proximity of signal equipment, etc.

Surprisingly, the train speeds in the set of passenger trains do not differ much from those of the set of freight trains, which is testified to by a relatively low variation coefficient of these two sets in Table 10.8.

For all trains, however, the histogram of velocity frequencies is rather flat (Fig. 10.12).

TABLE 10.8. Statistical characteristics of train velocities (steel plate girder bridge, No. 5, $l = 30$ m)

Characteristic	Unit	Trains		
		Passenger	Freight	Together
Number of measurements	1	31	44	75
Maximum velocity	km h^{-1}	104.0	73.7	104.0
Average velocity	km h^{-1}	75.8	47.5	59.2
Standard deviation	km h^{-1}	14.96	10.93	18.9
Variation coefficient	1	0.197	0.230	0.319

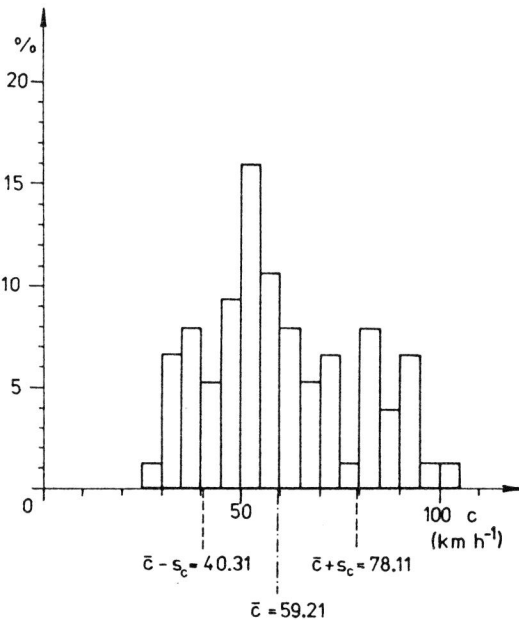

Fig. 10.12. Frequency histogram of velocities c, 100% = 75 trains, traffic load 20.1×10^6 tonnes per year, $\bar{c}$ – mean value, s_c – standard deviation (steel plate girder bridge, No 5, l = 30 m).

11. Statistical counting methods for the classification of random stress-time history

The stress-time history of railway bridges due to the passage of trains consisting of vehicles of different types is a more or less random process. The random component in the length coordinate (along the train) is usually higher than the dynamic component, because in mixed freight trains different axle forces follow with unequal spacings. An example of the stress-time history in a railway bridge is shown in Fig. 11.1.

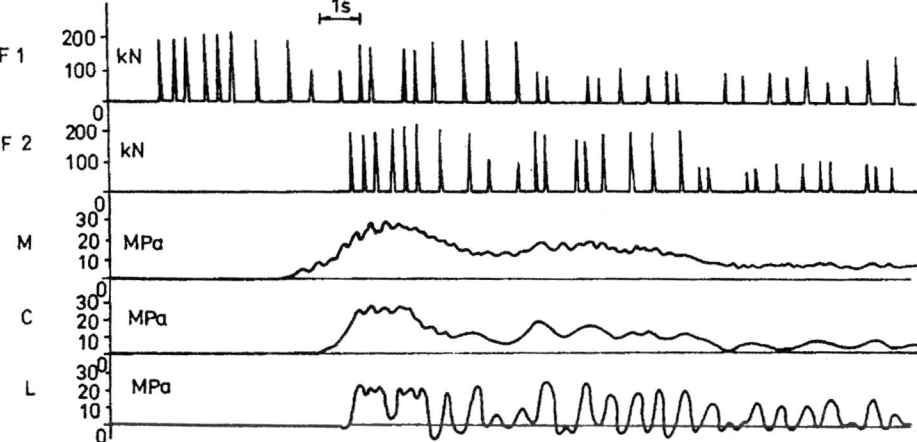

Fig. 11.1. Example of stress-time history in main girder (*M*), cross-girder (*C*) and stringer (*L*) of a 32 m span steel railway bridge (No 1) during the passage of a mixed freight train constaining 35 cars driven by an electric locomotive type E 699.1 at the velocity 30 km h^{-1}. Instantaneous dynamic axle forces 53.15 m in front of the bridge (*F1*) and at midspan (*F2*).

The analysis and classification of the stress-time histories represent a specific problem in structural dynamics.

The random processes in accordance with Sect. 1.4 are described by statistical characteristics of the first order (mean value, variance, standard deviation, probability density) and/or of the second order (correlation function, power spectral densities). For the problems of strength, fatigue and reliability of structures, however, special methods have been developed classifying the extreme values of random functions and the modes of their alternation. These methods, intended to characterize the random process with reference to fatigue damage, are of empirical character only. They endeavour to replace the random process with a certain number of equivalent complete stress cycles, often regardless of their sequence in time. Equivalence between these methods is achieved by making the fatigue life the same.

This chapter will give a survey of the most important methods of classification of stress-time histories. Particular attention will be paid to the rain-flow counting method which has been used for the analysis of structures subjected to fatigue load with ever increasing frequency. For this reason, this chapter describes in some detail both the method itself and the instrumentation for the evaluation; also a flow diagram and computer program are given.

11.1 Statistical counting methods

To clarify the concepts, Fig. 11.2 contains the terms most frequently used in the field of the statistical counting methods. There are a number of such methods, the most important of which are described in this chapter; for others see [49], [51], [93], [143]. All methods depend on the sensitivity of the respective devices, i.e. on the counting dead-zone in which the instruments are not able to record the vibrations (see Fig. 11.2).

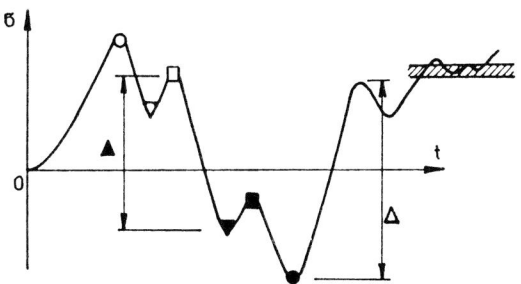

Fig. 11.2. Most important terms in counting methods:
○ – absolute maximum within the interval ⟨0, t⟩, ● – absolute minimum within the interval ⟨0, t⟩, □ – positive local maximum, ■ – negative local maximum, ▽ – positive local minimum, ▼ – negative local minimum, △ – positive stress range, ▲ – negative stress range.
The positive stress range and the negative stress range of equal magnitude form a complete cycle.
//// – insensitive dead-zone, 0 – t basic (mean) level.

11.1.1 Sampling method

In the sampling method, [49], according to Fig. 11.3, the stress values are counted at regular time intervals T. To characterize the stress-time history faithfully, it must hold that $T < 1/(2f)$, where f is the highest frequency. In some classification instruments, the counting is carried out during the short interval τ, which should be less than about $T/50$ to obtain sufficiently accurate results.

There is also a variant of the sampling method, [49], which, counts the maximum value in every interval T (see Fig. 11.4) The sampling methods depend on the position of the base (or zero) level.

11.1 Statistical counting methods

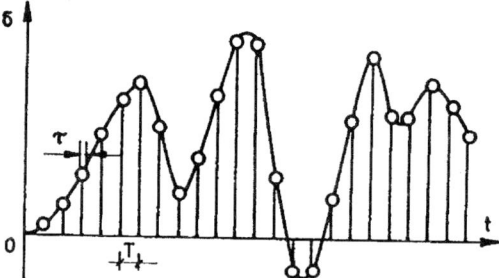

Fig. 11.3. Sampling method, $T < 1/(2f)$, $\tau \leqslant T/50$, $\bigcirc$ – count.

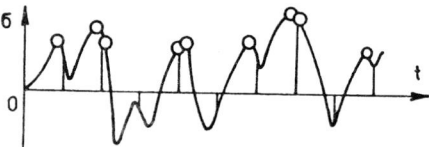

Fig. 11.4. Sampling method counting the maximum in every interval T, $\bigcirc$ – count.

11.1.2 Threshold method

The treshold method, [49], adds the time Σt_i for which the recorded stresses are within the ith class, i.e. between the thresholds $i - 1$ and i (see Fig. 11.5).

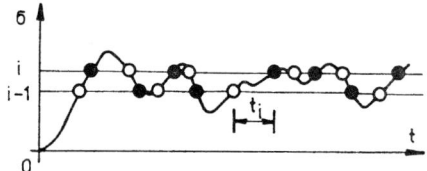

Fig. 11.5. Threshold method, Σt_i, $\bigcirc$ – start of counting in the i-th class, ● – stop of counting in the ith class.

11.1.3 Peak counting methods

a) Absolute peaks method

The absolute peaks method, [49], counts the absolute maxima within the individual intervals above the mean level and absolute minima within the intervals below the mean level. In Fig. 11.6, the absolute maxima are *3, 7, 11* and the absolute minima *6, 8, 12*. The result depends on the position of the mean level.

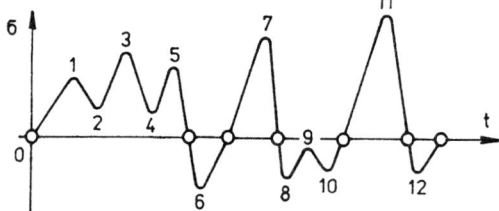

Fig. 11.6. Peak counting methods, $\bigcirc$ – transition of mean (basic) level.

b) Relative peaks method

The method of relative peaks, [49], counts all local maxima above the mean level (i.e. the points *1, 3, 5, 7, 11* in Fig. 11.6) and all local minima below the mean level (i.e. the points *6, 8, 10, 12* in Fig. 11.6). The result depends on the position of the mean level.

c) Local peaks method

The method of local peaks, [49], counts separately all local maxima (*1, 3, 5, 7, 9, 11* in Fig. 11.6) and all local minima (*2, 4, 6, 8, 10, 12* in Fig. 11.6). The result does not depend on the position of the mean level.

In addition to these principal methods, there are other peak counting methods, see [143].

11.1.4 Level crossing methods

a) Level crossing count

The level crossing method, [49], counts every signal transition across the class level separately for the positive transition on the rising parts of the process above the mean level and separately for the negative transition on the decaying parts of the process below the mean level (see Fig. 11.7). The transition across the mean level are usually included in the first positive class.

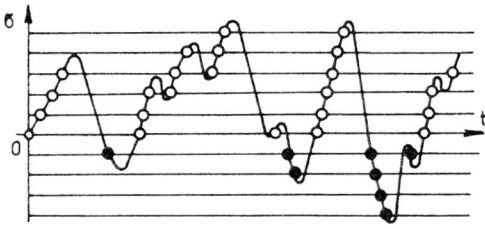

Fig. 11.7. Level crossing count,
○ – count in positive classes, ● – count in negative classes.

b) Fatigue meter method

This method, [51], is similar to the level crossing count, only every lower (higher) subsequent class within one interval above (below) the basic level is counted only once (see Fig. 11.8). The intermediate small stress cycles are lost in this method.

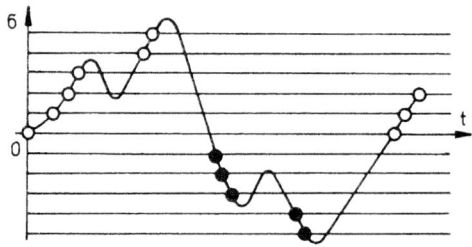

Fig. 11.8. Fatigue meter method,
○ – count in positive classes, ● – count in negative classes.

11.1.5 Stress range counting methods

For further methods see [143].

a) Range counting

The stress range method, [49], [143], counts the differences between two succeeding local extremes (peaks). Every stress range is considered as a half-cycle (see Fig. 11.9). The large dead zone will neglect small stress ranges.

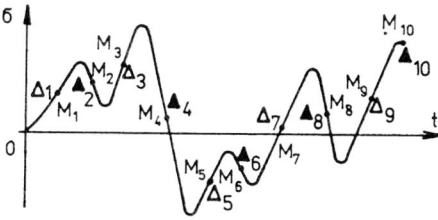

Fig. 11.9. Stress range and range-mean counting,
$\triangle_i$ – positive stress range, $\blacktriangle_i$ – negative stress range, M_i – mean value of stress range.

b) Range-pair counting

This method, [143], counts the stress ranges as in the stress range method. The positive and negative stress ranges of equal magnitude are arranged together and form one complete cycle regardless of their sequence in time. Unpaired stress ranges are lost (Fig. 11.10).

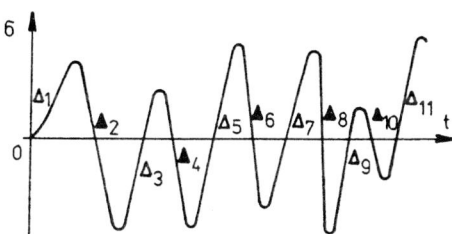

Fig. 11.10. Range-pair counting, complete cycles $\triangle_{1,10}$, $\triangle_{2,7}$, $\triangle_{3,4}$, $\triangle_{5,8}$, uncounted stress ranges in the interval $\langle 0, t \rangle$: $\triangle_{6,9,11}$.

c) Rain-flow counting method

The rain-flow method counts complete stress cycles corresponding to the closed loops in the $\sigma(\varepsilon)$ diagram; for details see Sect. 11.2.

11.1.6 Multiparametric methods

a) Range-mean counting

This method, [51], is identical to the stress range method (see Sect. 11.1.5a); moreover, it also records the mean values M_i, which are assigned to the respective stress range $\triangle_i$ or $\blacktriangle_i$ (see Fig. 11.9).

b) Correlation table

More complete information about the random process is afforded also by the correlation table [129] – see Table 11.1 – in which the occurrence rates of local maxima x_M are recorded vertically and the occurrence rates of local minima x_m horizontally, classified into a certain number of classes. In every field of the correlation table, a local maximum is assigned a local minimum. The correlation table makes it possible to derive the empirical probability density of maxima x_M (in vertical columns), of minima x_m (in horizontal lines), of stress ranges x_R (in the direction of the main diagonal) and of mean values x_S (in the direction of the secondary diagonal).

TABLE 11.1. Example of a correlation table of maxima and minima and the derived empiric probability densities of maxima x_M, minima x_m, stress ranges x_R and mean values x_S according to [129], n_i = number

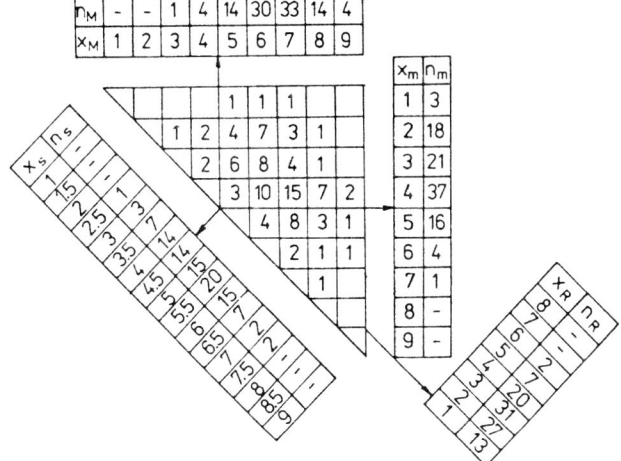

11.2 Rain-flow counting method

This method was reported orally in Japan in 1968 for the first time [139]; it first appeared in the literature in 1972 when it was described by N. E. Dowling, [51]. In the CIS, a very similar count was developed under the name of ‚full-cycle method' and to date there exist several variants of this count all over the world, [162], [220].

The method obtained its name from an idea that water is flowing along a pagoda-shaped roof (Fig. 11.13). In order to obtain such a pagoda-shaped roof the $\varepsilon(t)$ diagram must be rotated by 90° (time-axis downwards).

11.2.1 Justification of the counting method from the $\sigma(\varepsilon)$ diagram

The decisive component for the fatigue of building materials consists of the irreversible part (the so-called hysteresis) in the relation between the stress σ and the strain ε, because the fatigue damage is particularly affected by the alternate plastic deformations (see [119]). The evaluation of random processes with reference to the fatigue of structures, therefore, must be based not on the stress-time history (Fig. 11.11), which is only in the elastic region proportional to the strain $\varepsilon(t)$, but from the stress-strain diagram (see Fig. 11.12).

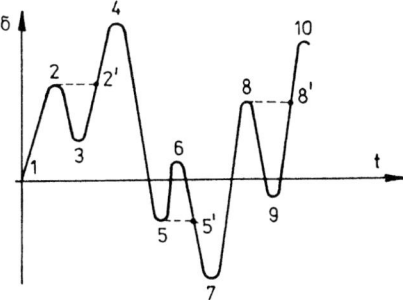

Fig. 11.11. Stress-time history.

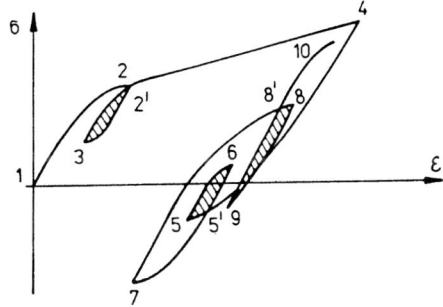

Fig. 11.12. Stress-strain diagram,
complete cycles: *2-3-2'*,
                             5-6-5',
                             8-9-8',
half cycles (stress ranges): *1-2-4*,
                             4-5-7,
                             7-8-10.

The stress-time history in Fig. 11.11 is converted into the stress-strain diagram of Fig. 11.12; on the other hand, Fig. 11.13 shows the strain-time history of the same random process. These three relations must be clearly distinguished.

The fundamental assumption of the rain-flow counting method is that the fatigue damage due to small induced stress cycles may be added to the fatigue damage due to large stress cycles. If the cycle *1–4* in Fig. 11.11 is interrupted by a small cycle *2–3–2'* the coordinate of point *2'* is very near to the point *2* in Fig. 11.12 and the material acts as if no interruption by an inserted cycle has taken place. Moreover, one complete cycle *2–3–2'* has remained at disposal.

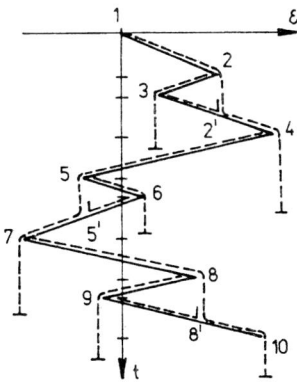

Fig. 11.13. Dependence of strain ε on time t,
complete cycles: *2-3-2'*,
   5-6-5',
   8-9-8',
half cycles (stress ranges): *1-2-4*,
   4-5-7,
   7-8-10.

The rain-flow counting method evaluates the strain-time history in the same way as the material reacts to a random loading process. It counts both the large amplitudes (half-cycles) and, separately, small inserted stress cycles (complete cycles), see Figs 11.12 and 11.13. This results in classified stress ranges which are paired into complete stress cycles after assessment, see Fig. 11.14.

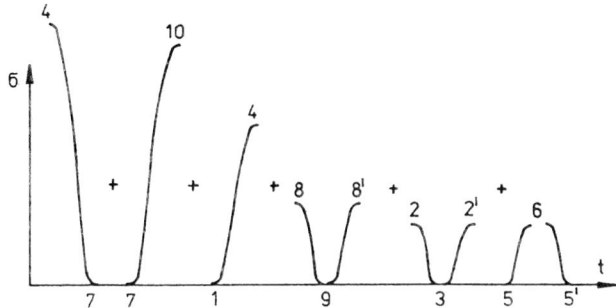

Fig. 11.14. Idealized stress-time history equivalent in respect of fatigue with stress-time history in Fig. 11.11 (after classification by the rain-flow counting method).

11.2 Rain-flow counting method

Among all the methods described only the rain-flow counting method faithfully reflects the behaviour of the material and characterizes its hysteresis. For this reason it is recommended for the evaluation of random stress-time histories with reference to the fatigue of materials. Every time interval of stress is counted only once and the same result is obtained, if the evaluation proceeded in the opposite direction.

The method, however, does not explain the fatigue process itself; it only describes it in a phenomenological way.

11.2.2 Counting rules

The progress of the stress-time history processing (Fig. 11.11) by the rain-flow method is shown in Fig. 11.13 in which the strain-time history has been rotated by 90° to enable the idea to be grasped of rain-flow along a pagoda-shaped roof. The following rules for the strain range counting are applied (one complete cycle = two strain ranges):

(1) One strain range is counted, when water flows from the initial strain to the maximum, from which it flows downwards along the next roof slope to the next maximum.

The flowing water is stopped, when

(1a) the strain minimum is lower than the initial strain,

(1b) it meets some former rain-flow, or when

(1c) it reaches the end of the strain record.

(2) The strain range is counted, when the water flows (from the highest maximum attained during the preceding range) downwards along the roof to the nearest minimum, from where it flows down to the nearest roof slope and to the next minimum.

It stops only, when

(2a) the strain maximum is higher than the initial maximum,

(2b) it meets some former rain-flow,

(2c) it reaches the end of the record.

(3) In this way the strain ranges are counted by the successive application of Rules (1) and (2).

(4) The roof parts which have not been covered by the rain-flow so far are counted as

(4a) the flow beginning in the maximum and flowing down across the following minima as described in Rule (2), or

(4b) the flow beginning in the minimum and flowing down accros the following maxima as described in Rule (1).

In either case the flow stops, when

(4c) it meets an earlier strain which is more extreme than the initial point of the flow, or when

(4d) it meets an earlier flow.

11.2.3 Algorithm for a computer

The rain-flow counting method is easy to program. There exist a number of programs for the evaluation of random processes by this method which provide slightly different results. One of the simplest programs is based on the following algorithm (Fig. 11.15):

(1) The local peaks $A(0)$, $A(1)$, ..., $A(k)$ are read from the stress-time history and digitalized.

(2) The set of local peaks obtained is decomposed into half-cycles and cycles.

(3) The condition for a cycle counting is defined by the relations

$$A(i-1) \leq A(i+1) < A(i) \leq A(i+2) \tag{11.1}$$

or

$$A(i-1) \geq A(i+1) > A(i) \geq A(i+2), \tag{11.2}$$

see Fig. 11.15a, b.

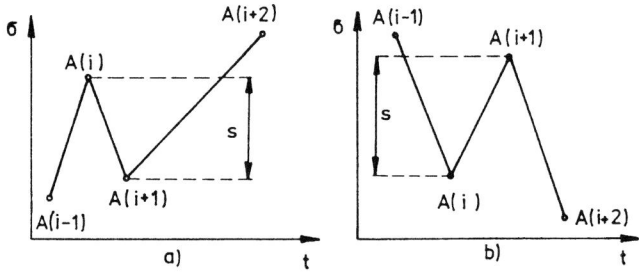

Fig. 11.15. Two basic cases of complete cycle counting in the rain-flow method according to Eq. (11.1) and Eq. (11.2):
a) rising, b) decaying part of the $\sigma(t)$ relation.
A – local extreme, i – order of local extreme, s – stress range between the $i+1$ and the ith local extremes.

(3a) The computation proceeds from the lowest $i = 1$ to the highest $i = k - 2$. If the conditions (11.1) or (11.2) have been complied with, one cycle = 2 half-cycles are counted with the range of

$$s = |A(i) - A(i+1)| \tag{11.3}$$

and with the mean value of

$$M = \frac{A(i) + A(i+1)}{2}. \tag{11.4}$$

(3b) The peaks $A(i)$ and $A(i+1)$ are eliminated from the sequence of extremes and the sequence is re-numbered.

(3c) The procedure according to (3a) and (3b) is repeated until at least one cycle has remained from the remaining sequence.

11.2 Rain-flow counting method

(4) If the decomposition into cycles has been completed, the stress ranges (i.e. the absolute values of differences of adjoining extremes) in the remaining sequence are called half-cycles. An example of completed decomposition into cycles is shown in Fig. 11.16; the remaining sequence is decomposed into half-cycles.

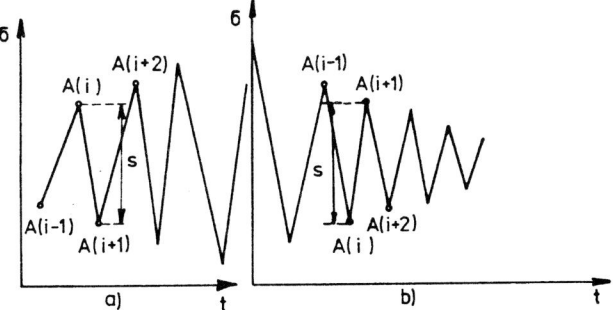

Fig. 11.16. Decomposition into cycles finished, counting of half-cycles to be made in the a) rising, b) decaying part of the $\sigma(t)$ relation.
Local extremes do not satisfy Eqs (11.1) and (11.2).
A – local extreme, i – order of local extreme, s – counted half cycle (stress range).

(5) The cycles and the remaining half-cycles of equal magnitudes are added in the course of the calculations; the result is a table of stress range frequencies, usually represented in the form of a frequency histogram and called the stress range spectrum. An example of such a spectrum obtained on a steel railway bridge during the passage of a train is shown in Table 11.2 (see also Figs. 12.4 and 12.5).

Table 11.2. Example of a stress range spectrum obtained on a cross-beam of the steel railway bridge No. 6 (Table 12.6) during the passage of a mixed freight train (diesel-electric locomotive T 478 + 10 cars + T 478), total weight 7420 kN, speed 38.5 km h^{-1}

Stress range $\Delta\sigma$ (MPa)	Number of complete stress cycles
0– 4	56
4– 8	4
8–12	6
12–16	3
16–20	5
20–24	5
24–28	3
28–32	0
32–36	1
Together	83

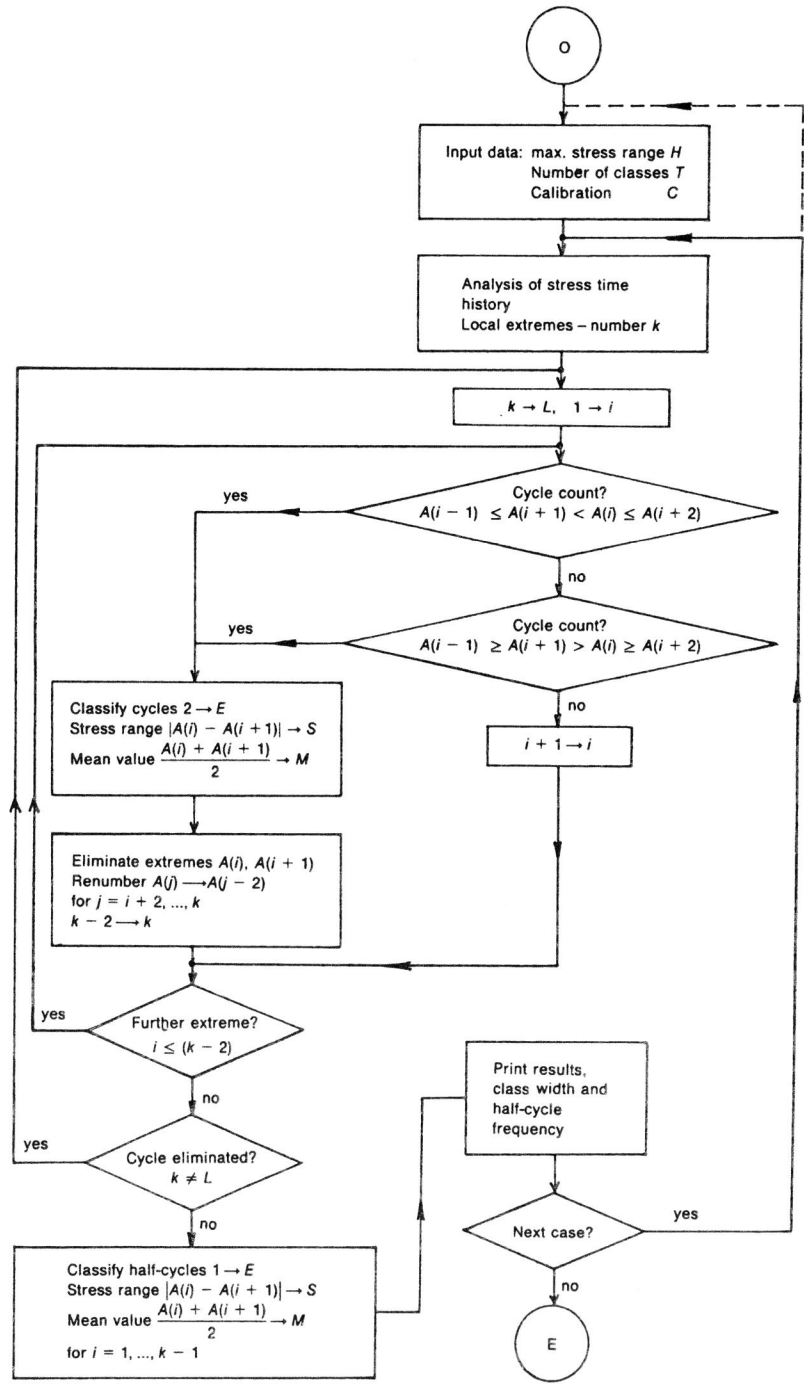

Fig. 11.17. Flow diagram of the computer program for the rain-flow counting method.

11.2 Rain-flow counting method

TABLE 11.3. Computer program for the rain-flow counting method (Hewlett Packard Language)

```
 0: spc 3; prt "RAIN FLOW METHOD"; prt "× × × × × × × × × × × × × × × ×"
 1: dim, X, Y, H, T, T[0:30], A[0:4100], D[0:30]
 2: spc; fxd 0; ent "MAX. RANGE = ?", H; prt "MAX. RANGE", H
 3: ent "NO OF CLASSES = ?", T; prt "NO OF CLASSES", T
 4: for I = 1 to T
 5: IH/T}D[I]
 6: next I
 7: spc; fxd 0; ent "STRAIN GAUGE?", X; prt "STRAIN GAUGE", X
 8: fxd 6; ent "CALIBRATION?", C; prt "CALIBRATION", C
 9: "TEST RUN": spc 3; fxd 0; ent "NO OF TEST RUN?", Y; prt "NO OF TEST RUN", Y; spc
10: for I = 1 to T
11: 0}T[I]
12: next I
13: fmt 2f3; time 6000; on err "ALL"
14: dsp "START OF ANALYZER"
15: for I = 1 to 4100
16: red 3, A[I], B
17: next I
18: "ALL": I}K; fxd 0; prt "NO OF EXTR", K
19: for I = 1 to K
20: otd A[I]}A[I]
21: if A[I]⟩ = 128; A[I]−255}A[I]
22: CA[I]}A[I]
23: next I
24: K}L;1}I
25: if A[I−1]⟨ = A[I+1] and A[I+1]⟨A[I] and A[I]⟨ = A[I+2]; gto +3
26: if A[I−1]⟩ = A[I+1] and A[I+1]⟩A[I] and A[I]⟩ = A[I+2]; gto +2
27: I+1}I; gto +6
28: abs (A[I]−A[I+1])}S; 2}E; gsb 44
29: for J=I+2 to K
30: A[J]}A[J−2]
31: next J
32: K−2}K
33: if I⟨ = K−2; gto −8
34: if K ≠ L; gto − 10
35: for I = 1 to K − 1
36: abs (A[I] −A[I+J])}S; 1}E; gsb 44
37: next I
38: prt "CLASS NUMBER"; fxd 0; fmt f3, "−", f3, f9
39: for I = 1 to T
40: wrt 16, D[I−1], D[I], T[I]
41: next I
42: gto "TEST RUN"
43: "CLASS":
44: for N = 1 to T
45: if S⟩ = D[N−1] and S<D[N]; T[N] +E} T[N]; ret
46: next N; ret
```

The described algorithm of the rain-flow counting method is represented by the flow diagram in Fig. 11.17. On this basis a program in the Hewlett Packard Language for a desk calculator Hewlett Packard 9825 was prepared. An excerpt from the program is presented in Table 11.3.

Figure 11.18 shows a schematic diagram of an evaluation instrument line for data processing by the rain-flow method. If the stress-time history has been obtained by calculation, it is most advantageous to evaluate it by the rain-flow counting method on the same computer. If the evaluation involves experimental data recorded on a measuring tape-recorder, it is necessary first to digitize them, feed them to the computer and eliminate local extrems. The feeding can be checked on an oscilloscope.

The method described may easily be extended by the evaluation of the mean values. In this way a biparametric method is obtained which classifies the stress ranges and their mean values and thus describes the random process in greater detail.

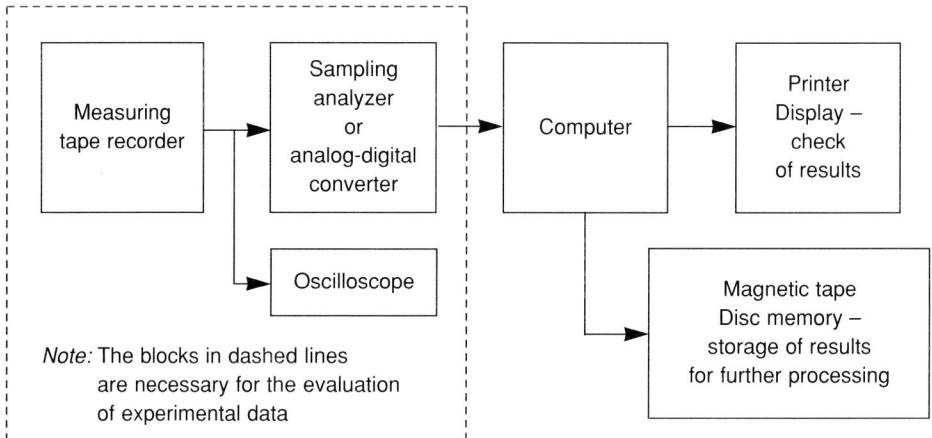

Fig. 11.18. Scheme of evaluation line for data processing by the rain-flow counting method.

11.3 Appreciation of counting methods

In the investigation of railway bridge vibrations, strength, fatigue and reliability, various statistical counting methods have been applied to classify the extreme or other stress values and the mode of their alternation in accordance with several requirements.

These clasification methods are of empirical character; the most widely known methods have been briefly described in Sect. 11.1.

The sampling method (Sect. 11.1.1) is used chiefly for the computation of statistical characteristics of the first and of the second order, as the numerical computation of these quantities operates with the data obtained in this way.

The probability of transition over a certain stress level can be ascertained by the threshold method (Sect. 11.1.2) or by the level crossing count (Sect. 11.1.4a).

The peak counting methods (Sect. 11.1.3) describe the random processes mostly with undue pessimism, while the fatigue meter method (Sect. 11.1.4b) with undue optimism in respect of the fatigue of structures.

The stress range counting methods (Sect. 11.1.5) are based on the knowledge that it is the stress range which is the statistically most significant factor affecting the fatigue of structures. The most frequently used method is the rain-flow counting method which faithfully characterizes the irreversible stress components of the stress-strain diagram that are decisive for the fatigue of structures. Moreover, this method also makes it possible to evaluate another parameter, e.g. the mean value of the vibration. The algorithm, flow diagram and computer program are easy to write.

11.4 Statistical evaluation of stresses

In selected places of the bridge maximum or minimum stresses can also be evaluated statistically. In this process two methods have been applied.

11.4.1 Stress extremes during train passage

From every recorded train passage, one extreme stress value is counted, i.e. max σ in the tensile bridge element or min σ in the compressed bridge element, or both, if the element is subjected to alternate loading. The set of these extreme values is evaluated statistically.

An example of statistical evaluation of maximum stresses in the tensile part of a two-span continuous girder is shown in Fig. 11.19. The histograms can only occasionally by smoothed by the Gauss curve.

This method of evaluation is used for the calculation of the probability that a certain stress threshold will not be exceeded. The threshold usually equals the mean value of the set $\bar{\sigma}$ and the γ– multiple of standard deviation s

$$P(\sigma < \bar{\sigma} + \gamma s) = \Phi(\gamma). \tag{11.5}$$

Here the function

$$\Phi(x) = \frac{1}{(2\pi)^{1/2}} \int_{-\infty}^{x} \exp(-y^2/2)\, dy = \frac{1}{2}\left[1 + \mathrm{erf}\left(\frac{x}{2^{1/2}}\right)\right] \tag{11.6}$$

represents the normal Gauss distribution. It is often assumed without any detailed verification.

Table 11.4 gives the coefficients of reliability γ and their respective probabilities $\Phi(\gamma)$ according to (11.5) for several practically used values for the one-sided probability of Gauss normal distribution, which is most frequently used in engineering practice.

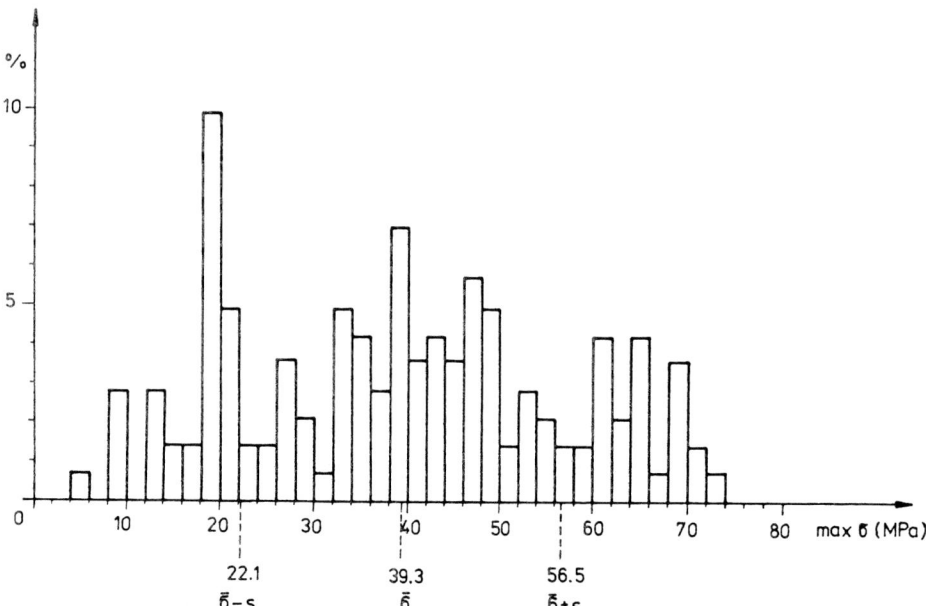

Fig. 11.19. Histogram of maximum stresses during the passage of 142 (100%) trains in the main continuous girder bridge (No 2), 2 × 16.2 m, $\bar{\sigma}$ = 39.3 MPa, s = 17.2 MPa.

Table 11.4. Probability that the mean stress value $\bar{\sigma}$ and the γ- multiple of the standard deviation s will not be exceeded (one-sided probability of Gaus normal distribution)

γ	$P(s < \bar{\sigma} + \gamma s) = \Phi(\gamma)$
0	0.5
0.842	0.8
1	0.841 31
1.282	0.9
1.5	0.933 19
1.645	0.95
2	0.977 235
2.326	0.99
3	0.998 648
3.090	0.999
4	0.999 968 7
5	0.999 999 7

In Table 11.5, the set of measurements from Fig. 11.19 is analyzed statistically and supplemented with the probability

$$P(\sigma > \max \max \sigma) \tag{11.7}$$

11.4. Statistical evaluation of stresses

that the stress will not exceed the max max of the whole set of max max σ, and the probability

$$P(\sigma > \sigma_n \delta) \tag{11.8}$$

that the stress will exceed the stress due to the standard load σ_n multiplied by the standard dynamic coefficient δ. This last named probability is usually very low in practice.

TABLE 11.5. Statistical evaluation of stresses in the main girder of the bridge No. 2 from Fig. 11.19

Quantity	Unit	Method	
		Peak counting	Sampling
Number of measurements	1	142	142
Maximum measured value	MPa	72.6	68.8
Mean value	MPa	39.3	34.1
Standard deviation	MPa	17.2	16.2
Variation coefficient	1	0.438	0.475
$\sigma_n \delta$	MPa	160.52	
$P(\sigma > \text{max max } \sigma)$	1	0.026	
$P(\sigma > \sigma_n \delta)$	1	5×10^{-13}	

11.4.2 Statistical evaluation of results of the sampling method

Some evaluation instruments evaluate the stress-time history by the sampling method (Sect. 11.1.1 and Fig. 11.3). The results of an example from Fig. 11.9 are tabulated in Table 11.5.

This method can be used for the computation of the *statistical dynamic coefficient*

$$\delta = 1 + \frac{\gamma s}{\bar{\sigma}}, \tag{11.9}$$

where $\bar{\sigma}$ – mean stress value obtained by the sampling method,
  s – standard deviation of this data set,
  γ – coefficient of reliability.

In bridge dynamics, the reliability coefficient is usually considered as $\gamma = 1.65$ which guarantees 95% reliability according to Table 11.4. It means that the mean stress value, multiplied by the statistical dynamic coefficient (11.9), will be exceeded in 5% of cases only.

Eq. (11.9) includes both the random load distribution along the length of the train and the actual dynamic component. The statistical dynamic coefficient (11.9) works well, when the dynamic coefficient during the passage of whole trains is to be found.

The classical *dynamic coefficient*

$$\delta = \frac{\max v(x, t)}{v_0}, \quad (11.10)$$

where max $v(x, t)$ – maximum dynamic deflection (or stress),
v_0 – the same value due to the static load,
can be used in practice only for the passage of a test vehicle, when the static test or the drive at a very low speed can be performed. This is not possible in the majority of cases for trains in every-day traffic.

12. Stress ranges in steel railway bridges

With references to the fatigue of steel structures it has appeared, [59] to [61], that the most important parameter influencing it is the stress range

$$\Delta\sigma = \sigma_{max} - \sigma_{min} \tag{12.1}$$

defined as the difference between the local maximum σ_{max} and the local minimum σ_{min} of stress. The values of σ_{max} and σ_{min} are considered algebraically, so that the stress range $\Delta\sigma$ is always a positive value. $\Delta\sigma$ is considered the most important parameter with respect to the response of the structure to current loading. Otherwise, fatigue is influenced substantially by stress concentrations, notches, weld details, and so on.

In case of complex stressing, e.g. stochastic stress-time history, the stress range is obtained by one of the statistical counting methods described in detail in Chapter 11.

For the fatigue of bridges, it is important to know the number of the individual stress ranges due to traffic loads per unit of time, e.g. per day, or per year. For this purpose, the stress ranges are classified into several (usually 20) classes and the number of stress ranges or stress cycles is ascertained in every class. In this way the histogram of stress range frequencies is obtained; it is called the stress spectrum which should not be confused with the frequency spectrum and the power spectral density.

The counting of stress ranges can be done either theoretically or experimentally.

12.1 Theoretical calculation of stress spectra

In the research program ORE D 128, [162], the problem was first solved theoretically: the loading schemes of typical trains of individual railway administrations were driven along a triangular influence line of the bending moment at midspan of a simply supported beam. The bending moment history thus obtained was classified by the rain-flow counting method to obtain the bending moment spectra. This procedure is represented schematically in Fig. 12.1. The principles of classification by the rain-flow counting method are briefly recapitulated in Fig. 12.2.

12.1.1 Characteristic trains

The following typical trains were selected as characteristic: passenger trains for suburban traffic, long-distance high-speed passenger trains, mixed freight trains with randomly arranged cars, and block freight trains carrying heavy substrates.

For every train, the total weight ΣF, velocity V, number of axles i and axle distances were determined.

For the CSD (former Czechoslovak State Railways), for instance, the following trains were selected as typical trains (Fig. 12.3; for other railway administrations see [162]):

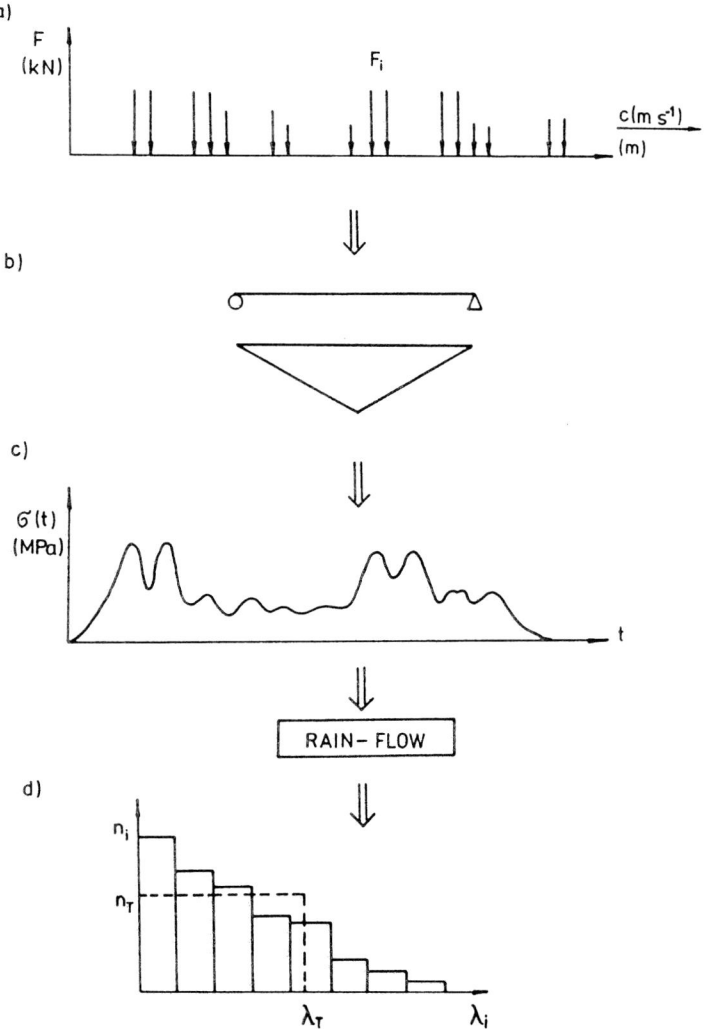

Fig. 12.1. a) Typical train with axle forces F_i travelling at the velocity c, b) Influence line of bending moment or stress at midspan of a simply supported beam, c) Computed stress-time history $\sigma(t)$ during the passage of a typical train, d) Histogram of numbers n_i of stress cycles (stress spectrum) after classification by rain-flow counting method; $\lambda_i = \Delta\sigma_i/(\delta \Delta\sigma_n)$ is the dimensionless stress range $\Delta\sigma_i$ in the ith class.

12.1 Theoretical calculation of stress spectra

1. Suburban train unit, velocity $V = 110$ km h^{-1}.
2. Express train with a four-axle locomotive, $V = 120$ km h^{-1}.
3. Mixed freight train with two four-axle locomotives L, 20 four-axle cars not fully loaded P and 20 cars fully loaded R, randomly assembled, $V = 90$ km h^{-1}.
3A. The same train as in 3, but with a different arrangement, the numbers of cars P and R remained the same (20 each), $V = 90$ km h^{-1}.
4. Very heavy block freight train with two six-axle locomotives and 47 four-axle fully loaded cars, $V = 90$ km h^{-1}.
5. Heavy block freight train with one six-axle locomotive and 36 fully loaded four-axle cars, $V = 90$ km h^{-1}.
6. Mixed freight train with two four-axle locomotives L, 8 empty four-axle cars P, 20 over loaded four-axle cars R, 4 not fully loaded four-axle cars T, 4 medium-loaded two-axle cars B and 4 medium-loaded four-axle cars W. The cars were randomly assembled, $V = 90$ km h^{-1}.

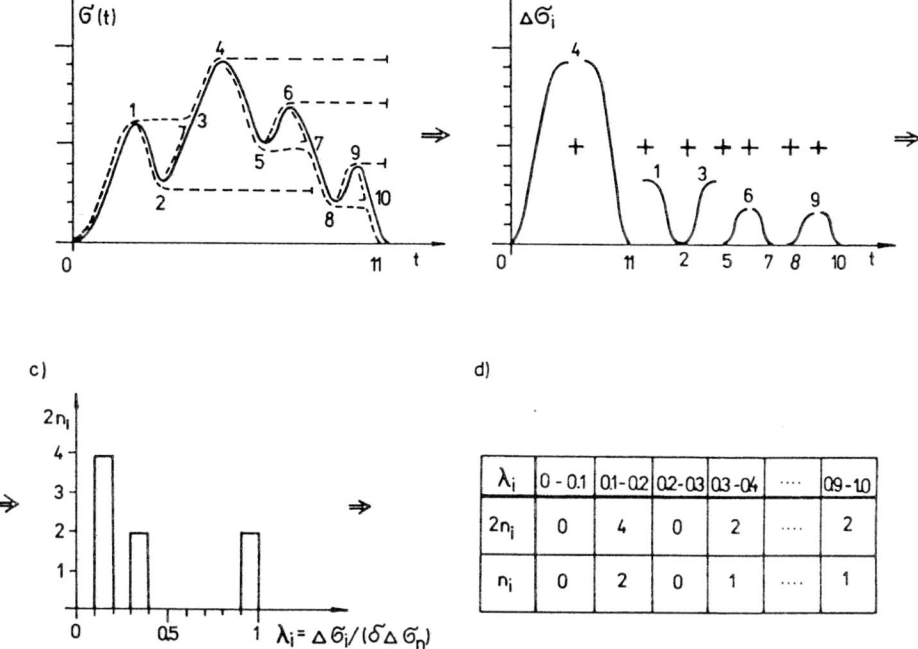

Fig. 12.2. Rain-flow counting method:
a) stress-time history $\sigma(t)$, b) stress ranges after classification, c) stress range spectrum in a histogram form, d) stress range spectrum in a tabular form.
$2n_i$ – number of stress ranges, n_i – number of stress cycles in the ith class.

220   *12. Stress ranges in steel railway bridges*

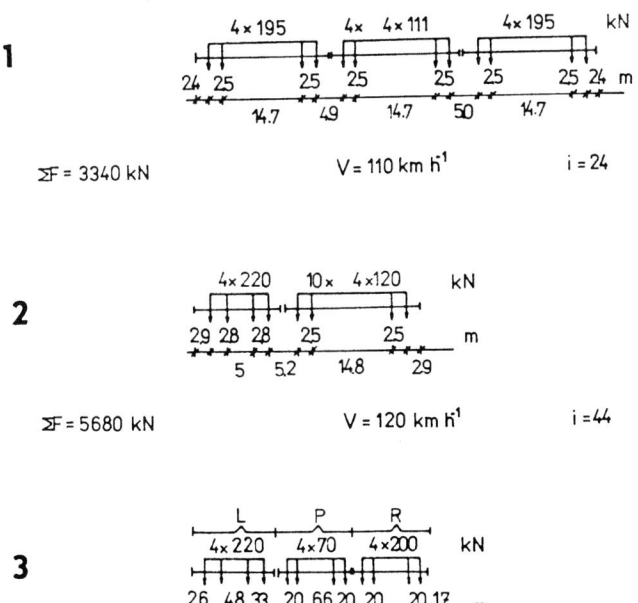

1

$\Sigma F = 3340$ kN $V = 110$ km h^{-1} $i = 24$

2

$\Sigma F = 5680$ kN $V = 120$ km h^{-1} $i = 44$

3

LL-PRPPPR-R-P-RR-P-PPPRRRPPR-PPPRPR-PRPRRRRRPPPRR

3A

LL-PRRPPRPRR-P-RRPPRRPRRPPPPR-PPRRR-R-PPRPPPPR-PR

$\Sigma F = 23\ 360$ kN $V = 90$ km h^{-1} $i = 168$

4

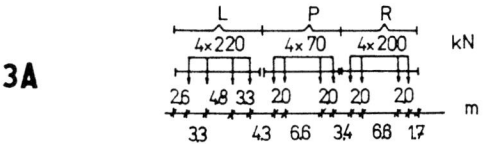

$\Sigma F = 40\ 000$ kN $V = 90$ km h^{-1} $i = 200$

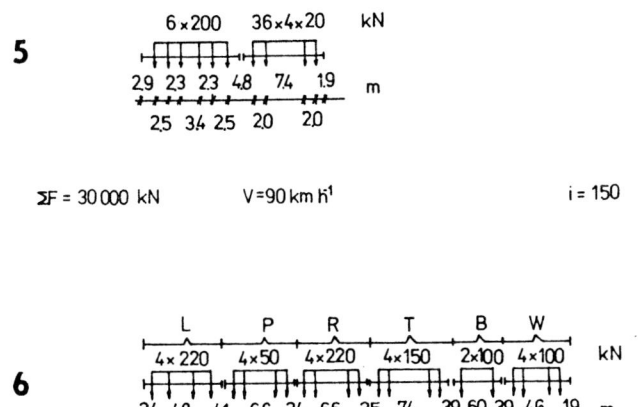

Fig. 12.3. Characteristic trains of ČSD:
1 – suburban train (multiple unit train), *2* – passenger train or express train (loco hauled passenger train). *3* – mixed freight train, *L* – locomotive, *P* – four-axle not fully loaded car, *R* – four-axle fully loaded car. *3A*, – mixed freight train with different car sequence in comparison to the train sub *3*, *4* – very heavy freight block train, *5* – heavy freight block train, *6* – mixed freight train, *L* – locomotive, *P* – four-axle empty car, *R* – four-axle overloaded car, *T* – four-axle not fully loaded car, *B* – two-axle medium loaded car, *W* – four-axle medium loaded car. $\sum F$ – total train weigth, *V* – velocity, *i* – number of axles.

12.1.2 Traffic loads

From the characteristic trains, the annual traffic load was composed for every railway administration [162]. This quantity is given in million tonnes per year transported over the given bridge. This traffic load is recorded in the statistics of railway administrations. According to conditions prevailing on main and secondary railway lines the traffic loads were classified into several groups. The example given in Table 12.1 gives an idea about the traffic loads on the ČSD lines:

T1 – Very heavy traffic load of 60×10^6 t/year prevailed on some sections of the main line of the ČSD.

T2 – Heavy traffic load of 25×10^6 t/year on main lines of the ČSD. (This traffic load was increased in Table 12.7 and later to 30×10^6 t/year.)

T3 – Medium traffic load of 10×10^6 t/year on other ČSD lines.

T4 – Light traffic load of 2×10^6 t/year on secondary ČSD lines.

The analysis of typical trains and traffic loads of the individual railway administrations has revealed that the very heavy traffic load *T1* of the CSD represents almost double the heavy traffic load of the BR, DB, NS, SBB or SNCF. A typical train for the railway traffic in the former Czechoslovakia (CSD) was a mixed freight train with randomly-arranged cars. The opposite extreme is represented for example by the trains of the NS, which consist mostly of block trains with constant axle forces and regular axle distances, see [162].

12.1.3 Bending moment spectra

The bending moment spectra were found by the method described above with the following assumptions:

Typical trains were considered as axle forces only, i.e. only their weight action was considered. Typical trains according to Fig. 12.3 were used for the generation of characteristic traffic according to Table 12.1.

The bridge was represented by a simply supported beam and for the calculation of its stresses it is assumed a triangular influence line of the bending moment at midspan. The bending moment was applied, because its use eliminates the influence of the cross section of a particular girder and the results, consequently, are of more general use.

The dynamic action of trains was considered only by the standard dynamic coefficient which depends, according to [212], on velocity and span.

The computations were carried out for the spans of 2, 3, 4, 5, 7, 10, 15, 20, 30 and 50 m. Examples of the tables of bending moment range for the spans of 3, 5, 30 and 50 m are shown in Tables 12.2 to 12.5. Tables 12.2 and 12.3 give the numbers $2n_i$ (see Fig. 12.2) in the individual classes of bending moments in kN m for all characteristic CSD trains. Tables 12.4 and 12.5 present analogous data for typical CSD railway traffic *T1–T4* according to Table 12.1.

However, it has appeared, [162], that the spectra obtained in this way afford pessimistic results in respect of fatigue. This is due to the following:

The characteristic trains are heavier than the actual trains; also their composition resulting in the traffic loads *Ti* is different.

The actual velocities are usually lower than those considered theoretically.

The dynamic action of whole trains is actually lower than that considered theoretically. The standard dynamic coefficient $\delta = 1 + \varphi$, [212], covers the highest dynamic actions of locomotives alone with 95% reliability, which are higher than the actions of whole trains.

The spatial interaction of all bridge elements makes the stresses in them lower than those considered theoretically.

Thus, the measured stresses are generally lower than the stresses determined by structural analysis.

TABLE 12.1. Theoretical traffic loads of CSD consisting of characteristic trains (according to Fig. 12.3)

Traffic load CSD		Characteristic train	Mass of one train (t)	Number of trains per year	Annual traffic load (10^6t)	Rounded-off traffic load value (10^6t)
T1 very heavy	Passenger	1 2	334 568	3 650 2 190	2.4	
	Freight	3 4 5 6	2 336 4 000 3 000 2 416	4 000 1 000 10 000 5 000	55.4	
Together				25 840	57.8	60
T2 heavy	Passenger	1 2	334 568	5 110 7 665	6.0	
	Freight	3 4 5 6	2 336 4 000 3 000 2 416	2 500 500 1 500 2 700	18.8	
Together				19 975	24.8	25
T3 medium	Passenger	1 2	334 568	3 650 1 825	2.2	
	Freight	3 4 5 6	2 336 4 000 3 000 2 416	1 500 – – 1 500	7.1	
Together				8 475	9.3	10
T4 light	passenger	1 2	334 568	3 650 –	1.2	
	freight	3 4 5 6	2 336 4 000 3 000 2 416	– – – 200	0.5	
Together				3 850	1.7	2

TABLE 12.2. Number of bending moment ranges of a simply supported beam during the passage of CSD characteristic trains according to Fig. 12.3

Span $l = 3$ m

CSD characteristic train	$\delta = 1 + \varphi$	Bending moment ranges (kN m) / Number of bending moment ranges $2n_i$

CSD train	δ	0–12	12–24	24–36	36–48	48–60	60–72	72–84	84–96	96–108	108–120	120–132	132–144	144–156	156–168	168–180	180–192	192–204	204–216	216–228	228–240	240–252
1	1.53																			8		
2	1.55																			4		
3	1.50			80				16	40		16			8						80	4	
4	1.50							160		8										196		
5	1.50							188		4										148		
6	1.50		32		16			144		32		40								80	16	16
3A	1.50	12		80		48		160							16					80		16

Span $l = 5$ m

CSD train	δ	0–29	29–58	58–87	87–116	116–145	145–174	174–203	203–232	232–261	261–290	290–319	319–348	348–377	377–406	406–435	435–464	464–493	493–522	522–551	551–580	580–609
1	1.52				4			4	12					8								
2	1.53	2	4			10			40						4		64					
3	1.50			62			16		16			2		4	4		96					
4	1.50						4		92						2		74					
5	1.50						2			70												
6	1.50		20	8	16			18	18					2	2		62					
3A	1.50	29		60		18	18		18			4		6	4		62				580	609

12.1 Theoretical calculation of stress spectra

TABLE 12.3. Number of bending moment ranges of a simply supported beam during the passage of CSD characteristic trains according to Fig. 12.3

Span $l = 30$ m

CSD characteristic train	$\delta = 1+\varphi$	Bending moment ranges (kN m)																				
		0	424	848	1272	1696	2120	2544	2968	3392	3816	4240	4664	5088	5512	5936	6360	6784	7208	7632	8056	8480
		–424	–848	–1272	–1696	–2120	–2544	–2968	–3392	–3816	–4240	–4664	–5088	–5512	–5936	–6360	–6784	–7208	–7632	–8056	–8480	–8904
		Number of bending moment ranges $2n_i$																				
1	1.20	4						2		2						2						2
2	1.22	4			18																	
3	1.16	84		10		2		6		2		4										
4	1.16	370	4																			
5	1.16	278											2						2			
6	1.16	74				4	2			2		2					2	2				
3A	1.16	80	6	4					14		6		4									

Span $l = 50$ m

CSD characteristic train	$\delta = 1+\varphi$	Bending moment ranges (kN m)																				
		0	1 100	2 200	3 300	4 400	5 500	6 600	7 700	8 800	9 900	11 000	12 100	13 200	14 300	15 400	16 500	17 600	18 700	19 800	20 900	22 000
		–1 100	–2 200	–3 300	–4 400	–5 500	–6 600	–7 700	–8 800	–9 900	–11 000	–12 100	–13 200	–14 300	–15 400	–16 500	–17 600	–18 700	–19 800	–20 900	–22 000	–23 100
		Number of bending moment ranges $2n_i$																				
1	1.15	24		2						2												2
2	1.17	68																				
3	1.12	62			4		2			4			2						2			
4	1.12	182									2											
5	1.12	140					2		2											2		
6	1.12	48	4		2	4				4										2	2	
3A	1.12	40			2	12			2	2			2								2	

TABLE 12.4. Number of bending moment ranges per year (in thousands) of a simply supported beam under CSD traffic loads according to Table 12.1

Span $l = 3$ m

CSD traffic load	Bending moment ranges (kN m)																				
	0	12	24	36	48	60	72	84	96	108	120	132	144	156	168	180	192	204	216	228	240
	–	–	–	–	–	–	–	–	–	–	–	–	–	–	–	–	–	–	–	–	–
	12	24	36	48	60	72	84	96	108	120	132	144	156	168	180	192	204	216	228	240	252

Number of bending moment ranges per year $2n_i$ ($\times 10^3$)

T1	160	320	80	240	–	2 726.4	87.6	–	208	–	58.4	87.6	77.2	80	–	–	–	–	2 434	–	152.8
T2	86.4	200	43.2	129.6	–	1 007.8	306.6	–	96.4	–	81.8	306.6	50.88	43.2	–	–	–	–	807.54	–	113.9
T3	48	120	24	72	–	418.4	73	–	48	–	58.4	73	29.2	24	–	–	–	–	276.5	–	55.3
T4	6.4	–	3.2	9.6	–	74.4	–	–	6.4	–	58.4	–	29.2	3.2	–	–	–	–	45.2	–	3.2

Span $l = 5$ m

CSD traffic load	Bending moment ranges (kN m)																				
	0	29	58	87	116	145	174	203	232	261	290	319	348	377	406	435	464	493	522	551	580
	–	–	–	–	–	–	–	–	–	–	–	–	–	–	–	–	–	–	–	–	–
	29	58	87	116	145	174	203	232	261	290	319	348	377	406	435	464	493	522	551	580	609

Number of bending moment ranges per year $2n_i$ ($\times 10^3$)

T1	2	108.76	288	100	120	88	104.6	377.4	–	700	–	22	–	95.2	28	18.76	1402	–	–	–	–
T2	1	84.66	176.6	54	68.2	45	69.04	502.5	–	105	–	11.8	–	70.1	16.4	36.06	486.4	–	–	–	–
T3	–	37.3	105	30	39	24	41.6	167.8	–	–	–	6	–	44.2	9	10.3	189	–	–	–	–
T4	–	4	1.6	4	3.2	–	18.2	47.4	–	–	–	0.8	–	30.4	0.4	0.4	12.4	–	–	–	–

TABLE 12.5. Number of bending moment ranges per year (in thousands) of a simply supported beam under CSD traffic loads according to Table 12.1

Span $l = 30$ m

CSD traffic load	Bending moment ranges (kN m)																			
	0–424	424–848	848–1272	1272–1696	1696–2120	2120–2544	2544–2968	2968–3392	3392–3816	3816–4240	4240–4664	4664–5088	5088–5512	5512–5936	5936–6360	6360–6784	6784–7208	7208–7632	7632–8056	8056–8480
	Number of bending moment ranges per year $2n_i$ ($\times 10^3$)																			
T1	3 879.4	34	21.9	79.42	8	20	17.3	24	30	—	25.3	26	10	—	—	4.38	—	18	—	2
T2	1 062.9	18.2	30.66	163	5	10.8	15.6	15	16.2	—	20.6	15.4	5.4	—	—	15.33	—	10.4	—	1
T3	258.9	9	21.9	47.85	3	6	10.3	9	9	—	13.3	9	3	—	—	3.65	—	6	—	—
T4	29.4	1.2	21.9	—	—	0.8	7.7	—	1.2	—	7.7	0.4	0.4	—	—	—	—	0.4	—	—

Span $l = 50$ m

CSD traffic load	Bending moment ranges (kN m)																			
	0–1100	1100–2200	2200–3300	3300–4400	4400–5500	5500–6600	6600–7700	7700–8800	8800–9900	9900–11000	11000–12100	12100–13200	13200–14300	14300–15400	15400–16500	16500–17600	17600–18700	18700–19800	19800–20900	20900–22000
	Number of bending moment ranges per year $2n_i$ ($\times 10^3$)																			
T1	2 306.5	20	7.3	26	20	18	10	10	43.3	8	—	—	—	—	—	4.38	—	38	—	2
T2	1 229.5	10.8	10.2	15.4	10.8	10.4	5.4	5.4	31.02	5	—	—	—	—	—	15.33	—	13.4	—	1
T3	376.7	6	7.3	9	6	6	3	3	19.3	3	—	—	—	—	—	3.65	—	6	—	—
T4	97.2	0.8	7.3	0.4	0.8	0.4	—	0.4	8.1	—	—	—	—	—	—	—	—	0.4	—	—

TABLE 12.6. Steel railway bridges of CSD subjected to experimental stress spectra measurements (experiments on bridges Nos. 7 and 8 were conducted by J. Sláma)

Bridge No.		1	2	3	4	5	6	7	8	Number of spectra
Structural type		Plate girder	Plate girder	Truss	Truss	Plate girder	Plate girder	Plate girder	Plate girder	
Deck type		Open	Box	Open	Open	Orthotropic	Orthotropic	Open	Open	
Traffic load (10^6 t/year)		39.6	23.7	12.75	9.0	20.1	2.1	19.1	19.1	
Ratio of annual and recorded traffic loads		227.61	365	427.5	371.3	365.22	569.32	476.05	578.97	
Main girder, simply supported	Span (m)	32.0		30.32	48.3	30.0	35.0	30.0	24.0	11
	No of spectra	1		1	1	1	1	4	2	
Main girder, continuous	Span (m)		2×16.2							3
	No. of spectra		3							
Diagonal	Length (m)			5.37	6.73					3
	No. of spectra			2	1					
Wind brace	Length (m)							4.17	4.29	5
	No. of spectra							3	2	
Cross-beam	Span (m)	3.2				4.9	5.65	5.8	6.12	11
	No. of spectra	1				1	2	4	3	
Stringer, midspan	Span (m)	2.7		3.79	3.45			3.0	3.0	19
	No. of spectra	1		1	1			8	8	
Stringer, support	Span (m)			3.79				3.0	3.0	9
	No. of spectra			1				4	4	
Longitudinal stiffener, midspan	Span (m)					1.24	1.75			3
	No. of spectra					1	2			
Longitudinal stiffener, support	Span (m)						1.75			2
	No. of spectra						2			
Deck plate, along	Span (m)					1.24	1.75			5
	No. of spectra					1	4			
Deck plate, across	Span (m)		0.2				0.375			8
	No. of spectra		1				7			
Number of spectra		3	4	5	3	4	18	23	19	79

Consequently, the data on the numbers of stress ranges in Tables 12.2 through 12.5 must be considered as maxima. However, their calculation corresponds with the methods currently used in structural analysis of bridges in accordance with the respective standards and national codes.

12.2 Experimental stress spectra

The stress spectra were recorded experimentally in 8 steel railway bridges of CSD, see Table 12.6. In the course of this investigation the response of the structures at selected points was measured under normal traffic conditions for a period of approximately 24 hours. The stress-time history was recorded and the stress ranges were classified by the rain-flow counting method according to Chapter 11 and Fig. 12.2.

In this way, histograms of stress ranges in all important elements of steel railway bridges were obtained (main girder, diagonal of a truss, wind brace, cross-beam, stringer, longitudinal stiffener and plate of orthotropic bridge deck) in the most varied traffic conditions on railway lines with traffic loads between 2.1×10^6 and 39.6×10^6 t/year. An example of an experimental stress spectrum is shown in Fig. 12.4a and in Table 11.2.

The analysis of all 79 stress spectra has enabled the following conclusions to be drawn:

a) The stress spectra include both the static and dynamic stress components due to running trains. In accordance with the rain-flow counting method these two components cannot be separated.

b) The number of stress cycles with large amplitudes is low. These cycles correspond to the static component of the trains, the groups of heavy cars and locomotives.

c) The number of stress cycles with small amplitudes is high. They characterize the dynamic action of running trains and their effects on the vibration of bridges.

d) The number of stress cycles in every class and the number and magnitude of classes depends on the intensity and composition of traffic loads. Roughly speaking, the number of cycles is approximately proportional to the magnitude of traffic load.

e) The number of stress cycles and the number of classes depend on the element of the bridge investigated, i.e. particularly on the length and the shape of the respective influence line as well as the cross section of the element. For this reason, all principal elements of steel bridges which have mutually different shapes of influence lines (main girder, cross-beam, stringer, etc.) were investigated. It has appeared that in shorter elements the number of stress cycles is usually higher than in longer elements. The stress spectra of less loaded and of secondary bridge elements are rather poor.

f) The histograms of stress ranges do not include any hypothesis of fatigue damage accumulation or any hypothesis of fracture mechanics. The stress spectra do not depend on Wöhler fatigue curves, either; they represent merely an evaluation of the response of the bridge to current traffic loads. From the available analysis methods only the rain-flow counting method is applied to the spectra.

g) A considerable number of oscillations arise usually in the lowest stress range class. It is difficult to distinguish, whether they are due to mechanical vibrations or to signal noise in the measuring instruments. For this reason, some authors, e.g. [101], recommend neglecting these data entirely. For the method of linear accumulation of fatigue damage, the number of stress cycles in the lowest class is of little significance; therefore, this problem loses its importance to a certain extent.

h) The variation in experimental results is considerable.

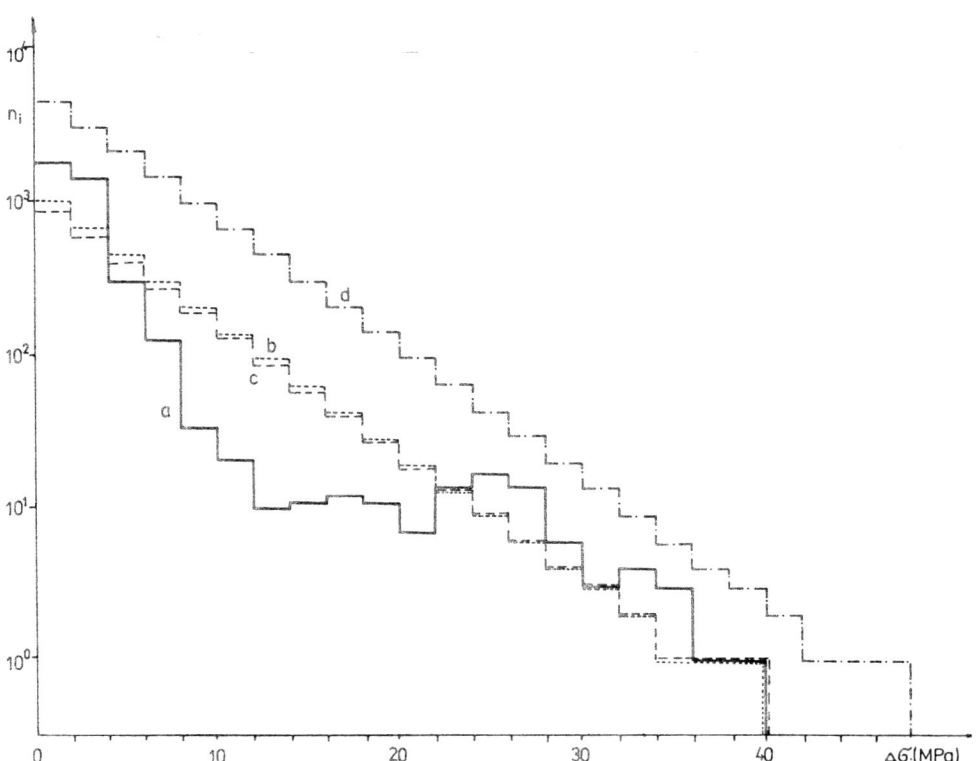

Fig. 12.4. Histogram of frequencies of complete stress cycles in the main girder of the bridge No. 5 during the 24 hours traffic:
a) experimental spectrum, b) theoretical spectrum according to equations (12.5) and (12.9), $\sigma_0 = 5.013$ MPa, c) theoretical spectrum according to equation (12.15), $\sigma_0 = 5.160$ MPa, d) theoretical spectrum according to equation (12.16), $\sigma_0 = 5.160$ MPa.

12.2.1 Empirical formula for the number of stress ranges

The first series of bridges investigated (Nos 1–5 in Table 12.6) provided 19 stress spectra which were statistically processed. To allow the comparison of results from the measurements on various bridges and their elements the stress ranges were written in the dimensionless form

$$\lambda_i = \frac{\Delta\sigma_i}{\delta \Delta\sigma_n} \tag{12.2}$$

where λ_i is a dimensionless stress range in the ith class,
  $\Delta\sigma_i$ – stress range,
  δ – standard dynamic coefficient,
  $\Delta\sigma_n$ – maximum stress range due to the standard load.

Theoretically λ_i can attain the value from 0 to 1 if the structure is not overstressed.

These standardized stress spectra were evaluated statistically by regression analysis. The following parameters were selected as independent variables:
  T – traffic load, i.e. weight of all trains passing along the bridge in a year (in million tonnes),
  l – length of the bridge element in metres,
  λ_i – 0; 0.1; 0.2; 0.3; 0.4; 0.5; 0.6; 0.7; 0.8; 0.9; 1 the dimensionless stress range (12.2).

Out of the four applied regression relations, the power regression

$$n_i = aT^b l^c \lambda_i^d e^{ks} \tag{12.3}$$

has appeared most satisfactory, where
  n_i – number of stress cycles in the ith class per year,
  a, b, c, d – regresion coefficients (12.4),
  s – standard deviation,
  $k = 1.65$ – coefficient guaranteeing 95% reliability.

Experimental data from the bridges Nos 1–5 (Table 12.6) have provided the following regression coefficients for railway bridges:

$$\begin{aligned} a &= 17.742, & c &= -0.354, \\ b &= 0.860, & d &= -4.464, \\ s &= 1.323, & k &= 1.65. \end{aligned} \tag{12.4}$$

The numbers of stress cycles n_i, in thousands per year, computed from (12.3) with the application of the data from (12.4), are given in Table 12.7.

12.2.2 Probability density of stress ranges

After the completion of experiments on eight bridges, all 79 histograms of stress cycles (Table 12.6) were subjected to statistical analysis. It has appeared that the exponential distribution of the probability density satisfies relatively well all the experimental conclusions a) to h) of Sect. 12.2.

TABLE 12.7. Number of stress cycles of steel railway bridges per year

Length l (m)	Traffic load CSD	λ_i									
		0.1	0.2	0.3	0.4	0.5	0.6	0.7	0.8	0.9	1
		n_i ($\times 10^3$/year)									
2	T1	120 000	5 500	900	250	92	41	20	11	6.7	4.2
	T3	67 000	3 000	490	140	51	22	11	6.2	3.7	2.3
	T3	26 000	1 200	190	53	20	8.7	4.4	2.4	1.4	0.9
	T4	6 500	290	48	13	4.9	2.2	1.1	0.61	0.36	0.22
3	T1	100 000	4 800	780	220	80	35	18	9.8	5.8	3.6
	T2	58 000	2 600	430	120	44	19	9.8	5.4	3.2	2.0
	T3	22 000	1 000	170	46	17	7.5	3.8	2.1	1.2	0.77
	T4	5 600	260	42	12	4.3	1.9	0.95	0.52	0.31	0.19
5	T1	88 000	4 000	650	180	66	29	15	8.1	4.8	3.0
	T2	48 000	2 200	360	99	37	16	8.1	4.5	2.7	1.7
	T3	19 000	850	140	38	14	6.3	3.2	1.7	1.0	0.64
	T4	4 700	210	35	9.6	3.6	1.8	0.79	0.44	0.26	0.16
30	T1	46 000	2 100	340	95	35	16	7.8	4.3	2.5	1.6
	T2	26 000	1 200	190	52	19	8.6	4.3	2.4	1.4	0.88
	T3	9 900	450	74	20	7.5	3.3	1.7	0.92	0.55	0.34
	T4	2 500	110	18	5.1	1.9	0.84	0.42	0.23	0.14	0.086
50	T1	39 000	1 800	290	79	29	13	6.5	3.6	2.1	1.3
	T2	21 000	970	160	44	16	7.2	3.6	2.0	1.2	0.73
	T3	8 300	370	61	17	6.3	2.8	1.4	0.77	0.46	0.29
	T4	2 100	94	15	4.3	1.6	0.70	0.35	0.19	0.11	0.071
100	T1	30 000	1 400	220	62	23	10	5.1	2.8	1.7	1.0
	T2	17 000	760	120	34	13	5.6	2.8	1.6	0.92	0.57
	T3	6 500	290	48	13	4.9	2.2	1.1	0.60	0.36	0.22
	T4	1 600	74	12	3.3	1.2	0.55	0.27	0.15	0.089	0.056

Traffic loads CSD per year

Notation	Mass	Number	Number of trains per hour
	of all trains passing along one line per year		
T1 Very heavy	60×10^6 t	35 040	4
T2 Heavy	30×10^6 t	26 280	3
T3 Medium	10×10^6 t	17 520	2
T4 Light	2×10^6 t	8 760	1

12.2 Experimental stress spectra

There are of course other possibilities, too. For instance, the author [76] considered earlier the chi-distribution and Sedlacek and Jacquemoud [189], supposed the beta-distribution. However, the exponential distribution has a great advantage in its simplicity.

The exponential is of the form

$$f(s) = \frac{1}{\sigma_0} e^{-s/\sigma_0} \tag{12.5}$$

where $s = \Delta\sigma$ are stress ranges,
σ_0 is parameter of exponential distribution.

The parameter σ_0 is usually determined from measured data by the least squares method. In this case, because of the conclusion g) from Sect. 12.2, it is necessary to suppress somewhat the influence of small stress ranges and accentuate the influence of high stress ranges. The least squares method, however, does not allow this modification. For this reason, the computation was based on the condition of equality of statistical moments of the kth order of theoretical and experimental results

$$n \int_0^\infty s^k f(s) \, ds = \sum_i n_i s_i^k \tag{12.6}$$

where $n = \Sigma n_i$ is the sum of the number of stress cycles,
n_i — number of stress cycles in the ith class,
$s_i = \Delta\sigma_i$ — stress range in the ith class.

The integral on the left-hand side of (12.6) is the statistical moment of the kth order which, after substitution of (12.5) and computation, yields

$$m_k = \int_0^\infty s^k f(s) \, ds = \sigma_0^k \, \Gamma(1 + k) \tag{12.7}$$

where $\Gamma(x)$ is the Gamma function.

Using (12.7) the unknown parameter σ_0 is then found from (12.6), i.e.

$$\sigma_0 = \left[\frac{\sum_i n_i s_i^k}{n \, \Gamma(1 + k)} \right]^{1/k} . \tag{12.8}$$

The condition (12.6) is equivalent to the condition of equality of the fatigue damage due the theoretical distribution of probability density and the fatigue damage due to experimental distribution of the number of stress cycles with the assumption of the validity of the linear theory of accumulation of fatigue damage (Sect. 13.2). For these reasons k was considered with the value of $k = 5$, because the measured stress ranges were generally within the range of the time fatigue limit, i.e. in the region in which the Wöhler fatigue curve has a minor slope $1/k = 1/5$.

All measured histograms of stress cycles were then evaluated according to Eq. (12.8). The values of σ_0 obtained were gathered into groups according to the individual elements of the steel railway bridges; in every group the maximum and the minimum values, the mean value of σ_0, the standard deviation and the variation coefficient were determined. The results of these calculations are given in Table 12.8.

Table 12.8 reveals that the principal and most heavily loaded elements of steel railway bridges, such as main girders, cross-beams and stringers, provide high vaues σ_0 while in secondary and slightly loaded members, such as wind braces, or elements of orthotropic bridge decks, the value of σ_0 remains low.

TABLE 12.8. Statistical evaluation of σ_0 in Eq. (12.8) for all investigated elements of steel railway bridges

Element of steel railway bridge	Number of measurements	max σ_0 (MPa)	min σ_0 (MPa)	mean σ_0 (MPa)	Standard deviation (MPa)	Variation coefficient (%)
Main girder simply supported	11	7.009	2.692	5.160	1.238	23.996
Main girder, continuous	3	7.060	5.379	6.464	0.941	14.557
Diagonal	3	5.805	3.835	4.779	0.988	20.665
Wind brace	5	1.692	1.508	1.588	0.085	5.353
Cross-beam	11	6.999	2.941	5.275	1.259	23.863
Stringer, midspan	19	8.285	3.980	5.883	1.247	21.199
Stringer, support	9	7.684	4.400	5.713	1.370	23.975
Longitudinal stiffener, midspan	3	4.605	2.092	3.304	1.259	38.106
Longitudinal stiffener, support	2	4.646	4.478	4.562	0.119	2.604
Deck plate, along	5	1.826	0.826	1.462	0.385	26.336
Deck plate, across	8	4.431	0.785	1.607	1.221	75.961

Figure 12.4. compares the measured (Fig. 12.4a) and theoretically calculated (Fig. 12.4b) stress spectra according to Eq. (12.5) with the value of $\sigma_0 = 5.013$ MPa, evaluated from Eq. (12.8) for the bridge No 5. The comparison reveals that the histograms are in good agreement in the region of large stress ranges, which are decisive for fatigue considerations.

The numbers of stress cycles n_i in the ith class are found from the probability density as follows:

$$n_i = n\,\Delta\sigma\,f(s_i) = \frac{n\,\Delta\sigma}{\sigma_0}\,e^{-\Delta\sigma_i/\sigma_0} \qquad (12.9)$$

12.2 Experimental stress spectra

where $n = \Sigma n_i$ is the total number of cycles in the spectrum under consideration,

$\Delta\sigma$ — division on the horizontal axis of the histogram (e.g. $\Delta\sigma = 2$ MPa in Fig. 12.4),

$\Delta\sigma_i$ — stress range in the ith class,

σ_0 — value of (12.8).

12.2.3 Number of stress cycles per year

The number of stress cycles calculated according to Eq. (12.9) was added across all classes i, giving the smoothed total number $n = \Sigma_i n_i$ cycles. This number was extrapolated by the ratio of the annual and measured traffic loads (Table 12.6) yielding the total number of stress cycles per year for every investigated bridge element.

This annual number of stress cycles depends on traffic load and on the investigated bridge element. According to previous experience (12.3) the power regression

$$n = aT^b l^c \tag{12.10}$$

was sought between the total number of stress cycles per year n, the traffic load T (in milions of tonnes per year) and the length of the bridge element l (in metres).

The regression coefficients a, b, c were found by the least squares method for the set of 79 triads of data n, T and l.

The attempt to divide this set of 79 data triads into subsets according to the individual elements of steel railway bridges and their analogous evaluation was not successful, because the subsets contained only few data.

The numerical computation resulted in the following values:

$$\begin{aligned} a &= 95\,642, \\ b &= 1.040, \\ c &= -0.165 \end{aligned} \tag{12.11}$$

with the coefficient of determination $R^2 = 0.558$. Standard deviation

$$s = \left[\frac{1}{79 - 2 - 1} \sum_i (\ln n_i - \ln a - b \ln T_i - c \ln l_i)^2 \right]^{1/2} \tag{12.12}$$

yielded the value of

$$s = 0.829. \tag{12.13}$$

The reliability zone is obtained by the multiplication of Eq. (12.10) by the factor

$$e^{t_{n-2}\cdot p\, s} = e^{1.99 \times 0.829} = 5.210 \tag{12.14}$$

which will ensure, [181], that the results have 95% reliability.

TABLE 12.9. Number of stress cycles in elements of steel railway bridges per year according to equations (12.15) and (12.16)

Bridge element	Main girder, simply supported		Main girder, continuous		Diagonal		Wind brace		Cross-beam	
Traffic load T (t/year)					30×10^6					
Span l (m)	30		30		6		5		5	
Mean σ_0 (MPa)	5.160		6.464		4.779		1.588		5.275	
$\Delta\sigma_i$ (MPa)	n_i	$n_i \, e^n$	n_i	$n_i \, e^n$	n_i	$n_i \, e^n$	n_i	$n_i \, e^n$	n_i	$n_i \, e^n$
	($\times 10^3$/year)									
5	690	3593	669	3487	899	4683	341	1774	926	4825
10	262	1363	309	1609	316	1645	14.6	76.1	359	1870
15	99.3	517	142	742	111	578	0.627	3.27	139	725
20	37.7	196	65.7	343	39.0	203	0.027	0.140	53.9	281
25	14.3	74.5	30.3	158	13.7	71.3	0.001	0.006	20.9	109
30	5.43	28.3	14.0	72.9	4.81	25.0			8.10	42.2
35	2.06	10.7	6.46	33.6	1.69	8.80			3.14	16.4
40	0.781	4.07	2.98	15.5	0.593	3.09			1.22	63.4
45	0.296	1.55	1.38	7.16	0.208	1.09			0.471	2.46
50	0.113	0.586	0.634	3.30	0.073	0.381			0.183	0.952
55	0.043	0.222	0.293	1.53	0.026	0.134			0.071	0.369
60	0.016	0.084	0.135	0.703	0.009	0.047			0.027	0.143
65	0.006	0.032	0.062	0.325	0.003	0.017			0.011	0.055
70	0.002	0.012	0.029	0.150	0.001	0.006			0.004	0.021
75	0.001	0.005	0.013	0.069		0.002			0.002	0.008
80		0.002	0.006	0.032		0.001			0.001	0.003
85		0.001	0.003	0.015						0.001
90			0.001	0.007						
95			0.001	0.003						
100			0.001	0.001						
105				0.001						
Σn_i ($\times 10^3$/year)	1111	5790	1243	6475	1386	7219	356	1854	1512	7878

12.2 Experimental stress spectra

TABLE 12.9. – continued

Bridge element	Stringer, midspan		Stringer, support		Longitudinal stiffener, midspan		Longitudinal stiffener, support		Deck plate, along		Deck plate, across	
Traffic load T (t/year)	30×10^6											
Span l (m)	3		3		2		2		2		0.4	
Mean σ_0 (MPa)	5.883		5.713		3.304		4.562		1.462		1.607	
$\Delta\sigma_i$ (MPa)	n_i	$n_i e^n$	n_i	$n_i e^n$	n_i	$n_i e^n$	n_i	$n_i e^n$	n_i	$n_i e^n$	n_i	$n_i e^n$
	($\times 10^3$/year)											
5	996	5191	1000	5212	977	5090	1074	5596	328	1709	530	2761
10	426	1422	417	2172	215	1121	359	1870	10.7	55.9	23.6	123
15	182	948	174	905	47.4	247	120	625	0.351	1.829	1.051	5.48
20	77.8	405	72.4	377	10.4	54.3	40.1	209	0.011	0.060	0.047	0.244
25	33.3	173	30.2	157	2.30	12.0	13.4	69.8		0.002	0.002	0.011
30	14.2	74.1	12.6	65.5	0.506	2.63	4.48	23.3				
35	6.08	31.7	5.24	27.3	0.111	0.580	1.50	7.80				
40	2.60	13.5	2.19	11.4	0.025	0.128	0.500	2.61				
45	1.11	5.79	0.911	4.75	0.005	0.028	0.167	0.871				
50	0.475	2.47	0.380	1.98	0.001	0.006	0.056	0.291				
55	0.203	1.06	0.158	0.824		0.001	0.019	0.097				
60	0.087	0.452	0.066	0.344			0.006	0.033				
65	0.037	0.193	0.027	0.143			0.002	0.011				
70	0.016	0.083	0.011	0.060			0.001	0.004				
75	0.007	0.035	0.005	0.025				0.001				
80	0.003	0.015	0.002	0.010								
85	0.001	0.006	0.001	0.004								
90	0.001	0.003		0.002								
95		0.001		0.001								
100		0.001										
Σn_i ($\times 10^3$/year)	1740	9066	1715	8936	1253	6527	1613	8404	339	1767	556	2890

TABLE 12.10. Number od stress cycles in steel railway bridges per year. Main girder, simply supported, influence of span according to equations (12.15) and (12.16)

Bridge element						Main girder, simply supported							
Traffic load T (t/year)						30×10^6							
Span l (m)	2		5		10		20		30		50		100	
Mean σ_0 (MPa)						5.160							
$\Delta\sigma_i$ (MPa)	n_i	$n_i e^n$	n_i	$n_i e^n$	n_i	$n_i e^n$	n_i	$n_i e^n$	n_i	$n_i e^n$	n_i	$n_i e^n$	n_i	$n_i e^n$
						($\times 10^3$/year)							
5	1078	5617	927	4829	827	4307	737	3842	690	3593	634	3303	565	2946
10	409	2132	352	1832	314	1634	280	1458	262	1363	241	1253	215	1118
15	155	809	133	695	119	620	106	553	99.3	517	91.3	476	81.4	424
20	58.9	307	50.6	264	45.2	235	40.3	210	37.7	196	34.6	180	30.9	161
25	22.4	116	19.2	100	17.1	89.3	15.3	79.7	14.3	74.5	13.1	68.5	11.7	61.1
30	8.48	44.2	7.29	38.0	6.50	33.9	5.80	30.2	5.43	28.3	4.99	26.0	4.45	23.2
35	3.22	16.8	2.77	14.4	2.47	12.9	2.20	11.5	2.06	10.7	1.89	9.86	1.69	8.80
40	1.221	6.36	1.05	5.47	0.937	4.88	0.835	4.35	0.781	4.07	0.718	3.74	0.641	3.34
45	0.464	2.42	0.398	2.08	0.355	1.85	0.317	1.65	0.296	1.55	0.273	1.42	0.243	1.27
50	0.176	0.916	0.151	0.788	0.135	0.703	0.120	0.627	0.113	0.586	0.103	0.539	0.092	0.481
55	0.067	0.348	0.057	0.299	0.051	0.267	0.046	0.238	0.043	0.222	0.039	0.204	0.035	0.182
60	0.025	0.132	0.022	0.113	0.019	0.101	0.017	0.090	0.016	0.084	0.015	0.078	0.013	0.069
65	0.010	0.050	0.008	0.043	0.007	0.038	0.007	0.034	0.006	0.032	0.006	0.029	0.005	0.026
70	0.004	0.019	0.003	0.016	0.003	0.015	0.002	0.013	0.002	0.012	0.002	0.011	0.002	0.010
75	0.001	0.007	0.001	0.006	0.001	0.006	0.001	0.005	0.001	0.005	0.001	0.004	0.001	0.004
80	0.001	0.003		0.002		0.002		0.002		0.002		0.002		0.001
85		0.001		0.001		0.001		0.001		0.001		0.001		0.001
$\Sigma n_i (\times 10^3$/year)	1737	9052	1494	7782	1332	6941	1188	6191	1111	5790	1022	5322	911	4747

12.2 Experimental stress spectra

After the substitution of (12.10) in (12.9), the mean number of cycles n_i in the stress range class $\Delta\sigma_i$ of the width $\Delta\sigma$ is obtained as

$$n_i = \frac{a\,\Delta\sigma}{\sigma_0}\,T^b l^c\,e^{-\Delta\sigma_i/\sigma_0}, \tag{12.15}$$

or the number of stress cycles n_i with 95% reliability

$$n_i = \frac{a\,\Delta\sigma}{\sigma_0}\,T^b l^c\,e^{-\Delta\sigma_i/\sigma_0}\,e^{t_{n-2,P}\,s}. \tag{12.16}$$

12.2.4 Number of stress cycles

The number of stress cycles in the elements of steel railway bridges according to equations (12.15) and (12.16), using the data from (12.11) to (12.14) was calculated in Tables 12.9 to 12.11. The tables give the mean numbers of stress cycles n_i according to (12.15) or the same data with 95% reliability according to (12.16), denoted briefly by $n_i e^{ts}$, in thousands per year. The width of the stress range classes is $\Delta\sigma_0 = 5$ MPa.

TABLE 12.11. Number of stress cycles in steel railway bridges per year. Main girder, simply supported, influence of traffic load T according to equations (12.15) and (12.16)

Bridge element	Main girder, simply supported							
Traffic load T (t/year)	2×10^6		10×10^6		30×10^6		60×10^6	
Span l (m)	30							
Mean σ_0 (MPa)	5.160							
$\Delta\sigma_i$ (MPa)	n_i	$n_i e^{ts}$	n_i	$n_i e^{ts}$	n_i	$n_i e^{ts}$	n_i	$n_i e^{ts}$
	($\times 10^3$/year)							
5	41.3	215	220	1146	690	3593	1418	7388
10	15.7	81.6	83.5	435	262	1363	538	2804
15	5.94	31.0	31.7	165	99.3	517	204	1064
20	2.25	11.7	12.0	62.6	37.7	196	77.5	404
25	0.855	4.46	4.56	23.8	14.3	74.5	29.4	153
30	0.325	1.69	1.73	9.02	5.43	28.3	11.2	58.1
35	0.123	0.642	0.657	3.42	2.06	10.7	4.23	22.1
40	0.047	0.244	0.249	1.30	0.781	4.07	1.61	8.37
45	0.018	0.092	0.095	0.493	0.296	1.55	0.610	3.18
50	0.007	0.035	0.036	0.187	0.113	0.586	0.231	1.21
55	0.003	0.013	0.014	0.071	0.043	0.222	0.088	0.457
60	0.001	0.005	0.005	0.027	0.016	0.084	0.033	0.174
65		0.002	0.002	0.010	0.006	0.032	0.013	0.066
70		0.001	0.001	0.004	0.002	0.012	0.005	0.025
75				0.001	0.001	0.005	0.002	0.009
80				0.001		0.002	0.001	0.004
85						0.001		0.001
90								0.001
Σn_i ($\times 10^3$/year)	66.5	346	355	1847	1111	5790	2285	11906

Table 12.9 gives the numbers of stress cycles in the individual elements of steel railway bridges differing from the value of σ_0 for a heavy traffic load $T = 30 \times 10^6$ t/year and for the spans l typical for the individual elements. The table shows that the principal bridge elements endure a higher number of stress cycles per year and also include stress cycles in higher stress range classes. The secondary bridge elements and the orthotropic bridge deck endure lower numbers of stress cycles and the stress ranges are lower. It should be noted, however, that the stress magnitude in an element depends on its cross section and, additionally, on the designer's distribution of material.

Table 12.10 shows the influence of the span l on the number of stress cycles in the main girder in identical conditions. It is obvious that this number increases slightly with decreasing span – see conclusion e) in Sect. 12.2.

Table 12.11 shows the influence of the traffic load T on the number of stress cycles in the main girder. The number of stress cycles increases with increasing load, which is in agreement with the conclusion d) in Sect. 12.2.

It should be noted in Tables 12.9 to 12.11 that the calculation of the number of stress cycles stops when the value after rounding drops below one unit per year.

Figure 12.4 shows a comparison of the measured and calculated stress spectra for the main girder of the bridge No. 5 in the course of 24 hours. Fig. 12.4c shows the theoretical spectrum according to Eq. (12.15), while Fig. 12.4d has been found with 95% reliability from Eq. (12.16). In both cases the mean value for the main girder in the form of a simply supportead beam $\sigma_0 = 5.160$ Mpa (Table 12.6) was applied. The mean theoretical spectrum in Fig. 12.4c corresponds well with the measured spectrum in Fig. 12.4a, especially in the region of higher stress ranges which are decisive for fatigue. Fig. 12.4d furnishes a very safe spectrum.

12.3 Growing traffic loads

Figures 10.1 and 10.2 have illustrated the increase of the traffic loads in the past, while all data on the number of vibrations are ascertained at the time when the bridge has been in operation for m years.

Let us assume that the increase of traffic loads is linear and that its annual increment is b. According to the data from Figs 10.1 and 10.2 the estimated increments b are given in Table 12.12.

Table 12.12. Annual increments of trafic loads on CSD railway lines

Trafic load	Annual increment b (10^6 t)
T1 very heavy	0.8
T2 heavy	0.4
T3 medium	0.2
T4 light	0.1

Let us assume that the increase of the traffic load is approximately due to the higher loading of the cars, so that the number of trains per year, n_T, remains the same. During the bridge life L the bridge is traversed by N_T trains:

$$N_T = n_T L . \tag{12.17}$$

According to the conclusion d) from Sect. 12.2 the number of stress cycles is roughly proportional to the traffic load passing along the bridge. Consequently, during the bridge life L every class contains

$$T n_i \tag{12.18}$$

stress cycles. Here n_i is the number of stress cycles in the ith class per year per unit traffic load T (i.e. per 1 million tonnes), and T is the sum of traffic loads traversing the bridge in the course of its life L. It is calculated by means of the summation of L terms of the arithmetic series from the equation

$$T = L \left[T_m + b \left(\frac{1 + L}{2} - m \right) \right] \tag{12.19}$$

where T_m is the traffic load at present, when the bridge has been in operation for m years,
    b – the annual traffic load increment according to Table 12.12,
    L – the total life of the bridge.

Should we wish also to characterize the increase of the traffic loads in the future, we would have to replace in (13.11), (13.14) and (13.17) n_i with (12.18) and n_T with (12.17).

12.4 Influence of overloading

Across the bridge No. 7, a pressure vessel for a nuclear power plant was transported during the experiments. The vessel was loaded on a special Krupp car Uaai 839 with 32 axles and axle loads of up to 160 kN.

The stress spectra recorded and evaluated during current one day traffic, during ten travels of the heavy train with the Krupp car and during one travel of the same train are shown in Fig. 12.5.

Figure 12.5 reveals that the maximum stress range due to the heavy train is almost twice as large as that due to current traffic. The numbers of stress cycles in the individual classes are higher than those of the current traffic in the majority of cases. It should be noted that the speed of the heavy train was very low (less than 30 km h^{-1}).

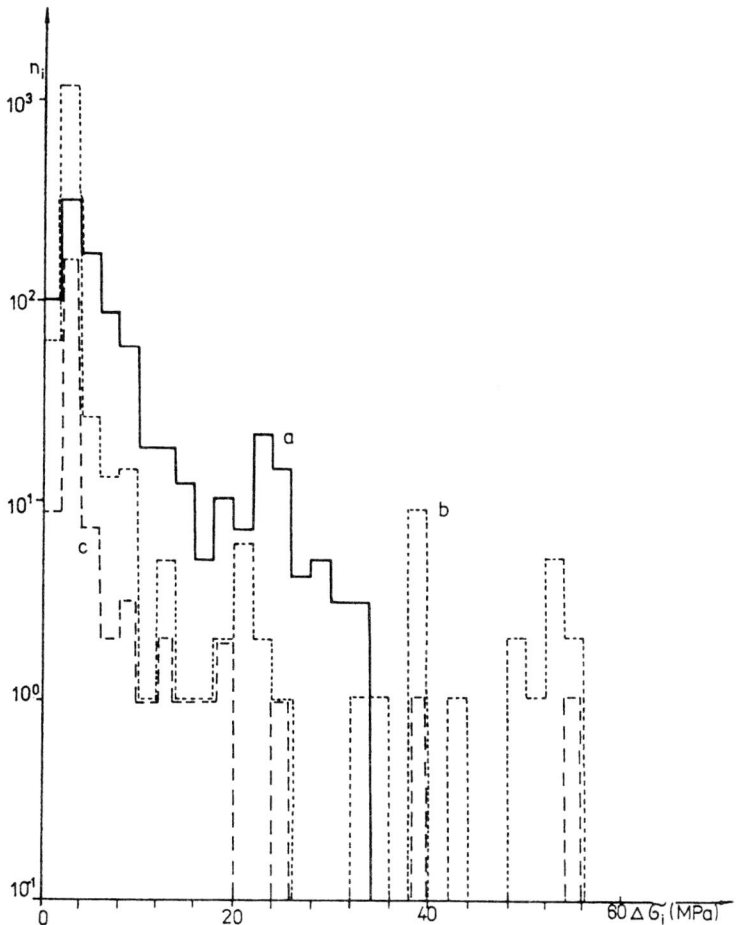

Fig. 12.5. Histogram of frequencies of complete stress cycles in the main girder of the bridge No. 7: a) daily traffic, 65 trains per 24 hours, b) ten passages of the Krupp very heavy train, c) one passage of the Krupp very heavy train.

12.5 Other factors

The length and shape of the influence line of the investigated element influences the form and magnitude of the stress spectrum. This influence was studied theoretically in [162] on a number of examples (bending moment at midspan of a simply supported beam, shear force above support and at midspan of a simply supported beam, bending moment at midspan of the first field and above the first intermediate support of a continuous beam of four equal spans and the support reaction of the same continuous beam). The load was represented by four characteristic trains NS 1, 2, 3 and 6 and the spans considered were 3, 4, 5, 7 and 10 m. The theoretical procedure described in Sect. 12.1 was used.

In comparison with the spectrum of bending moments at midspan of a simply supported beam the results were both less and more favourable, see [162].

In the quantity σ_0 (Table 12.8) the experimental procedure described in Sect. 12.2.2 incorporates the influence of the length and shape of the influence line, because the individual elements of steel railway bridges provide characteristic shapes of influence lines and their lengths are variable only within narrow limits. Therefore, the problem is characterized with sufficient accuracy.

The number of further, so far unsolved, problems include secondary stresses, stress concentrations, spatial stresses, shear stresse, etc., which can significantly influence the form and magnitude of stress spectra and, consequently, the fatigue of the bridge.

12.6 Appreciation of stress spectra

The stress spectra in steel railway bridges were obtained in this chapter by three methods:

1. The theoretical spectra of bending moments given in Tables 12.2 to 12.5 are very conservative with respect to fatigue. A minor increase of accuracy could result from the consideration of the effect of inertia of moving trains.

2. The spectra shown in Table 12.7, found from Eq. (12.3), are suitable for the fatigue assessment of new bridges. With regard to the long life of bridges they guarantee a sufficient safety margin.

3. The number of stress cycles according to Eq. (12.15) affords realistic values for contemporary traffic; therefore, it is suitable for the calculation of stress spectra in the fatigue assessment of existing bridges. Eq. (12.16) guarantees sufficient reliability. Some spectra found from these equations are given in Tables 12.9 to 12.11, they follow the real shape of influence lines of standard elements of steel railway bridges.

The advantage of the representation of the bridge response to usual traffic loads in the form of a stress spectrum consists of its independence of any hypothesis of fatigue damage cumulation, fracture mechanics or Wöhler fatigue curve. In practical applications, however, one of these hypotheses must be assumed, see Chapter 13.

13. The assessment of steel railway bridges for fatigue

Two approaches have been developed for the fatigue assessment of steel railway bridges:

The first is based on the hypothesis of fatigue damage accumulation and is well suited for the structures which are in the design phase. Hence this approach has become part of bridge design standards in many countries. The design based on the fatigue damage accumulation hypothesis results in a structure with individual elements which are more or less balanced with reference to fatigue, although their life expectancy cannot be estimated with full reliability.

The second approach is based on the principles of fracture mechanics which assumes an initial crack of a certain length and envisages its propagation until it has attained a critical length. For this reason, the second approach is applied to existing structures in which fatigue cracks have originated under traffic load. The methods of fracture mechanics make it possible to estimate the time of periodic inspections of bridges and, possibly, the residual life expectancy of the structure.

Apart from these two principal approaches to fatigue assessment of steel railway bridges there is a number of further variants and combinations. Nevertheless, even fracture mechanics and particularly the theory of fatigue damage accumulation do not completely clarify the causes of fatigue and remain theories of phenomenological character.

In the fatigue assessment of railway bridges it is assumed that the loading process and, consequently, its response in the form of stresses in the bridge is of random character (stochastic process) and that the probability, that after a small stress range a large stress range follows, is identical with the probability of the phenomena appearing in the opposite sequence. For this reason the counting methods described in Chapter 11 should be used.

13.1 Theory of fatigue damage accumulation

The research of fatigue of steel structures, [59] to [61], has yielded the following most important results:

1. Fatigue damage is substantially influenced by the stress ranges

$$\Delta\sigma = \sigma_{max} - \sigma_{min} \tag{13.1}$$

where σ_{max} is the local maximum, and

σ_{min} – the local minimum, see also equation (12.1) and Fig. 11.2.

Further parameters, such as the mean stress value, steel quality, etc. have been found to be statistically less significant for large civil engineering structures.

13.1 Theory of fatigue damage accumulation

2. The relation between the number of cycles preceding damage N and the stress range $\Delta\sigma$ (Wöhler fatigue curve) may be represented, on the bilogarithmic scale, by one, two or several straight lines of the type (Fig. 13.1)

$$\Delta\sigma = (C/N)^{1/k}, \qquad (13.2)$$

where $1/k$ is the slope of the Wöhler curve on a bilogarithmic scale and
C – a constant depending on the investigated structural detail, on stress concentration, weld type, required reliability, etc.

Figure 13.2 shows the fatigue curves due to ORE, [162], for one slope and for two slopes, and Fig. 13.3 shows the Wöhler curve with two slopes according to Czechoslovak standards ČSN 73 1401 and ČSN 73 6205, where the respective constants and the point of deviation depend on the investigated structural detail.

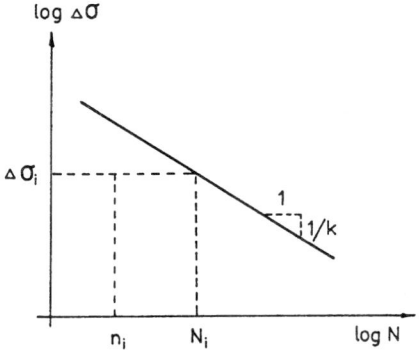

Fig. 13.1. Wöhler fatigue curve and Palmgren–Miner theory of fatigue damage accumulation.

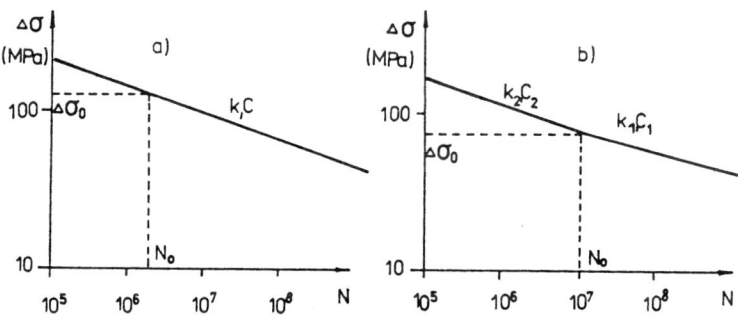

Fig. 13.2. Wöhler fatigue curve according to ORE [162]
a) with one slope
  SN1: $k = 3$, $C = 4 \times 10^{12}$, $N_0 = 2 \times 10^6$, $\Delta\sigma_0 = 126$ MPa, $\log N = 12.6 - 3 \log \Delta\sigma$,
  SN3: $k = 3.75$, $C = 1.5 \times 10^{14}$, $N_0 = 2 \times 10^6$, $\Delta\sigma_0 = 126$ MPa, $\log N = 14.18 - 3.75 \log \Delta\sigma$,
b) with two slopes
  SN2: $k_1 = 4$, $C_1 = 2.95 \times 10^{14}$, $k_2 = 3$, $C_2 = 4 \times 10^{12}$, $N_0 = 10^7$, $\Delta\sigma_0 = 73.56$ MPa, $\log N = 12.6 - 3 \log \Delta\sigma$,
    $N \leq 10^7$, $\log N = 14.47 - 4 \log \Delta\sigma$, $N > 10^7$,
  SN4: $k_1 = 6.5$, $C_1 = 9 \times 10^{19}$, $k_2 = 3.75$, $C_2 = 1.5 \times 10^{14}$, $N_0 = 2 \times 10^6$, $\Delta\sigma_0 = 126$ MPa,
    $\log N = 14.8 - 3.75 \log \Delta\sigma$, $N \leq 2 \times 10^6$, $\log N = 19.95 - 6.5 \log \Delta\sigma$, $N > 2 \times 10^6$.

Figure 13.4 shows the fatigue curve for normal nominal stress according to Eurocode 3, [56], which has three parts with points of deviation at $N = 5 \times 10^6$ and $N = 1 \times 10^8$ and the slopes $k_1 = 5$ and $k_2 = 3$. For $N > 10^8$, the limiting value of $\Delta\sigma_0$ is considered to act; stresses below this are not taken into account. The categorization of structural details given in [56] is quantified by a reference value $\Delta\sigma_C$ for $N = 2 \times 10^6$. The fatigue limit at a constant amplitude corresponds to the fatigue strength $\Delta\sigma_1$ at $N = 5 \times 10^6$. The set of fatigue curves in Fig. 13.4 represents mean values minus twice the standard deviation of test results with constant stress range. When the test results are used for the categorization of a certain detail, the 95% confidence interval for the 95% survival is calculated at $N = 2 \times 10^6$ cycles to determine the standard deviation and the number of samples (the minimum number being 10 samples).

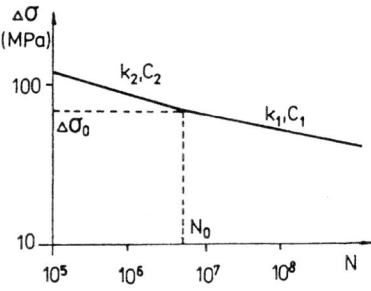

Fig. 13.3. Wöhler fatigue curve according to Czechoslovak standards ČSN 73 1401 and ČSN 73 6205 with two slopes,
$k_1 = 5.45$, $k_2 = 3.45$, $\Delta\sigma = (C_1/N)^{1/k_1}$ for $N > N_0$, $\Delta\sigma = (C_2/N)^{1/k_2}$ for $N \leq N_0$, for structural details:

Detail	$\Delta\sigma_0$ (MPa)	N_0	log C_1	log C_2
A	170	2×10^6	18.457	13.996
B	123	2.78×10^6	17.834	13.654
C	87	3.96×10^6	17.168	13.289
D	69	5.01×10^6	16.722	13.044
E	53	6.51×10^6	16.211	12.762
F	43	8.11×10^6	15.811	12.544
G	35	1×10^7	15.415	12.327

3. The fatigue damage D is described sufficiently by the Palmgren–Miner theory of linear fatigue damage, [168], [146],

$$D = \sum_i \frac{n_i}{N_i},\qquad(13.3)$$

where n_i is the number of stress cycles of the range $\Delta\sigma_i$, Fig. 13.1,
N_i – the number of stress cycles on the Wöhler fatigue curve, and
i – the ordinal number of the stress range level.
4. According to this theory fatigue failure occurs, when
$$D = 1.\qquad(13.4)$$

13.2 Method of equivalent damage

For practical application of stress spectra the method of equivalent damage (also known as the λ_T method) has proved useful. This method is based on the comparison of fatigue damage due to two stress spectra:

1. A real or a theoretical spectrum $(n_i, \Delta\sigma_i)$ measured or computed for n_T trains.
2. A fictitious single-step spectrum with the maximum stress range $\delta \Delta\sigma_n$ and the number of cycles n_T.

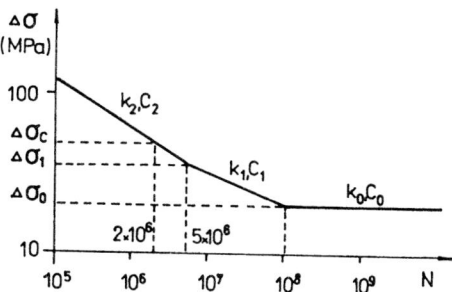

Fig. 13.4. Wöhler fatigue curve for normal nominal stresses according to Eurocode 3 [56] with three slopes, $k_0 = \infty$, $k_1 = 5$, $k_2 = 3$, $\Delta\sigma = \Delta\sigma_0$ for $N > 1 \times 10^8$, $\Delta\sigma = (C_1/N)^{1/k_1}$ for $5 \times 10^6 \leq N \leq 1 \times 10^8$, $\Delta\sigma = (C_2/N)^{1/k_2}$ for $N < 5 \times 10^6$:

Detail category (for $N = 2 \times 10^6$) $\Delta\sigma_c$ (MPa)	Fatigue limit at constant amplitude for $N = 5 \times 10^6$ $\Delta\sigma_1$ (MPa)	Limit value for $N = 10^8$ $\Delta\sigma_0$ (MPa)	log C_1	log C_2
160	117	65	17.036	12.901
140	104	57	16.786	12.751
125	93	51	16.536	12.601
112	83	45	16.286	12.451
100	74	40	16.036	12.301
90	66	36	15.786	12.151
80	59	32	15.536	12.001
71	52	29	15.286	11.851
63	46	25	15.036	11.701
56	41	23	14.786	11.551
50	37	20	14.536	11.401
45	33	18	14.286	11.251
40	29	16	14.036	11.101
36	26	15	13.786	10.951

Here the quantity $\delta \Delta\sigma_n$ indicates the maximum stress range at the investigated point due to the standard load $\Delta\sigma_n$ multiplied by the standard dynamic coefficient δ (but without the load factor).

If the equation of the Wöhler curve (13.2) in the form of

$$N_i = C \, \Delta\sigma_i^{-k} \tag{13.5}$$

is substituted in Eq. (13.3), the damage due to actual traffic load is obtained

$$D = \frac{1}{C} \sum_i n_i \, \Delta\sigma_i^k . \tag{13.6}$$

The single-step stress spectrum $(n_T, \Delta\sigma)$ yields the damage

$$D = \frac{n_T}{N} = \frac{1}{C} \, n_T \, \Delta\sigma^k . \tag{13.7}$$

The equivalent damage method compares the damage (13.6) and the damage (13.7):

$$\frac{1}{C} \, n_T \, \Delta\sigma^k = \frac{1}{C} \sum_i n_i \, \Delta\sigma_i^k \tag{13.8}$$

and calculates $\Delta\sigma$

$$\Delta\sigma^k = \frac{1}{n_T} \sum_i n_i \, \Delta\sigma_i^k . \tag{13.9}$$

The individual classes of stress ranges $\Delta\sigma_i$ are converted into a dimensionless form by dividing by $\delta \Delta\sigma_n$:

$$\lambda_i = \frac{\Delta\sigma_i}{\delta \Delta\sigma_n} . \tag{13.10}$$

Therefore, both sides of equation (13.9) are divided by the same quantity $\delta \Delta\sigma_n$ so that equation (13.9) yields the parameter λ_T (see also Fig. 12.1d):

$$\lambda_T = \frac{\Delta\sigma}{\delta \Delta\sigma_n} = \left[\frac{1}{n_T} \sum_i n_i \, \lambda_i^k \right]^{1/k} . \tag{13.11}$$

λ_T is called the traffic load factor and is found from the data with numbers of stress cycles n_i in the individual dimensionless stress clasess λ_i and the number of trains n_T given in Table 12.7 or Tables 12.9 to 12.11.

The practical application of the λ_T method is very simple: in the case of the single-slope Wöhler curve it depends, according to equation (13.11), on k only; it does not depend on the contant C because in the derivation only equation (13.3), and not equation (13.4), has been used.

The traffic load factor λ_T for Wöhler curves with two slopes $k_1 > k_2$ can be found approximately as follows (assuming that the majority of stress ranges is in the sector of the k_1 slope):

13.2 Method of equivalent damage

In this case the damage is

$$D = \sum_i \frac{n_i \, \Delta\sigma_i^{k_1}}{C_1} + \sum_j \frac{n_j \, \Delta\sigma_j^{k_2}}{C_2} = \frac{n_T \, \Delta\sigma^k}{C}, \quad (13.12)$$

where the indices i, j and the constants C_1, C_2, k_1 and k_2 pertain to sectors *1* and *2*, respectively, of the Wöhler curve. A fictitious Wöhler curve with a uniform slope of $1/k$ has the constant C and the number of cycles n_T which equals the number of trains per year.

Equation (13.12) can be divided by $(\delta \Delta\sigma_n)^k$ so that from equation (13.11),

$$\lambda_T = \left\{ \frac{C}{n_T} \left[\frac{(\delta \Delta\sigma_n)^{k_1-k}}{C_1} \sum_i n_i \, \lambda_i^{k_1} + \frac{(\delta \Delta\sigma_n)^{k_2-k}}{C_2} \sum_j n_j \, \lambda_j^{k_2} \right] \right\}^{1/k}. \quad (13.13)$$

Assuming that the number of cycles n_T appears in the sector *1* of Wöhler curve, it holds true that $k = k_1$, $C = C_1$, and

$$\lambda_T = \left\{ \frac{1}{n_T} \left[\sum_i n_i \, \lambda_i^{k_1} + \frac{C_1}{C_2} \frac{1}{(\delta \Delta\sigma_n)^{k_1-k_2}} \sum_j n_j \, \lambda_j^{k_2} \right] \right\}^{1/k_1}. \quad (13.14)$$

If n_T appears in the sector *2*, the derivation is similar.

When the stress ranges $\Delta\sigma_i$ appear in both the first and the second sectors of the Wöhler curve and the length of the service life of the structure L is prescribed, λ_T must be calculated iteratively from the data of n_i, $\Delta\sigma_i$ and the number of trains n_T per year:

The number of trains N_T for the whole service life of the bridge is

$$N_T = L n_T. \quad (13.15)$$

The prolonged first sector of Wöhler curve yields, according to equation (13.2) and Fig. 13.3

$$\Delta\sigma_T = (C_1 / N_T)^{1/k_1} = \Delta\sigma_0 (N_0 / N_T)^{1/k_1} \quad (13.16)$$

and, according to equation (13.11),

$$\lambda_T = \left[\frac{1}{n_T} \sum_i n_i \, \lambda_i^{k_1} \right]^{1/k_1}. \quad (13.17)$$

The comparative stress range $\Delta\sigma_{com}$ is defined by means of equation (13.16) and equation (13.17) as

$$\Delta\sigma_{com} = \Delta\sigma_T / \lambda_T. \quad (13.18)$$

The damage (13.12) is modified into the form of

$$D = \sum_i \frac{1}{N_0} \left(\frac{\Delta\sigma_{com}}{\Delta\sigma_0} \right)^{k_1} n_i \, \lambda_i^{k_1} + \sum_j \frac{1}{N_0} \left(\frac{\Delta\sigma_{com}}{\Delta\sigma_0} \right)^{k_2} n_j \, \lambda_j^{k_2} \quad (13.19)$$

and the service life in years from equation (13.19)
$$L = 1/D \tag{13.20}$$
is compared with the assumed service life. If both data differ, $\Delta\sigma_{com}$ is corrected and the computation of D and L according to equations (13.19) and (13.20), respectively, is repeated. When the assumed and calculated periods of service life are the same, λ_T is found from equation (13.18)
$$\lambda_T = \Delta\sigma_T / \Delta\sigma_{com} . \tag{13.21}$$

In the case of Wöhler curves with a horizontal sector (Fig. 13.4) the process is generally the same, but stress cycles n_i are neglected for which $\Delta\sigma_i < \Delta\sigma_0$.

13.3 Limit state for fatigue

The limit state for fatigue is defined for bridges as folows:
The bridge must guarantee the passage of a minimum number of trains N_{min} without being damaged due to fatigue.

When the fatigue service life of the bridge is expressed by the number of trains during its service life N_T, then – according to the limit states theory – the bridge must be so designed as to ensure that
$$N_T \geq N_{min} . \tag{13.22}$$
This is the condition of the limit state of fatigue.

The fatigue life N_T, expressed by the number of trains, can be found from equations (13.11) and (13.2):
$$N_T = \frac{1}{\lambda_T^k} \frac{C}{(\delta\Delta\sigma_n)^k} . \tag{13.23}$$
Therefore, it is necessary that
$$\frac{1}{\lambda_T^k} \frac{C}{(\delta\Delta\sigma_n)^k} \geq N_{min} \tag{13.24}$$
or
$$\lambda_T \delta\Delta\sigma_n \leq \left(\frac{C}{N_{min}}\right)^{1/k} . \tag{13.25}$$

The constant C can be expressed, according to equation (13.2) as follows:
$$C = N_0 \Delta\sigma_0^k , \tag{13.26}$$
where N_0, $\Delta\sigma_0$ indicates a point on the Wöhler curve. After the introduction of equation (13.26) into the inequality (13.25) we obtain
$$\lambda_T \delta\Delta\sigma_n \leq \left(\frac{N_0}{N_{min}}\right)^{1/k} \Delta\sigma_0 \tag{13.27}$$

13.3 Limit state for fatigue

or

$$\lambda_T \delta \Delta\sigma_n \leq \gamma_4 \Delta\sigma_0 \tag{13.28}$$

with the notation

$$\gamma_4 = \left(\frac{N_0}{N_{min}}\right)^{1/k} . \tag{13.29}$$

In a limit case it may happen that

$$N_T = N_{min} \tag{13.30}$$

and the coefficient of the service life of the bridge γ_4 is

$$\gamma_4 = \left(\frac{N_0}{N_T}\right)^{1/k} . \tag{13.31}$$

In the majority of cases the number of trains N_T is expressed as the product of the number of trains per year n_T and the assumed service life of the bridge L in years, i.e.

$$N_T = n_T L . \tag{13.32}$$

For the usual number $N_0 = 2 \times 10^6$ the coefficient of service life of the bridge is

$$\gamma_4 = \left(\frac{2 \times 10^6}{n_T L}\right)^{1/k} . \tag{13.33}$$

In the left-hand side of inequality (13.28) the coefficient of probability of load occurrence on multi-track bridges γ_3 is included. It has appeared [162] that in the case of railway bridges this coefficient varies within the limits of

$$q^{-1+(1/k)} \leq \gamma_3 \leq 1 , \tag{13.34}$$

where q is the number of tracks on the bridge. The upper limit in (13.34) corresponds to the case when the train on the bridge always meets the trains on all other tracks. The lower limit

$$\gamma_3 = q^{-1+(1/k)} \tag{13.35}$$

represents the opposite case, when the train on the bridge never meets any train on other tracks.

Detailed calculations in [162] have shown that the probability γ_3 approaches the lower limit of (13.35); therefore, it is taken into account in fatigue assessment.

13.4 Fatigue assessment of bridges according to limit states theory

The derived condition (13.28) can appear in the assessment of steel railway bridges for fatigue according to the limit states theory in the following form:

$$\gamma_1 \gamma_2 \gamma_3 \delta \Delta\sigma_n \leq \gamma_4 \gamma_5 \Delta\sigma_R . \tag{13.36}$$

The left-hand side of inequality (13.36) represents the response to the load, while the right-hand side represents the fatigue strength of the investigated structural detail. In equation (13.36) the symbols have the following meaning:

$\gamma_1 = \lambda_T$ is the fatigue load factor; it is found from the stress spectra in Tables 12.7, 12.9 to 12.11 according to equations (13.11), (13.14) or (13.21),

γ_2 is the factor characterizing the influence of the form of influence line, based on the data of Table 12.7. If the data for typical elements of steel railway bridges from Tables 12.9 to 12.11 are used, $\gamma_2 = 1$,

γ_3 is the factor characterizing the probability of load intensity on multi-track bridges (13.35),

γ_4 is the factor of planned service life of the bridge (13.33),

γ_5 is the factor of notch effects, which depends on the categorization of the structural detail of the investigated element,

$\Delta\sigma_R$ is the basic design fatigue limit. Naturally, the product of $\gamma_5 \Delta\sigma_R$ may be combined into a single limit value of fatigue strength [56].

The factor γ_3 according to equation (13.35) is calculated for several values of q and k in Table 13.1. In practice, the value of $\gamma_3 = 0.60$ is considered, which corresponds, in the case of two-track bridges, approximately with any value of k.

The coefficient γ_4 according to equation (13.33) is given in Table 13.2 for four traffic loads of CSD from Table 12.7 and for two values of k. This coefficient is used to characterize the planned service life the bridge. For instance, in the case of a temporary bridge, the permitted fatigue strength limit may be increased in this way.

The purpose of the assessment condition (13.36) is to compare the greatest stress range due to the standard load $\Delta\sigma_n$ multiplied by the standard dynamic coefficient δ, reduced by the factor γ_1 according to traffic load intensity and its probability on multi-track bridges γ_3 with the fatigue strength of the investigated structural detail $\gamma_5 \Delta\sigma_R$ with reference to the planned service life of the bridge γ_4.

TABLE 13.1. Factor γ_3 for multi-track bridges

Number of tracks on the bridge q	γ_3	
	$k = 3$	$k = 5$
1	1	1
2	0.63	0.57
3	0.48	0.42
4	0.40	0.33
5	0.34	0.28

13.4 Fatigue assessment of bridges according to limit states theory

Table 13.2. Factor γ_4 of the planned service life of bridges

Coefficient k	Bridge service life L (years)	CSD traffic load			
		T1	T2	T3	T4
			γ_4		
3	30	1.24	1.36	1.56	1.97
	60	0.98	1.08	1.24	1.56
	90	0.86	0.95	1.08	1.36
	120	0.78	0.86	0.98	1.24
5	30	1.14	1.20	1.31	1.50
	60	0.99	1.05	1.14	1.31
	90	0.91	0.97	1.05	1.20
	120	0.86	0.91	0.99	1.14

13.5 Propagation of fatigue cracks

According to [119], fracture mechanics distinguishes three phases of the fatigue process which follow one another or even overlap:

a) the phase of changes of mechanical properties,
b) the phase of nucleation of fatigue cracks,
c) the phase of crack propagation for which the stress conditions at the head of the crack are decisive. This phase terminates with fatigue fracture, which is represented for example by the Wöhler curve.

For civil engineering structures operating in normal traffic and temperature conditions the third phase lasts a relatively long time. From practice it is generally known that a bridge with a fatigue crack can serve for a long time in normal conditions. Therefore, the phase of crack propagation is given great attention.

Paris-Erdogan [169] derived a differential equation for the velocity of fatigue crack propagation in the third phase

$$\frac{da}{dN} = C_0 (\Delta K)^m, \qquad (13.37)$$

where a – crack length,
N – number of stress cycles,
C_0, m – material constants,
ΔK – range of stress intensity factor which characterizes the stress conditions at the head of the crack. The quantity ΔK depends on the direction of applied force with reference to the direction of crack opening and is found from

$$\Delta K = \Delta \sigma_{ev} (\pi a)^{1/2} f(a), \qquad (13.38)$$

where $\Delta\sigma_{ev}$ – equivalent stress range under random load (for stochastic stress time history),

$f(a)$ – dimensionless function depending on stress and geometric conditions of crack environs, crack length and structure dimensions (e.g. plate width b etc.).

For ΔK or $f(a)$, manuals have been produced, giving these quantities for the most varied cases. In Figs. 13.5 to 13.9 the functions $f(a)$ have been taken from [96] for the cases of cracks most often occurring in steel railway bridges. Further cases can be found in [179] and [190].

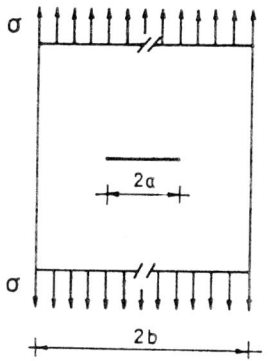
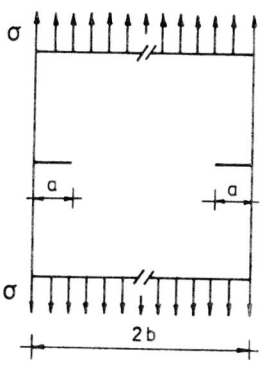
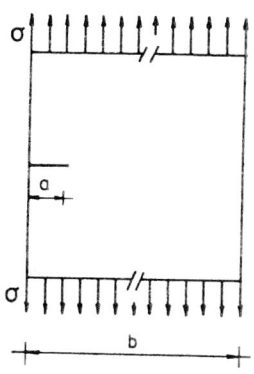

Fig. 13.5. Symmetrical crack in a plate at normal stress, valid for all a/b:
$f(a) = [1 - 0.5a/b + 0.326$
$\cdot (a/b)^2]/(1 - a/b)^{1/2}$.

Fig. 13.6. Two cracks in a plate at normal stress, valid for all a/b:
$f(a) = [1.12 - 0.61a/b + 0.13$
$\cdot (a/b)^3]/(1 - a/b)^{1/2}$.

Fig. 13.7. Asymmetrical crack in a plate at normal stress, valid for $a/b < 0.7$:
$f(a) = 1.12 - 0.23a/b + 10.6$
$\cdot (a/b)^2 - 21.7(a/b)^3 + 30.4(a/b)^4$.

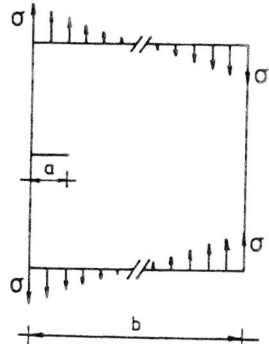
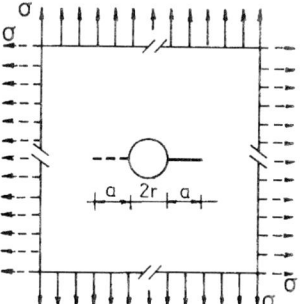

Fig. 13.8. Asymmetrical crack in a plate at flexural stresses, valid for $a/b < 0.7$:
$f(a) = 1.12 - 1.39a/b + 7.3$
$\cdot (a/b)^2 - 13(a/b)^3 + 14(a/b)^4$.

Fig. 13.9. Plate with a circular hole and with one or two cracks in uniaxial and biaxial stress state. Explanations, see the Table on page 255.

Explanations to the Fig. 13.9

	One crack		Two cracks	
	Stress			
a/r	Uniaxial	Biaxial	Uniaxial	Biaxial
	$f(a)$			
0	3.36	2.24	3.36	2.24
0.1	2.73	1.98	2.73	1.98
0.2	2.30	1.82	2.41	1.83
0.3	2.04	1.67	2.15	1.70
0.4	1.86	1.58	1.96	1.61
0.5	1.73	1.49	1.83	1.57
0.6	1.64	1.42	1.71	1.52
0.8	1.47	1.32	1.58	1.43
1.0	1.37	1.22	1.45	1.38
1.5	1.18	1.06	1.29	1.26
2.0	1.06	1.01	1.21	1.20
3.0	0.94	0.93	1.14	1.13
5.0	0.81	0.81	1.07	1.06
10	0.75	0.75	1.03	1.03
∞	0.707	0.707	1.00	1.00

The Paris–Erdogan equation was derived for harmonic stresses in which

$$\Delta\sigma_{ev} = \Delta\sigma. \tag{13.39}$$

In the case of stochastic stress time history an equivalent stress range $\Delta\sigma_{ev}$ causing the same crack propagation, must be substituted in equation (13.38). This is a difficult requirement which can only be complied with approximately. Therefore, Rolfe and Barsom [178], recommend the calculation of the quadratic mean from random stress ranges

$$\Delta\sigma_{ev} = \left(\frac{1}{n}\sum_i \Delta\sigma_i^2\right)^{1/2}, \tag{13.40}$$

where $n = \sum_i n_i$ is the number of all stress cycles.

Maarschalkerwaart [137], bases his considerations on the method of equivalent damage (Sect. 13.2), applied to a Wöhler fatigue curve of one slope, arriving at

$$\Delta\sigma_{ev} = \left(\frac{1}{n}\sum_i n_i\, \Delta\sigma_i^k\right)^{1/k} \tag{13.41}$$

and recommends $k = 3$.

The method of single-step equivalent damage (Sect. 13.2) can also be applied to a Wöhler curve of two or three slopes (Fig. 13.3 and Fig. 13.4). The method is based on the damage (13.12) and the assumption that $\Delta\sigma_{ev}$ appears in the sector of Wöhler curve with the slope k_1. Then equation (13.12) yields

$$\Delta\sigma_{ev} = \left[\frac{1}{n}\left(\sum_i n_i\, \Delta\sigma_i^{k_1} + \frac{C_1}{C_2}\sum_j n_j\, \Delta\sigma_j^{k_2}\right)\right]^{1/k_1}. \tag{13.42}$$

In the case of a Wöhler curve with three slopes (Fig. 13.4), the values $\Delta\sigma_i \leq \Delta\sigma_0$ are omitted and only the values $\Delta\sigma_0 < \Delta\sigma_i \leq \Delta\sigma_1$ and $\Delta\sigma_1 < \Delta\sigma_j$ are used.

If the numbers of random stress ranges n_i have an exponential distribution of probability density (12.9), the equivalent stress range (13.41) can be found as (after the substitution of (12.9) and integration):

$$\Delta\sigma_{ev} = \left[\sigma_0^k \Gamma(1 + k)\right]^{1/k}. \tag{13.43}$$

In this process $k = 5$.

The Paris–Erdogan differential equation is non-linear with reference to the presence of a in equation (13.38). Its approximate solution consists of the introduction of its initial crack length a_0 and is left constant in the course of integration. Equation (13.37) is thus modified into the form of

$$dN = \frac{da}{C_0\left[\Delta\sigma_{ev}(\pi a_0)^{1/2} f(a_0)\right]^m (a/a_0)^{m/2}} \tag{13.44}$$

and integrated from a_0 to a (while N_0 corresponds with a_0):

$$N - N_0 = \int_{a_0}^{a} \frac{da}{C_0\left[\Delta\sigma_{ev}(\pi a_0)^{1/2} f(a_0)\right]^m (a/a_0)^{m/2}}. \tag{13.45}$$

The integration in (13.45) presents the number of cycles N which increases the fatigue crack from a_0 to a:

$$N - N_0 = \frac{a_0}{C_0\left[\Delta\sigma_{ev}(\pi a_0)^{1/2} f(a_0)\right]^m}\, \frac{2}{m-2}\left[1 - \left(\frac{a_0}{a}\right)^{(m/2)-1}\right] \quad \text{for } m > 2 \tag{13.46}$$

or

$$N - N_0 = \frac{a_0}{C_0\left[\Delta\sigma_{ev}(\pi a_0)^{1/2} f(a_0)\right]^2}\, \ln\frac{a}{a_0} \quad \text{for } m = 2. \tag{13.47}$$

As an example let us follow the propagation of a crack of initial length $a_0 = 10;\ 15;\ 20;\ 25;\ 30$ mm at midspan of a stringer, span $l = 3$ m, under traffic load $T = 30 \times 10^6$ t/year. The crack is asymmetric according to Fig. 13.7, $b = 200$ mm, $f(a_0)$ is in Table 13.4. From Table 12.8 $\Delta\sigma_0 = 5.883$ MPa, so that for $k = 5$ we obtain from equation (13.43) the equivalent stress range

$$\Delta\sigma_{ev} = 15.33 \text{ MPa} \quad \text{or} \quad \Delta\sigma_{ev} = 21.32 \text{ MPa} \tag{13.48}$$

for traffic load with 95% reliability, see equation (12.16).

13.5 Propagation of fatigue cracks

Further we consider

$$C_0 = 4 \times 10^{-13} \text{ N}^{-3} \text{ mm}^{11/2}, \quad m = 3. \quad (13.49)$$

The results of the computation of the crack length a as a function of the number of cycles N according to equation (13.46) are shown in Fig. 13.10. The values for traffic loads with 95% reliability are shown in dashed lines. The horizontal axis also shows the scales of L and L' for the service life in years according to equation (13.52) with the assumption of $n = 2.742 \times 10^6$ for L or $n = 1.429 \times 10^7$ for L', found from equation (12.10).

Figure 13.10 shows the velocity of crack propagation which increases with increasing initial crack length.

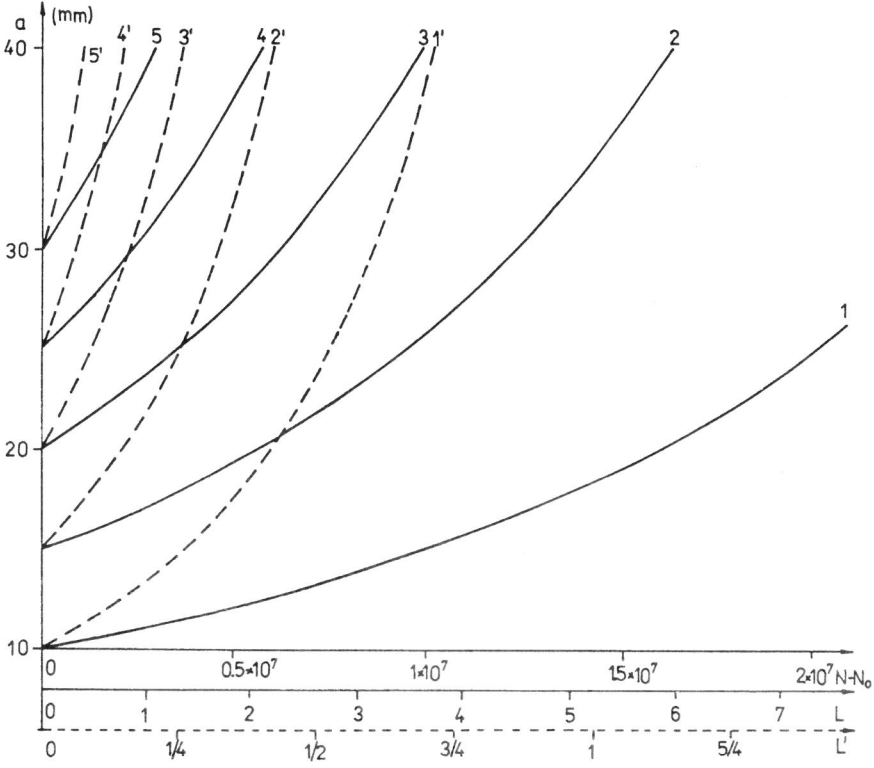

Fig. 13.10. Dependence of crack length a on the number of stress cycles N for the initial crack length a_0
$1 - a_0 = 10$ mm, $4 - a_0 = 25$ mm,
$2 - a_0 = 15$ mm, $5 - a_0 = 30$ mm,
$3 - a_0 = 20$ mm,
—— $\Delta\sigma_{ev} = 15.33$ MPa, - - - $\Delta\sigma_{ev} = 21.32$ MPa (traffic load with 95% reliability), L, L' in years.

13.6 Fatigue life and interval of railway bridge inspections

13.6.1 Residual service life for initial crack a_0

Fatigue fracture of the structure occurs, when a crack of initial length a_0 propagates to attain its critical length a_{cr}. In this proces $a_{cr} \gg a_0$ is considerably longer than the initial length a_0. This means that in equation (13.46) the value of $(a_0/a)^{(m/2)-1}$ can be neglected in comparison to 1, so that the residual service life expressed in terms of the number of cycles is

$$N - N_0 = \frac{a_0}{C_0 [\Delta\sigma_{ev} (\pi a_0)^{1/2} f(a_0)]^m [(m/2) - 1]} . \qquad (13.50)$$

Let us denote

$$n = \sum_i n_i \qquad (13.51)$$

as the total number of cycles of all amplitudes per year, which can also be found from the empirical formula (12.10) with the coefficients (12.11).

In the following, n is found from equation (12.10). However, more realistic results can be obtained with $\sum n_i$ data from the last line of Table 12.9, which differ from (12.10) by their being rounded and terminated, if the number of cycles drops below one per year.

The residual service life in years is then found from equation (13.50) and equation (13.51):

$$L = \frac{N - N_0}{n} . \qquad (13.52)$$

The expression (13.52) has been used in Table 13.3 for the estimation of the residual service life of some elements of steel railway bridges from Table 12.8. The initial length of fatigue crack was considered as $a_0 = 1, 2, 3, 4$ mm; the calculation done for the typical span l, number of cycles per year according to equation (12.10), stress σ_0 from Table 12.8, equivalent stress $\Delta\sigma_{ev}$ from equation (13.43), $f(a_0)$ was estimated, $C_0 = 4 \times 10^{-13}$ N^{-3} mm$^{11/2}$, $m = 3$, $k = 5$, $T = 30 \times 10^6$ t/year.

Table 13.3 shows that the service life of a bridge element is shortened by the increasing crack length a_0 and the increasing equivalent stress $\Delta\sigma_{ev}$ or σ_0.

13.6.2 Service life assuming fatigue damage accumulation

Considering the hypothesis of fatigue damage accumulation, the fatigue damage per year D is found first from equation (13.3). After L years the damage is LD. According to the Palmgren–Miner theory (13.4) fatigue failure occurs, when $LD = 1$. Hence the service life

$$L = 1/D . \qquad (13.53)$$

13.6 Fatigue life and interval of railway bridge inspections

TABLE 13.3. Residual service life of some elements of steel railway bridges for initial crack length a_0 under traffic load $T = 30 \times 10^6$ t/year

Element of a steel railway bridge	Span	No. of cycles per year, eq. (12.10)	Stress from Table 12.8, eq. (12.8)	Equivalent stress, eq. (13.43)	Initial crack length a_0 (mm)			
					1	2	3	4
					$f(a_0)$			
					1.12	1.12	1.13	1.14
	l (m)	n ($\times 10^6$)	σ_0 (MPa)	$\Delta\sigma_{ev}$ (MPa)	Service life L (years)			
Main beam, simply supported	30	1.876	5.160	13.44	140	99	79	67
Main beam, continuous	30	1.876	6.464	16.84	71	50	40	34
Diagonal	6	2.446	4.779	12.45	135	96	76	64
Wind bracing	5	2.521	1.588	4.14	3537	2528	2009	1694
Cross-beam	5	2.521	5.275	13.74	98	69	55	46
Stringer, midspan	3	2.742	5.883	15.33	65	46	36	31
Stringer, above support	3	2.742	5.713	14.88	71	50	40	34
Longitudinal stiffener midspan	2	2.932	3.304	8.61	342	242	192	162
Longitudinal stiffener above support	2	2.932	4.562	11.88	130	92	73	62
Deck plate, along	2	2.932	1.462	3.81	3941	2788	2216	1869
Deck plate, across	0.4	3.824	1.607	4.19	2272	1607	1277	1077

For Wöhler curve with one slope and exponential distribution of cycle number probability (12.5) the damage is found as

$$D = \int_0^\infty \frac{n(s)\,ds}{N(s)}, \qquad (13.54)$$

where we substitute ($s = \Delta\sigma$)

$$n(s) = \frac{nL}{\sigma_0} e^{-s/\sigma_0}, \quad N(s) = Cs^{-k}. \qquad (13.55)$$

After integration of equation (13.54), equation (13.53) yields the service life

$$L = \frac{C}{n\sigma_0^k \Gamma(1+k)}. \qquad (13.56)$$

For the stringer midspan, $l = 3$ m, $T = 30 \times 10^6$ t/year, $\sigma_0 = 5.883$ MPa, $n = 2.742 \times 10^6$, $C = 3.436 \times 10^{14}$ N^5 mm^{-10}, $k = 5$ the service life L is computed from equation (13.56) as

$$L = 148 \text{ years}. \qquad (13.57)$$

For a Wöhler curve with three slopes (Fig. 13.4), equation (13.54) must be integrated by parts while omitting $\Delta\sigma_i < \Delta\sigma_0$:

$$D = \frac{nL}{\sigma_0}\left[\frac{1}{C_1}\int_{\Delta\sigma_0}^{\Delta\sigma_1} s^{k_1} e^{-s/\sigma_0}\,ds + \frac{1}{C_2}\int_{\Delta\sigma_1}^\infty s^{k_2} e^{-s/\sigma_0}\,ds\right].$$

In this way the service life is obtained for the exponential distribution of the probability density of stress ranges (for $k_1 = 5$, $k_2 = 3$)

$$L = \frac{\sigma_0}{n} (I_1 + I_2)^{-1}, \qquad (13.58)$$

where

$$I_1 = \frac{\sigma_0}{C_1} [e^{-\Delta\sigma_0/\sigma_0}(\Delta\sigma_0^5 + 5\sigma_0 \Delta\sigma_0^4 + 20\sigma_0^2 \Delta\sigma_0^3 + 60\sigma_0^3 \Delta\sigma_0^2$$
$$+ 120\sigma_0^4 \Delta\sigma_0 + 120\sigma_0^5) - e^{-\Delta\sigma_1/\sigma_0}(\Delta\sigma_1^5 + 5\sigma_0 \Delta\sigma_1^4$$
$$+ 20\sigma_0^2 \Delta\sigma_1^3 + 60\sigma_0^3 \Delta\sigma_1^2 + 120\sigma_0^4 \Delta\sigma_1 + 120\sigma_0^5)],$$

$$I_2 = \frac{\sigma_0}{C_2} e^{-\Delta\sigma_1/\sigma_0}(\Delta\sigma_1^3 + 3\sigma_0 \Delta\sigma_1^2 + 6\sigma_0^2 \Delta\sigma_1 + 6\sigma_0^3). \qquad (13.59)$$

For the same stringer, $l = 3$ m, $T = 30 \times 10^6$ t/year, $n = 2.742 \times 10^6$, $\sigma_0 = 5.883$ MPa, $\Delta\sigma_0 = 20$ MPa, $\Delta\sigma_1 = 37$ MPa, $k_1 = 5$, $k_2 = 3$, $C_1 = 3.436 \times 10^{14}\,\text{N}^5\,\text{mm}^{-10}$, $C_2 = 2.518 \times 10^{11}\,\text{N}^3\,\text{mm}^{-6}$ equation (13.58) yields

$$L = 206 \text{ years}. \qquad (13.60)$$

Finally for the experimental number n_i of stress ranges $\Delta\sigma_i$ equations (12.12) and (13.53) present

$$L = \left[\frac{1}{C_1} \sum_i n_i \Delta\sigma_i^{k_1} + \frac{1}{C_2} \sum_j n_j \Delta\sigma_j^{k_2}\right]^{-1}. \qquad (13.61)$$

The use of the respective columns of n_i or $n_i\,e^{ts}$ in Table 12.9 for stringer midspan in equation (13.61) yields the life of

$$L = 192 \text{ years} \quad \text{or} \quad L = 37 \text{ years, respectively}, \qquad (13.62)$$

for traffic load with 95% realibility.

Should we not neglect $\Delta\sigma_i < \Delta\sigma_0$ we would obtain

$$L = 175 \text{ years} \quad \text{or} \quad L = 34 \text{ years, respectively}. \qquad (13.63)$$

13.6.3 Interval of railway bridge inspections

The estimation of the service life of steel railway bridges given in Table 13.3 and in equations (13.57), (13.60), (13.62) and (13.63) show a great variation and, consequently, a considerable unreliability of results. For this reason the estimation of fatigue life calculated either on the basis of fracture mechanics or by means of the fatigue damage accumulation methods cannot be recommended.

Of more practical importance is the estimation of the interval of bridge inspections and thus the prevention of unexpected propagation of a fatigue crack. This is possible by the methods of fracture mechanics.

13.6 Fatigue life and interval of railway bridge inspections

The particular consideration is based on the assumption that during a current bridge inspection a fatigue crack of a length a_0 has been discovered. The objective is to estimate the period (time interval) L_i until the next inspection during which the crack will increase by a certain measurable length Δa (in practice about $\Delta a = 5$ mm).

Once again, the value of L_i in years will be found from equation (13.52)

$$L_i = \frac{N - N_0}{n}, \qquad (13.64)$$

where, according to equation (13.46),

$$N - N_0 = \frac{a_0}{C_0[\Delta\sigma_{ev}(\pi a_0)^{1/2} f(a_0)]^m [(m/2) - 1]} \left[1 - \left(\frac{a_0}{a_0 + \Delta a} \right)^{(m/2)-1} \right] \qquad (13.65)$$

and $n = \sum_i n_i$ is the total number of all cycles per year.

For the same example as that used in Sect. 13.5 and in Fig. 13.10 the intervals L_i in years or months were calculated and given in Table 13.4., which also gives the intervals found for traffic loads with 95% reliability. Table 13.4 shows that the intervals of inspections are growing shorter, if the initial crack length is greater.

TABLE 13.4. Interval of inspections of a stringer with a crack of a length a_0 under traffic load of 30×10^6 t/year

Crack length			Inspection interval			
Initial	At the end of inspection period	$f(a_0)$ Fig. 13.7			Traffic load with 95% reliability	
			L_i		L_i	
a_0 (mm)	$a = a_0 + 5$ (mm)		years	= months	years	= months
10	15	1.13	3.7	44	0.26	3.1
15	20	1.15	2.1	25	0.15	1.8
20	25	1.18	1.3	16	0.09	1.1
25	30	1.22	0.9	10	0.06	0.8
30	35	1.27	0.6	7	0.04	0.4

When the inspection, carried out after the interval L_i reaches an increased crack length, this new crack length is considered as a_0 and a further inspection interval L_i is found from equations (13.64) and (13.65). If the fatigue crack increases, the inspection interval will decrease. Such safeguarding of railway traffic represents the biggest contribution of fracture mechanics to bridge engineering practice.

Appendix

14. Thermal interaction of long-welded rail with railway bridges

Railway bridges induce additional forces into long-welded rail, thereby limiting its use. For this reason, the problem is given great attention, since there is a general tendency to use long-welded rail wherever possible. Its use results in a significant reduction of dynamic effects and a smoother vehicle ride (and consequent ecological and economic advantages). In addition, it facilitates the mechanized maintenance of railway tracks.

Thermal stress state does not form part of the dynamics of structures, but temperature is, naturally, a function of time. For these reasons, this chapter on thermal interaction is added as a supplement to describe in detail the behaviour of bridges and rails during temperature changes.

14.1 Theoretical model of long-welded rail and bridge

14.1.1 Basic assumptions

The theoretical solution of the problem of mutual interaction between a long-welded rail and a bridge during temperature changes is based on the following assumptions:

(a) The bridge girder and the long-welded rail are considered as a system of i bars as in Chapter 8. The bridge has one fixed and one or several movable bearings, in which friction is neglected. The long-welded rail is of infinite length. The simplest case is shown in Fig. 14.1.

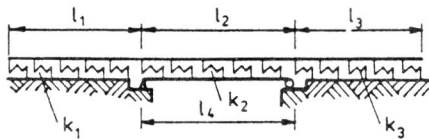

Fig. 14.1. Theoretical model of a bridge with long-welded rail.

(b) The connection between the rails and bridge in a horizontal direction is idealized by an elastic layer, i.e. by a system of infinitely close horizontal springs (Fig. 14.1). With the assumption of linear action of these springs their spring constant (stiffness) per unit length in the ith field is k_i (N mm^{-2}). It is a force per unit length which causes a unit displacement of the track (i.e. two rails) with reference to the bridge or the rail bed. Linearity is used here to simplify a highly complex relation between the resistance of the rail against displacement, which is generally non-linear and depends on many factors such as the individual elements of permanent way and its maintenance, on temperature, moisture, degree of soiling, vertical load, and so on.

(c) It is assumed that the bridge is not influenced by the rails during temperature changes and that it can move freely. The opposite effect, i.e. the influence of the bridge on the stresses in the rail is, naturally, considerable.

(d) According to Hooke's law and physical laws governing thermal extension, the longitudinal force N in a bar with the modulus of elasticity E and cross section area A is

$$N = EA(u' - \alpha \Delta t), \tag{14.1}$$

where u' – relative elongation (strain) of the bar,
$\quad \alpha$ – coefficient of thermal extension of the bar,
$\quad \Delta t = t - t_0$ – temperature change with reference to the initial (basic) temperature t_0,
$\quad t$ – temperature (°C).

(e) Longitudinal stresses in the bridge and in the rails are independent of bending stresses and, therefore, they can be investigated separately [70].

(f) It can be assumed in many cases that the rails on the bridge and beyond the bridge are indentical and that their fastening is similar. In computer applications, the omission of this assumption does not bring any difficulties.

14.1.2 Basic equations and their solution

If an element of the bar, length dx, is considered (see Fig. 14.2), it is possible to write the condition of equilibrium of forces in a horizontal direction according to the assumptions (e) and (b) as:

$$-N + N + N' \, dx - ku \, dx = 0, \tag{14.2}$$

where $N = N(x)$ – longitudinal normal force in the bar at the point x,
$\quad u = u(x)$ – horizontal longitudinal displacement of the bar, where the dash indicates differentiation with respect to x.

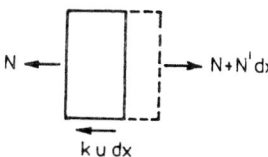

Fig. 14.2. Forces affecting an element of the bar.

Equation (14.2) can be written

$$N' - ku = 0 \tag{14.3}$$

or, after the substitution $N' = EAu''$ from equation (14.1), according to assumption (d)

$$-EAu'' + ku = 0. \tag{14.4}$$

Equation (14.4) is the basic differential equation valid for the bar subjected to longitudinal forces.

a) Fixed bearing on one bridge end

Let us investigate the basic case, when the long-welded rail passes over a bridge provided with a fixed bearing on its left-hand side (either a simply supported or continuous beam) – Fig. 14.3 and assumption (a). The origin of coordinates of every bar is at its left-hand side with the exception of the bar $i = 1$, where the origin is at its right-hand end.

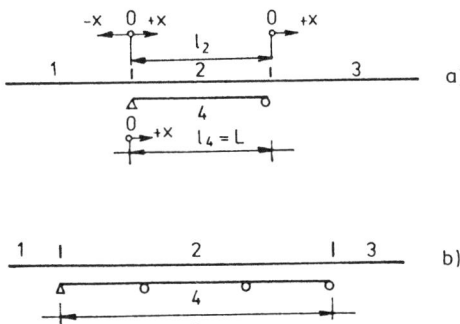

Fig. 14.3. Fixed bearing on a one bridge end
a) simply supported beam, b) continuous beam.

The differential equations (14.4) for the bars, denoted in Fig. 14.3 with numbers $i = 1, 2, 3, 4$ and with lengths of l_i, are as follows

$$-E_i A_i u_i'' + k_i u_i = 0, \quad i = 1,3,$$
$$-E_2 A_2 u_2'' + k_2 (u_2 - u_4) = 0,$$
$$-E_4 A_4 u_4'' + k_4 (u_4 - u_2) = 0 \qquad (14.5)$$

and the forces in the bars are found from equation (14.1) as

$$N_i = E_i A_i (u_i' - \alpha_i \Delta t_i), \qquad (14.6)$$

where $\Delta t_i = t_i - t_{i0}$ is the temperature change in the ith bar from its basic temperature t_{i0}, and
α_i – coefficient of thermal extension.

According to (c) it is assumed that the bridge (bar 4) may expand freely and that, consequently, the solution for the bar 4, according to the law of thermal extension, is

$$u_4(x) = \alpha_4 \Delta t_4 x, \qquad u_4'(x) = \alpha_4 \Delta t_4,$$
$$u_4''(x) = 0, \qquad N_4(x) = 0 \qquad (14.7)$$

and does not depend on the behaviour of the bar 2.

The system of differential equations (14.5) and (14.7) is constrained by boundary conditions indicating that the displacements in bars 1 and 3 in

14.1 Theoretical model of long-welded rail and bridge

Fig. 14.3 at a sufficient long distance from the bridge equal zero and that the displacements and the forces in adjoining bars at their joints are identical:

$$u_1(-\infty) = 0, \qquad u_3(\infty) = 0,$$
$$u_1(0) = u_2(0), \qquad N_1(0) = N_2(0),$$
$$u_2(l_2) = u_3(0), \qquad N_2(l_2) = N_3(0),$$
$$u_4(0) = 0, \qquad N_4(l_4) = 0, \qquad (14.8)$$

where $l_2 = l_4$ is the length of the bridge, see Fig. 14.3.

The assumed solution (14.7) for bar 4 satisfies the boundary conditions (14.8) for this bar.

The general solution of the system of differential equations (14.5) is

$$u_i(x) = B_i e^{\lambda_i x} + C_i e^{-\lambda_i x}, \qquad i = 1, 3,$$
$$u_2(x) = B_2 e^{\lambda_2 x} + C_2 e^{-\lambda_2 x} + \alpha_4 \Delta t_4 x, \qquad (14.9)$$

where

$$\lambda_i^2 = \frac{k_i}{E_i A_i}, \qquad (14.10)$$

B_i and C_i are constants found after the substitution of equation (14.9) in the boundary conditions (14.8).

The assumption (f) considerably simplifies the calculation of the constants B_i and C_i because $EA = E_i A_i$, $i = 1, 2, 3$, where E and A are modulus of elasticity and area of the cross-section of the rail, respectively, so

$$\lambda^2 = \lambda_i^2 = \frac{k_i}{E_i A_i} = \frac{k}{EA}, \qquad i = 1, 2, 3, \qquad (14.11)$$

$$\Delta t = \Delta t_i = t_i - t_{i0} = t - t_0, \qquad i = 1, 2, 3 \qquad (14.12)$$

where Δt is the difference in temperature t of the rails from the initial temperature t_0 of the rail fixing,

$$\Delta T = \Delta T_4 = t_4 - t_{40} \qquad (14.13)$$

is the difference between the mean temperature of the bridge t_4 and the mean temperature of the bridge t_{40} when the rails were fixed to it,

$$\alpha = \alpha_i, \qquad i = 1, 2, 3 \qquad (14.14)$$

is the coefficient of thermal extension of the rails, and

$$\alpha_0 = \alpha_4 \qquad (14.15)$$

is the coefficient of thermal extension of the bridge.

This notation yields expressions for the displacements of the individual bars in the following form:

$$u_1(x) = \frac{\alpha_0 \Delta T}{2\lambda} [1 - (1 + \lambda l_2) e^{-\lambda l_2}] e^{\lambda x},$$

$$u_2(x) = \frac{\alpha_0 \Delta T}{2\lambda} [2\lambda x + e^{-\lambda x} - (1 + \lambda l_2) e^{-\lambda(l_2 - x)}],$$

$$u_3(x) = \frac{\alpha_0 \Delta T}{2\lambda} (\lambda l_2 - 1 + e^{-\lambda l_2}) e^{-\lambda x},$$

$$u_4(x) = \alpha_0 \Delta T x \tag{14.16}$$

and for the forces in the bars

$$N_1(x) = -EA \alpha \Delta t \left\{ 1 + \frac{\alpha_0 \Delta T}{2\alpha \Delta t} [-1 + (1 + \lambda l_2) e^{-\lambda l_2}] e^{\lambda x} \right\},$$

$$N_2(x) = -EA \alpha \Delta t \left\{ 1 + \frac{\alpha_0 \Delta T}{2\alpha \Delta t} [-2 + e^{-\lambda x} + (1 + \lambda l_2) e^{-\lambda(l_2 - x)}] \right\},$$

$$N_3(x) = -EA \alpha \Delta t \left[1 + \frac{\alpha_0 \Delta T}{2\alpha \Delta t} (\lambda l_2 - 1 + e^{-\lambda l_2}) e^{-\lambda x} \right]$$

$$N_4(x) = 0. \tag{14.17}$$

The expression factored out in equations (14.7), i.e.,

$$N = -EA \alpha \Delta t \tag{14.18}$$

is the force in the long-welded rail due to the temperature change Δt and the expressions in brackets in equations (14.17) give the contribution of the bridge to the force in the long-welded rail.

b) Fixed bearing in the middle of the bridge

Further practical cases are shown in Fig. 14.4. The common feature for all these cases is the possibility that the bridge may expand in both directions from the fixed bearing. These cases can be described by differential equations in accordance with the notation of the individual bars from Fig. 14.4.

$$-E_i A_i u_i'' + k_i u_i = 0, \quad i = 1, 4,$$

$$-E_2 A_2 u_2'' + k_2 (u_2 - u_5) = 0,$$

$$-E_3 A_3 u_3'' + k_3 (u_3 - u_6) = 0,$$

$$-E_5 A_5 u_5'' + k_5 (u_5 - u_2) = 0,$$

$$-E_6 A_6 u_6'' + k_6 (u_6 - u_3) = 0. \tag{14.19}$$

14.1 Theoretical model of long-welded rail and bridge

The boundary conditions are:

$$u_1(-\infty) = 0, \qquad u_4(\infty) = 0,$$
$$u_1(0) = u_2(0), \qquad N_1(0) = N_2(0),$$
$$u_2(l_2) = u_3(0), \qquad N_2(l_2) = N_3(0),$$
$$u_3(l_3) = u_4(0), \qquad N_3(l_3) = N_4(0),$$
$$u_5(l_5) = 0, \qquad N_5(0) = 0,$$
$$u_6(0) = 0, \qquad N_6(l_6) = 0. \tag{14.20}$$

The system of equations (14.19) with the boundary conditions (14.20) is solved by the same procedure as in Sect. a) and with the same simplification:

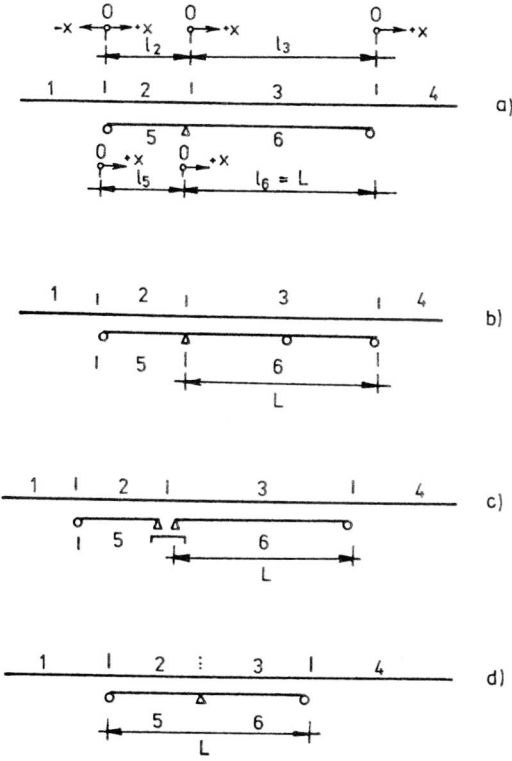

Fig. 14.4 Fixed bearing in the middle of the bridge
a) two-span continuous beam,
b) three-span continuous beam,
c) two fixed bearings on central pier,
d) bridge without fixed bearings.

$$u_1(x) = -\frac{\alpha_0 \Delta T}{2\lambda} [\lambda l_2 - 1 + (1 + \lambda l_3) e^{-\lambda(l_2 + l_3)}] e^{\lambda x},$$

$$u_2(x) = -\frac{\alpha_0 \Delta T}{2\lambda} [2\lambda (l_5 - x) - (1 + \lambda l_2) e^{-\lambda x} + (1 + \lambda l_3) e^{-\lambda(l_2 + l_3 - x)}],$$

$$u_3(x) = \frac{\alpha_0 \Delta T}{2\lambda} [2\lambda x + (1 + \lambda l_5) e^{-\lambda(l_2 + x)} - (1 + \lambda l_3) e^{-\lambda(l_3 - x)}],$$

$$u_4(x) = \frac{\alpha_0 \Delta T}{2\lambda} [\lambda l_3 - 1 + (1 + \lambda l_5) e^{-\lambda(l_2 + l_3)}] e^{-\lambda x},$$

$$u_5(x) = -\alpha_0 \Delta T (l_5 - x),$$

$$u_6(x) = \alpha_0 \Delta T x, \tag{14.21}$$

$$N_1(x) = -EA\alpha\Delta t \left\{ 1 + \frac{\alpha_0 \Delta T}{2\alpha \Delta t} [\lambda l_5 - 1 + (1 + \lambda l_3) e^{-\lambda(l_2 + l_3)}] e^{\lambda x} \right\},$$

$$N_2(x) = -EA\alpha\Delta t \left\{ 1 + \frac{\alpha_0 \Delta T}{2\alpha \Delta t} [(1 + \lambda l_3) e^{-\lambda(l_2 + l_3 - x)} \right.$$
$$\left. + (1 + \lambda l_5) e^{-\lambda x} - 2] \right\},$$

$$N_3(x) = -EA\alpha\Delta t \left\{ 1 + \frac{\alpha_0 \Delta T}{2\alpha \Delta t} [(1 + \lambda l_3) e^{-\lambda(l_3 - x)} \right.$$
$$\left. + (1 + \lambda l_5) e^{-\lambda(l_2 + x)} - 2] \right\},$$

$$N_4(x) = -EA\alpha\Delta t \left\{ 1 + \frac{\alpha_0 \Delta T}{2\alpha \Delta t} [\lambda l_3 - 1 + (1 + \lambda l_5) e^{-\lambda(l_2 + l_3)}] e^{-\lambda x} \right\},$$

$$N_5(x) = 0, \quad N_6(x) = 0. \tag{14.22}$$

c) Alternate bearings on one pier

A bridge of two spans with alternate fixed and movable bearings on an intermediate pier is shown in Fig. 14.5. This case is characterized by the differential equations (14.19) similar to the above. However, their boundary conditions are:

$$\begin{aligned}
u_1(-\infty) &= 0, & u_4(\infty) &= 0, \\
u_1(0) &= u_2(0), & N_1(0) &= N_2(0), \\
u_2(l_2) &= u_3(0), & N_2(l_2) &= N_3(0), \\
u_3(l_3) &= u_4(0), & N_3(l_3) &= N_4(0), \\
u_5(0) &= 0, & N_5(l_5) &= 0, \\
u_6(0) &= 0, & N_6(l_6) &= 0.
\end{aligned} \tag{14.23}$$

14.1 Theoretical model of long-welded rail and bridge

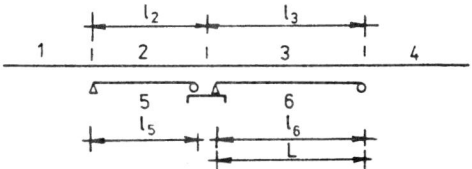

Fig. 14.5. Alternate (fixed and movable) bearings on intermediate pier.

The solution of the system of equations (14.19) with the boundary conditions (14.23) has the following form:

$$u_1(x) = \frac{\alpha_0 \Delta T}{2\lambda}\left\{1 - \left[\lambda l_2 + (1 + \lambda l_3)e^{-\lambda l_3}\right]e^{-\lambda l_2}\right\}e^{\lambda x},$$

$$u_2(x) = \frac{\alpha_0 \Delta T}{2\lambda}\left\{2\lambda x + e^{-\lambda x} - \left[\lambda l_2 + (1 + \lambda l_3)e^{-\lambda l_3}\right]e^{-\lambda(l_2-x)}\right\},$$

$$u_3(x) = \frac{\alpha_0 \Delta T}{2\lambda}\left[2\lambda x + (\lambda l_2 + e^{-\lambda l_2})e^{-\lambda x} - (1 + \lambda l_3)e^{-\lambda(l_3-x)}\right],$$

$$u_4(x) = \frac{\alpha_0 \Delta T}{2\lambda}\left[\lambda l_3 - 1 + (\lambda l_2 + e^{-\lambda l_2})e^{-\lambda l_3}\right]e^{-\lambda x},$$

$$u_5(x) = \alpha_0 \Delta T x,$$

$$u_6(x) = \alpha_0 \Delta T x, \tag{14.24}$$

$$N_1(x) = -EA\alpha\Delta t\left[1 + \frac{\alpha_0 \Delta T}{2\alpha\Delta t}\left\{-1 + \left[\lambda l_2 + (1 + \lambda l_3)e^{-\lambda l_3}\right]e^{-\lambda l_2}\right\}e^{\lambda x}\right],$$

$$N_2(x) = -EA\alpha\Delta t\Big[1 + \frac{\alpha_0 \Delta T}{2\alpha\Delta t}\{-2 + e^{-\lambda x}$$
$$+ \left[\lambda l_2 + (1 + \lambda l_3)e^{-\lambda l_3}\right]e^{-\lambda(l_2-x)}\}\Big],$$

$$N_3(x) = -EA\alpha\Delta t\left\{1 + \frac{\alpha_0 \Delta T}{2\alpha\Delta t}\left[-2 + (\lambda l_2 + e^{-\lambda l_2})e^{-\lambda x} + (1 + \lambda l_3)e^{-\lambda(l_3-x)}\right]\right\},$$

$$N_4(x) = -EA\alpha\Delta t\left\{1 + \frac{\alpha_0 \Delta T}{2\alpha\Delta t}\left[\lambda l_3 - 1 + (\lambda l_2 + e^{-\lambda l_2})e^{-\lambda l_3}\right]e^{-\lambda x}\right\},$$

$$N_5(x) = 0, \qquad N_6(x) = 0. \tag{14.25}$$

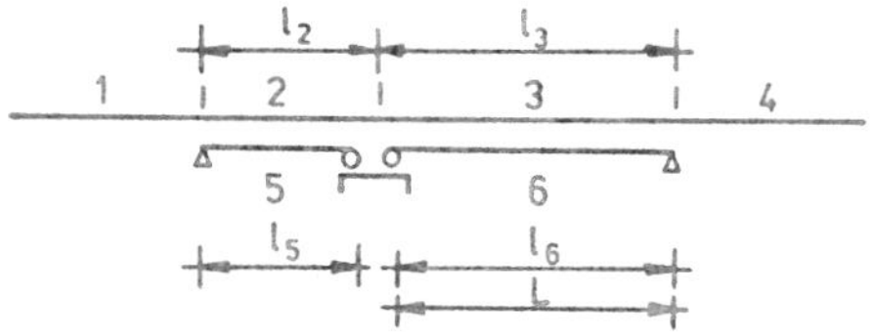

Fig. 14.6. Two movable bearings on a pier.

d) Two movable bearings on one pier

The case of a two span bridge with two movable bearings on the intermediate pier is shown in Fig. 14.6. In this case, the differential equations (14.19) are applicable, but the boundary conditions are:

$$\begin{aligned}
u_1(-\infty) &= 0, & u_4(\infty) &= 0, \\
u_1(0) &= u_2(0), & N_1(0) &= N_2(0), \\
u_2(l_2) &= u_3(0), & N_2(l_2) &= N_3(0), \\
u_3(l_3) &= u_4(0), & N_3(l_3) &= N_4(0), \\
u_5(0) &= 0, & N_5(l_5) &= 0, \\
u_6(l_6) &= 0, & N_6(0) &= 0.
\end{aligned} \qquad (14.26)$$

The solution of the system of equations (4.19) with boundary conditions (14.26) has the following form:

$$u_1(x) = \frac{\alpha_0 \Delta T}{2\lambda} \left[1 - \lambda(l_2 + l_3)\,e^{-\lambda l_2} - e^{-\lambda(l_2+l_3)}\right] e^{\lambda x},$$

$$u_2(x) = \frac{\alpha_0 \Delta T}{2\lambda} \left\{2\lambda x + e^{-\lambda x} - \left[e^{-\lambda(l_2+l_3)} + \lambda(l_2 + l_3)\,e^{-\lambda l_2}\right] e^{\lambda x}\right\},$$

$$u_3(x) = -\frac{\alpha_0 \Delta T}{2\lambda} \left[2\lambda(l_6 - x) + e^{-\lambda(l_3-x)} - e^{-\lambda(l_2+x)} - \lambda(l_2 + l_3)\,e^{-\lambda x}\right],$$

$$u_4(x) = \frac{\alpha_0 \Delta T}{2\lambda} \left[\lambda(l_2 + l_3)\,e^{-\lambda l_3} - 1 + e^{-\lambda(l_2+l_3)}\right] e^{-\lambda x},$$

$$u_5(x) = \alpha_0 \Delta T x,$$

$$u_6(x) = -\alpha_0 \Delta T (l_6 - x), \qquad (14.27)$$

$$N_1(x) = -EA\alpha \Delta t \left\{1 + \frac{\alpha_0 \Delta T}{2\alpha \Delta t}\left[\lambda(l_2 + l_3)\,e^{-\lambda l_2} - 1 + e^{-\lambda(l_2+l_3)}\right] e^{\lambda x}\right\},$$

$$N_2(x) = -EA\alpha \Delta t \left[1 + \frac{\alpha_0 \Delta T}{2\alpha \Delta t}\left\{-2 + e^{-\lambda x} + \left[e^{-\lambda(l_2+l_3)} + \lambda(l_2 + l_3)\,e^{-\lambda l_2}\right] e^{\lambda x}\right\}\right],$$

$$N_3(x) = -EA\alpha \Delta t \left[1 + \frac{\alpha_0 \Delta T}{2\alpha \Delta t}\left\{-2 + e^{-\lambda(l_3-x)} + \left[e^{-\lambda l_2} + \lambda(l_2+l_3)\right] e^{-\lambda x}\right\}\right],$$

$$N_4(x) = -EA\alpha \Delta t \left\{1 + \frac{\alpha_0 \Delta T}{2\alpha \Delta t}\left[-1 + \lambda(l_2 + l_3)\,e^{-\lambda l_3} + e^{-\lambda(l_2+l_3)}\right] e^{-\lambda x}\right\},$$

$$N_5(x) = 0, \qquad N_6(x) = 0. \qquad (14.28)$$

14.1 Theoretical model of long-welded rail and bridge

e) Three-span bridge with alternate bearings

The case of a three-span bridge with alternate fixed and movable bearings is shown in Fig. 14.7. It is characterized by the following differential equations:

$$-E_i A_i u_i'' + k_i u_i = 0, \quad i = 1,5,$$
$$-E_2 A_2 u_2'' + k_2 (u_2 - u_6) = 0,$$
$$-E_3 A_3 u_3'' + k_3 (u_3 - u_7) = 0,$$
$$-E_4 A_4 u_4'' + k_4 (u_4 - u_8) = 0. \quad (14.29)$$

The pertinent boundary conditions are:

$$u_1(-\infty) = 0, \qquad u_5(\infty) = 0,$$
$$u_1(0) = u_2(0), \qquad N_1(0) = N_2(0),$$
$$u_2(l_2) = u_3(0), \qquad N_2(l_2) = N_3(0),$$
$$u_3(l_3) = u_4(0), \qquad N_3(l_3) = N_4(0),$$
$$u_4(l_4) = u_5(0), \qquad N_4(l_4) = N_5(0),$$
$$u_6(0) = 0, \qquad N_6(l_6) = 0,$$
$$u_7(0) = 0, \qquad N_7(l_7) = 0,$$
$$u_8(0) = 0, \qquad N_8(l_8) = 0. \quad (14.30)$$

With the same simplifying assumptions as in the case at a) the solution of the system of equations (14.29) with boundary conditions (14.30) results in:

$$u_1(x) = \frac{\alpha_0 \Delta T}{2\lambda} \left[1 - \left\{\lambda l_2 + \left[\lambda l_3 + (1 + \lambda l_4)\, e^{-\lambda l_4}\right] e^{-\lambda l_3}\right\} e^{-\lambda l_2}\right] e^{\lambda x},$$

$$u_2(x) = \frac{\alpha_0 \Delta T}{2\lambda} \left[2\lambda x + e^{-\lambda x} - \left\{\lambda l_2 + \left[\lambda l_3 + (1 + \lambda l_4)\, e^{-\lambda l_4}\right] e^{-\lambda l_3}\right\} e^{-\lambda(l_2 - x)}\right],$$

$$u_3(x) = \frac{\alpha_0 \Delta T}{2\lambda} \left\{2\lambda x + (\lambda l_2 + e^{-\lambda l_2})\, e^{-\lambda x} - \left[\lambda l_3 + (1 + \lambda l_4)\, e^{-\lambda l_4}\right] e^{-\lambda(l_3 - x)}\right\},$$

$$u_4(x) = \frac{\alpha_0 \Delta T}{2\lambda} \left\{2\lambda x + \left[\lambda l_3 + (\lambda l_2 + e^{-\lambda l_2})\, e^{-\lambda l_3}\right] e^{-\lambda x} - (1 + \lambda l_4)\, e^{-\lambda(l_4 - x)}\right\},$$

$$u_5(x) = \frac{\alpha_0 \Delta T}{2\lambda} \left\{\lambda l_4 - 1 + \left[\lambda l_3 + (\lambda l_2 + e^{-\lambda l_2})\, e^{-\lambda l_3}\right] e^{-\lambda l_4}\right\} e^{-\lambda x},$$

$$u_i(x) = \alpha_0 \Delta T x, \quad i = 6, 7, 8. \quad (14.31)$$

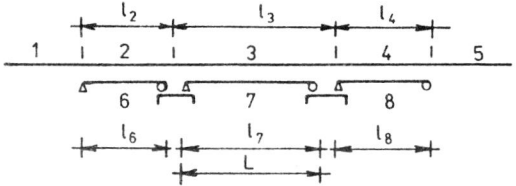

Fig. 14.7. Three-span bridge with alternate fixed and movable bearings on intermediate piers.

$$N_1(x) = -EA\alpha\Delta t \left[1 - \frac{\alpha_0 \Delta T}{2\alpha\Delta t}\left\{1 - \{\lambda l_2 + [\lambda l_3 \right.\right.$$
$$\left.\left. + (1+\lambda l_4)e^{-\lambda l_4}]e^{-\lambda l_3}\}e^{-\lambda l_2}\right\}e^{\lambda x}\right],$$

$$N_2(x) = -EA\alpha\Delta t\left[1 + \frac{\alpha_0 \Delta T}{2\alpha\Delta t}\left\{-2 + \{\lambda l_2 + [\lambda l_3 \right.\right.$$
$$\left.\left. + (1+\lambda l_4)e^{-\lambda l_4}]e^{-\lambda l_3}\}e^{-\lambda(l_2-x)} + e^{-\lambda x}\right\}\right],$$

$$N_3(x) = -EA\alpha\Delta t\left\{1 + \frac{\alpha_0 \Delta T}{2\alpha\Delta t}\left[-2 + [\lambda l_3 + (1+\lambda l_4)e^{-\lambda l_4}]e^{-\lambda(l_3-x)}\right.\right.$$
$$\left.\left. + (\lambda l_2 + e^{-\lambda l_2})e^{-\lambda x}\right]\right\},$$

$$N_4(x) = -EA\alpha\Delta t\left\{1 + \frac{\alpha_0 \Delta T}{2\alpha\Delta t}\left[-2 + (1+\lambda l_4)e^{-\lambda(l_4-x)}\right.\right.$$
$$\left.\left. + [\lambda l_3 + (\lambda l_2 + e^{-\lambda l_2})e^{-\lambda l_3}]e^{-\lambda x}\right]\right\},$$

$$N_5(x) = -EA\alpha\Delta t\left[1 + \frac{\alpha_0 \Delta T}{2\alpha\Delta t}\{\lambda l_4 - 1 \right.$$
$$\left. + [\lambda l_3 + (\lambda l_2 + e^{-\lambda l_2})e^{-\lambda l_3}]e^{-\lambda l_4}\}e^{-\lambda x}\right],$$

$$N_i(x) = 0, \quad i = 6, 7, 8. \tag{14.32}$$

f) Bridge with n-spans and with alternate bearings

An analogous procedure may also be used for the solution of the case of a bridge with n spans and with alternate fixed and movable bearings; for $n = 3$ see Fig. 14.7. It can be shown that the highest stresses in the rail are above the last movable bearing. In case of the bridge with equal spans $l_i = L$, therefore, this stress has the form

$$\sigma = \frac{N_{n+1}(L)}{A} = -E\alpha\Delta t\left[1 + \frac{\alpha_0 \Delta T}{2\alpha\Delta t}(\lambda L - 1 + C_n)\right], \tag{14.33}$$

where the constant C_n depends on the number of spans n

$$C_1 = e^{-\lambda L},$$
$$C_n = (\lambda L + C_{n-1})e^{-\lambda L}, \quad n \geq 2. \tag{14.34}$$

The greatest relative displacement u between the bridge and the rail can be found above the last movable bearing as:

$$u = u_{2(n+1)}(L) - u_{n+1}(L) = \frac{\alpha_0 \Delta T}{2\lambda}(1 + \lambda L - C_n). \tag{14.35}$$

For $n \geq 3$ the expressions (14.33) and (14.35) differ very little.

14.1.3 Examples

Figures 14.8 and 14.9 show the distribution of the longitudinal force $N_i(x)$ and the longitudinal displacement $u_i(x)$ in the long-welded rail on a single-, two- and three-span bridge with various bearing arrangement. It is the case of a steel bridge with the long-welded rail fastened directly on the bridge deck, without ballast, in extreme winter conditions. The following input data were substituted in the respective equations given in Sect. 14.1.2:

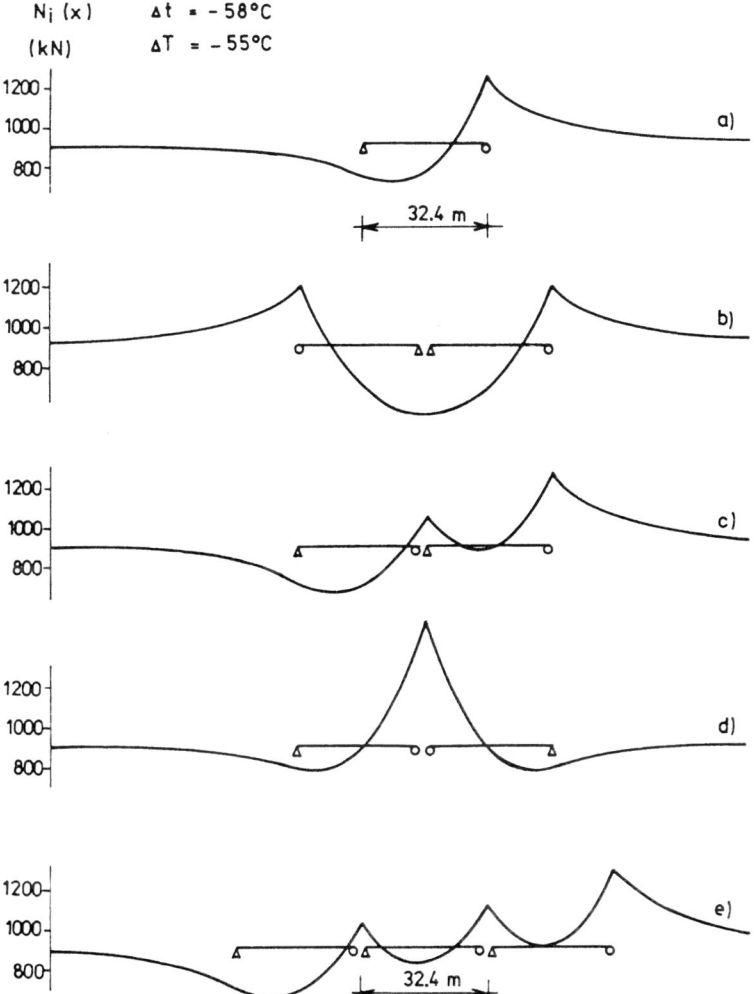

Fig. 14.8. Theoretical distribution of the force in the long-welded rail on a steel bridge with direct fastening of rails in extreme winter conditions:
a) simply supported beam, b) two fixed bearings on central pier, c) alternate bearings on central pier, d) two movable bearings on central pier, e) three-span bridge with alternate bearings on intermediate piers.

$\Delta t = -58°$ C, $\quad\Delta T = -55°$ C,
$k = 8$ N mm^{-2}, $\quad E = 2.1 \times 10^5$ MPa,
$\alpha = 12 \times 10^{-6}/°$C, $\quad \alpha_0 = 9 \times 10^{-6}/°$C,
$A = 6250$ mm^2, $\quad L = 32.4$ m.

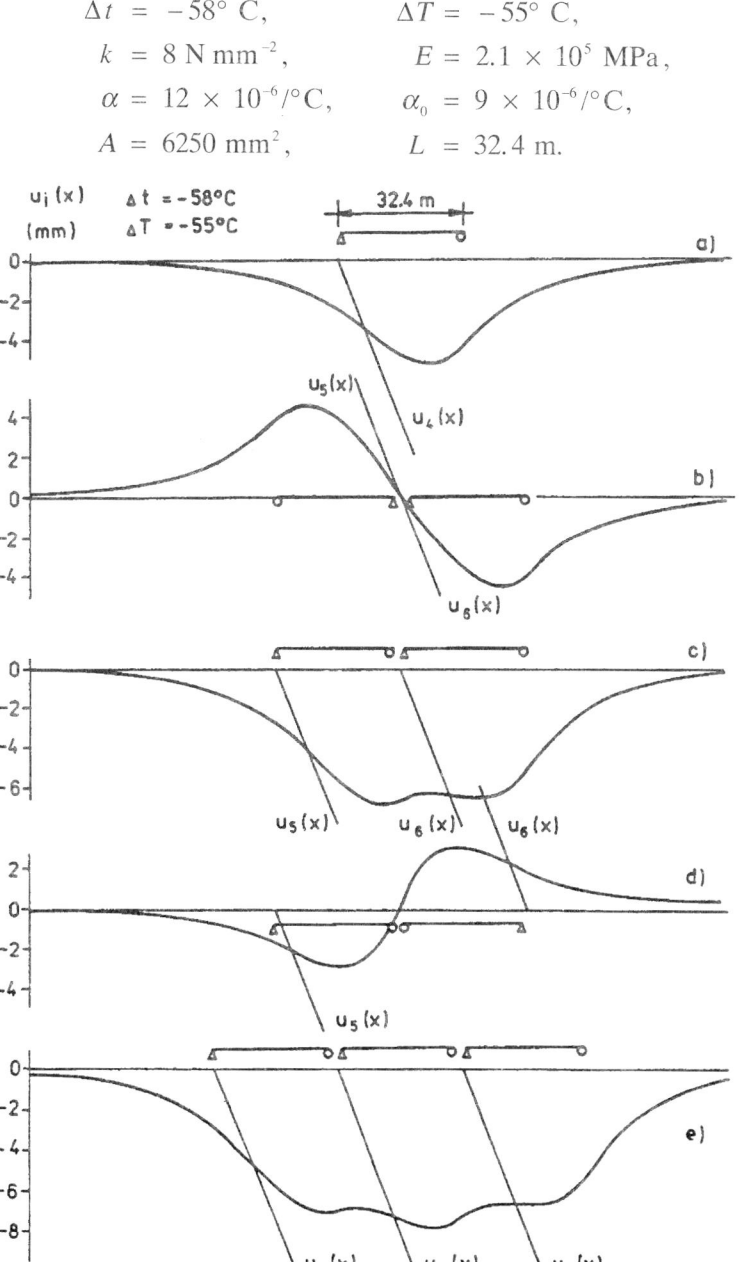

Fig. 14.9. Theoretical distribution of long-welded rail displacements on a steel bridge with direct fastening of rail in extreme winter conditions:
a) simply supported beam, b) two fixed bearings on central pier, c) alternate bearings on central pier, d) two-movable bearings on central pier, e) three-span bridge with alternate bearings on intermediate piers.

14.2 Comparison of theory with experiments

Experimental research was carried out on three steel bridges of the late Czechoslovak State Railways with characteristic fastening of long-welded rail (for a detailed description of these bridges see [78]):

1. Bridge with direct fastening of rails on to bridge deck, without ballast. Steel plate girder construction of span 32.4 m, simply supported, skew, with orthotropic floor system below; the rail is fixed by means of ribbed sole-plates and rubber pads.

2. Bridge with continuous ballast and wooden sleepers. Steel plate girder bridge, continuous with three spans 25 + 55 + 25 m and with a stiffening arch in the central span, orthotropic floor system carrying ballast with wooden sleepers and sole-plates.

3. Bridge with free fastening without ballast, three span continuous steel plate girder of spans 42.9 + 54.6 + 42.9 m. The open deck lying above consists of stringers and cross-beams. The long-welded rails are attached to wooden sleepers with ribbed sole-plates, however, the sleepers together with rails can move along the stringers transferring only vertical and horizontal transverse forces.

The experimental method was taken from [160] and it is described in detail in [78]. As it was not possible to free the long-welded rail on the bridge before experiments, all measured data refer to the time, when the rail temperature had attained the fixing temperature. It is assumed that at that moment there are no stresses in the rail.

During the tests the following quantities were measured:
- the temperature of the ambient air, of the bridge in one cross section and of the rail,
- the longitudinal stresses in several cross sections of the rail in front of, on and beyond the bridge,
- the displacement of the bridge with respect to the abutment, of the rails with respect to the bridge and of the rails with respect to the abutment of the permanent way.

The most important results of the experiments are summarized in Table 14.1.

For the purpose of comparison of theory with experiments it is necessary to determine all constants occurring in equations (14.1) and (14.4). Their determination is easy with the exception of the coefficient of rigidity k which depends on numerous factors and it also varies along the track. For these reasons the stress distribution along the long-welded rail was found for several values of k. Those values of k which approached most closely the experimental values were considered as experimental values in Table 14.1.

The distribution of forces along the long-welded rail on the three investigated bridges is shown in Figs 14.10 to 14.12. Equations (14.17) and (14.22) were used

TABLE 14.1. The most important results of experiments on steel bridges of CSD and their comparison with theory

Bridge		Unit	1	2	3
Type of superstructure			Direct fastening	Ballast, wooden sleepers	Free fastening
Static system			simple	continuous with arch	continuous
Span		m	32.4	25+55+25	42.9+54.6+42.9
Expansion length		m	32.4	80+25	97.5+42.9
Number of days of measurements		day	12	8	14
Thermal delay of the bridge to temperature of air		minute			25
Thermal delay of lower chord to temperature of air		minute	100	40	40
Maximum thermal differences in a day of	Air	°C	21.0	17.0	23.0
	Upper chord		38.1	31.8	33.0
	Web, (1) of bridge deck		(1) 29.1	(1) 30.9	22.8
	Lower chord in sunshine		15.5	15.5	30.7
	Lower chord in shade				15.6
	Rail		33.3	24.0	34.4
Maximum instantaneous thermal differences of	Upper and lower chord in shade	°C	21.7	16.7	16.0
	Upper and lower chord in sushine				3.9
	Left and right hand side of lower chord		0.9	5.4	13.6
	Left and right hand side of bridge deck		12.4	5.7	
Force in a rail over movable bearing $S_M/\Delta t$	(2) $\dfrac{T}{E}$	kN/°C	21.2 20.4	19.5 17.4	23.8 19.5
Force in a rail over fixed bearing $S_F/\Delta t$	(2) $\dfrac{T}{E}$		12.7 16.2	13.7 13.7	11.1 10.5
$(S_M - S_F)/\Delta t$	(2) $\dfrac{T}{E}$		8.5 5.6	5.8 5.1	12.7 9.1
Force in a fixed bearing $X_F/\Delta T$	(3) $\dfrac{T}{E}$		8.8 7.6	4.1 3.9	9.5
Mutual displacement of rail to abutment $/\Delta t$	(2) $\dfrac{T}{E}$	mm/°C	0.076 0.107	0.140 0.107	0.220 0.133
Mutual displacement of rail to bridge $/\Delta t$	(2) $\dfrac{T}{E}$		0.208 0.142	0.441	(5) 0.266 0.275
Coefficient of thermal extension of the bridge/ΔT	(3) E	10^{-6}/°C	9.03	6.64	9.79
Coefficient of thermal extension of the bridge	(4) E		11.79	9.66	10.25
Coefficient of elastic fastening k	E	N/mm²	8	1	2

Notes: (1) bridge deck,
(2) related to temperature of rail,
(3) related to mean temperature of bridge,
(4) related to temperature of ambient air,
(5) displacement of rail to stringer,
T theory,
E experiment.

14.2 Comparison of theory with experiments 277

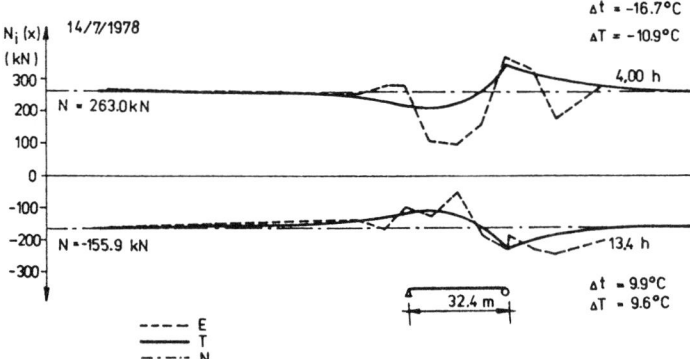

Fig. 14.10. Theoretical (T) and experimental (E) force distribution in the long-welded rail in the morning (0400 h) and afternoon (1340 h) on a steel bridge of span 32.4 m with direct fastening of rails; for N – see equation (14.18).

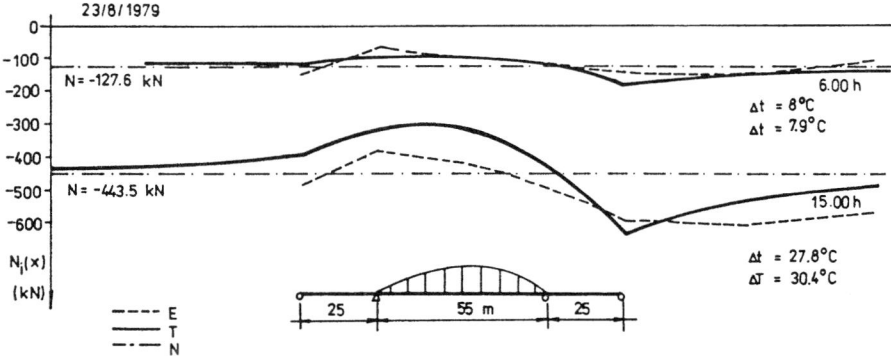

Fig. 14.11. Theoretical (T) and experimental (E) force distribution in the long-welded rail in the morning (0600 h) and afternoon (1500 h) on a steel bridge of spans 25 + 55 + 25 m with ballast and timber sleepers; for N – see equation (14.18).

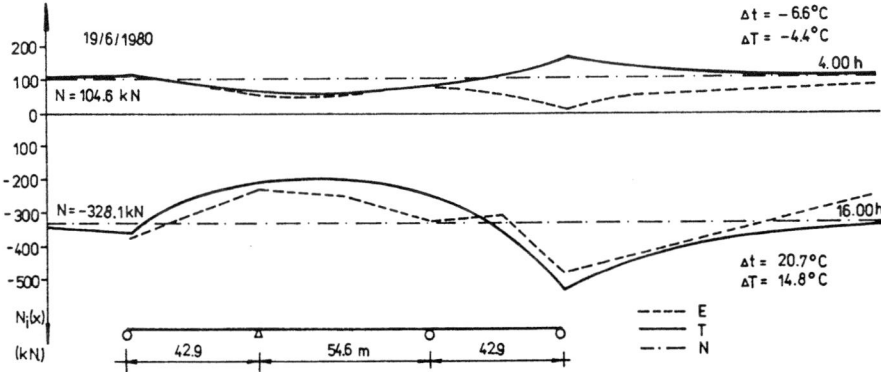

Fig. 14.12. Theoretical (T) and experimental (E) force distribution in the long-welded rail in the morning (0400 h) and afternoon (1600 h) on a steel bridge of spans 42.9 + 54.6 + 42.9 m with free fastening of rails; for N – see equation (14.18).

for the comparison of theory with experiments. Table 14.1 also contains the data of the horizontal longitudinal forces below the fixed bearing X_F and of the forces in the rail above the fixed (S_F) and movable (S_M) bearings, referred to the temperature range of the bridge and/or rail by 1°C.

Table 14.1 and Figs 14.10 to 14.12 show relatively good agreement between the theory from Sect. 14.1 and the experimental results.

14.3 Expansion length of bridges with long-welded rail

It follows from the theoretical and experimental analysis of thermal interaction between long-welded rails and the bridge that the bridge increases axial forces particularly above movable bearings. This increase depends on the expansion length of the bridge.

The *expansion length of* the bridge L is defined as the greatest distance from the fixed bearing to the farthest end of the structure that may freely extend or move. In the case of bridges with no fixed bearings, which have only movable bearings, such as rubber, teflon, sliding (friction) bearings etc., the expansion length of the bridge is the total length of the structure.

The expansion bridge length is, consequently, the most important factor limiting the use of long-welded rail. It must be determined so as to comply with four conditions:
– the strength of rail material,
– the gap in rail after its fracture,
– the mutual displacement between rail and bridge,
– the stability of the long-welded rail.

All considerations must be based on maximum and minimum temperatures which can occur in the local climatic conditions.

14.3.1 Maximum and minimum temperatures

The formulae given in Sect 14.1 contain temperature ranges Δt in the rail (14.12) and ΔT in the bridge (14.13), which depend on the maximum and/or the minimum temperatures and on the fixing temperatures. These temperatures, naturally, depend on local climatic conditions and should be estimated by every railway administration.

The determination of Δt and ΔT must be based on extreme air temperatures which may occur in a period longer than 50 years with a 95% probability. These temperatures were assessed on the occasion of the preparation of the Czechoslovak Standard ČSN 73 6203 and are tabulated in Table 14.2.

On the basis of these data the extreme temperatures of insulated bridges of various material with or without ballast which give a mean bridge temperature were estimated. The studies of UIC and BR on this problem were taken also into account.

TABLE 14.2. Maximum and minimum temperatures of air, of bridges and of rails

			Temperature		ΔT		Δt	
			max.	min.	max.	min.	max.	min.
			(°C)					
Ambient air (1)			+36	−30				
Bridge (2)	Steel	With ballast	+30	−30	+20	−50		
		Without ballast	+35	−35	+25	−55		
	Composite	With and without balast	+30	−20	+20	−40		
	Concrete, prestressed	With and without ballast	+30	−15	+20	−35		
Rail			+60	−30				
Fixing temperature of rails		Wooden sleepers	+28	+15			+45	−58
		Prestressed sleepers	+28	+10			+50	−58
Fixing temperature of bridges (2)			+20	+10				

Notes: (1) extreme temperatures that can appear with probability of 95% in the time period longer than 50 years,
(2) mean temperature of the bridge.

The extreme rail temperatures, i.e., +60°C and −30°C, are based on the values traditionally valid in Central Europe.

Also the fixing temperatures of rails were taken from a CSD regulation. With the assumption of these fixing temperatures the mean bridge temperature may be about +20°C for the maximum fixing temperature and about +10°C for the minimum fixing temperature.

On the basis of these values the difference ΔT of the bridge and Δt of the rails were found (see Table 14.2) and used for the calculation of the expansion bridge lengths. Table 14.2 shows that, with regard to the fixing temperature, the maximum temperature ranges occur during winter in this particular case.

14.3.2 Strength condition

The first condition for the calculation of the maximum admissible expansion bridge length is that the rail stress must not exceed a certain value. Since the rail stresses are found according to the theory of admissible stresses and not on the theory of limit states, the admissible stress σ_{adm} which can be used for thermal stresses in the long-welded rail was derived in accordance with [127], so that the sum of the stresses did not exceed the yield stress σ_{kt} = 600 MPa attained by rail steels of the quality 95 CSD-Vk and higher. The individual stress components are given in detail in Table 14.3.

TABLE 14.3. Stresses in long-welded rail

Type of rail			unit	R 65	S 49, T
Cross-section area of weared rail A			mm²	6823	5088
Stress components due to	Traffic		MPa	90	130
	Bad track conditions, additional			15	20
	Bending in a curve			20	20
	Production			20	20
	Flat wheels			100	100
	Temperature	σ_{adm}		255	210
	Reserve			100	100
	Total stress			600	600

Thus, for the calculation of thermal stresses in a long-welded rail the values of σ_{adm} = 255 MPa may be used for the type R 65 and σ_{adm} = 210 MPa for types S 49 and T of the rails, respectively. In this approach the fully worn rail of the cross section area A (see Table 14.3) and the modulus of elasticity $E = 2.1 \times 10^5$ MPa are considered. It is also safely assumed that $\alpha_0 = \alpha = 12 \times 10^{-6}/°C$. The coefficient of horizontal longitudinal fixing of rails depends chiefly on the presence of ballast and the type of fixing (see Table 14.4). Also the results of our experiments were taken into account.

With these assumptions the maximum admissible expansion bridge length L was found from the condition

$$\sigma \leq \sigma_{adm}, \qquad (14.36)$$

where σ is the maximum stress in a long-welded rail on the bridge. This stress appears above the movable bearing according to the individual cases shown

14.3 Expansion length of bridges with long-welded rail

below. The condition (14.36) will be applied in that season when ΔT and Δt achieve their greatest values.

With respect to the larger cross section area of the rail of type R 65 and its higher admissible stress the expansion bridge length L with rails of type R 65, found from the strength condition, is always greater than for rail types S 49 and T.

Table 14.4. Coefficient of horizontal longitudinal fastening of rails

Railway bed	Fastening and sleepers		k (N mm^{-2})
With ballast	Wooden sleepers		1
	Prestressed sleepers		2
Without ballast	Free fastening	(1)	2
	Indirect fastening	(2)	7
	Direct fastening	(3)	8
Notes: (1) rail, sole-plate, wooden sleepers – can slide along stringers (2) rail, sole-plate, wooden sleepers, stringers, (3) rail, sole-plate, bridge deck.			

a) Fixed bearing on one bridge end

In the case represented by Fig. 14.3 the equations (14.17) present the highest stress

$$\sigma = \frac{N_3(0)}{A} = -E\alpha\Delta t \left[1 + \frac{\alpha_0 \Delta T}{2\alpha \Delta t}(\lambda L - 1 + e^{-\lambda L})\right]. \quad (14.37)$$

b) Fixed bearing in the middle of the bridge.

In the cases shown in Fig. 14.4, see equations (14.22), the highest stress appears above the movable bearing of greater expansion length

$$\sigma = \frac{N_4(0)}{A} = -E\alpha\Delta t \left\{1 + \frac{\alpha_0 \Delta T}{2\alpha \Delta t}\left[\lambda L - 1 + \left(1 + \frac{\lambda L}{1+r}\right)e^{-\lambda L(2+r)/(1+r)}\right]\right\} \quad (14.38)$$

where $l_3 = l_6 = L$, and in the case of three-span continuous girders the ratio of individual spans $1 : r : 1$ is considered. (In the summary Table 14.8 the following notation has been used: Case 2: $r = 0$, Case 3: $r = 0$, but $L = 2l_3$, Case 6: $r = 1$, and Case 7: $r = 1.3$, which is the most frequent case for three-span continuous bridges.)

c) Alternate bearings on one pier

In the case of the two-span bridge with alternate bearings (shown in Fig. 14.5, see equations (14.25)) the highest stress in the rail occurs also above the second movable bearing

$$\sigma = \frac{N_4(0)}{A} = -E\alpha\Delta t \left\{ 1 + \frac{\alpha_0 \Delta T}{2\alpha \Delta t} \left[\lambda L - 1 + (\lambda L + e^{-\lambda L}) e^{-\lambda L} \right] \right\}. \tag{14.39}$$

d) Two movable bearings on one pier

In the case of the bridge of two equal spans and two movable bearings above the pier (Fig. 14.6, equations (14.28)) the highest stress in the rail is above the pier:

$$\sigma = \frac{N_3(0)}{A} = -E\alpha\Delta t \left[1 + \frac{\alpha_0 \Delta T}{2\alpha \Delta t} (\lambda L - 1 + e^{-\lambda L}) \right]. \tag{14.40}$$

e) Three-span bridge with alternate bearings

For the bridge of three equal spans (Fig. 14.7, equations (14.32)) the highest stress in the rails is above the last movable bearing:

$$\sigma = \frac{N_5(0)}{A} = -E\alpha\Delta t \left[1 + \frac{\alpha_0 \Delta T}{2\alpha \Delta t} \left\{ \lambda L - 1 + \left[\lambda L + (\lambda L + e^{-\lambda L}) e^{-\lambda L} \right] e^{-\lambda L} \right\} \right] \tag{14.41}$$

and for n spans equation (14.33) is applicable.

14.3.3 Gap condition

The safety of railway traffic must be guaranteed even when long-welded rail cracks on the bridge. The gap in the rail is further increased by the deflection of the bridge, particularly in the case of bridges with floor systems lying above. The gap can occur only during low temperature conditions.

This requirement is very severe and extremely disadvantageous for bridges, althought it should also be applied to the long-welded rail of open track.

According to the bibliography and experience, the admissible width of the gap varies between 30 and 50 mm (see [205]) which results in the drop of the wheel centre by 0.78 mm, provided the rail is supported rigidly. Actually, an impact occurs during the movement of traffic which, however, does not depend on the gap width. For this reason the calculation of the expansion length is based on the condition

$$a \leq 50 \text{ mm}. \tag{14.42}$$

14.3 Expansion length of bridges with long-welded rail

a) Width of the gap in long-welded rail

If the long-welded rail cracks, two semi-infinite beams are formed, the displacements of which are described by the differential equation (14.4), see Fig. 14.13. Let us apply to equation (14.4) the following boundary conditions for the right-hand side of the beam

$$u(\infty) = 0, \quad N(0) = 0. \tag{14.43}$$

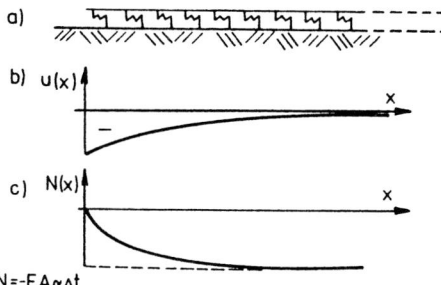

Fig. 14.13. a) Semi-infinite beam, b) distribution of horizontal displacement $u(x)$, c) distribution of longitudinal force $N(x)$.

With these conditions the solution of equation (14.4) results in

$$u(x) = -\frac{\alpha \Delta t}{\lambda} e^{-\lambda x}, \quad N(x) = -EA\alpha \Delta t (1 - e^{-\lambda x}). \tag{14.44}$$

The rupture of the long-welded rail, consequently, gives rise to the gap

$$u_3 - u_2 \tag{14.45}$$

(see Fig. 14.14), where

$$u_2 = \frac{\alpha \Delta t}{\lambda_2}, \quad u_3 = -\frac{\alpha \Delta t}{\lambda_3} \tag{14.46}$$

and λ_2 and λ_3 are found from equations (14.11) for the left and right-hand side rail sectors, respectively.

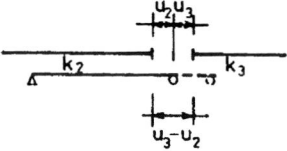

Fig. 14.14. Gap after the rupture of long-welded rail above movable bearing of a bridge.

b) Influence of deflection and depth of the bridge.

The gap in the rail is increased by the deflection of the bridge and it also depends on the bearing depth of the bridge. The width c for this case is shown in Fig. 14.15; it is calculated with the assumption that the ratio $m_1 = l/h$ of the span l to the bearing depth h and the ratio $m_2 = l/f$ of the span l to the deflection f is due to half the standard load attained in practice.

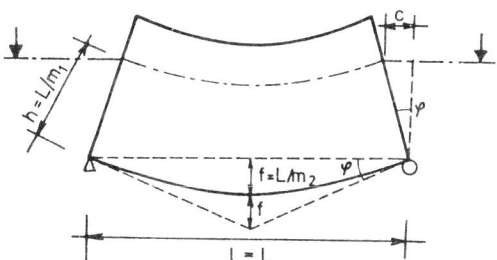

Fig. 14.15. Influence of bridge deflection and girder depth on the gap in the rail.

The gap width then

$$c = \frac{l}{m_1} \sin \varphi = \frac{4l}{m_1 m_2}, \qquad (14.47)$$

where $\operatorname{tg} \varphi = 4 f/l$ and $\sin \varphi \approx \operatorname{tg} \varphi$ (parabolic deflection is assumed). In the case of two movable bearings on a pier the double value of c should be taken into account, because the highest stress is above the central pier.

c) Computation of expansion length

The condition of the maximum permitted gap width is, consequently

$$u_3 - u_2 + c \leqslant a. \qquad (14.48)$$

After the substitution of (14.46) and (14.47), equation (14.48) can be written

$$L \leqslant \left[a + \alpha \Delta t \left(\frac{1}{\lambda_2} + \frac{1}{\lambda_3} \right) \right] \frac{m_1 m_2}{4}, \qquad (14.49)$$

where it is assumed that $l = L$ (safe).

In the calculation of the expansion length L according to equation (14.49) the following values were considered: $a = 50$ mm, $\alpha = 12 \times 10^{-6}/°C$, $\Delta t = -58°C$ and the coefficients k_2 and k_3 were from Table 14.5, because winter conditions and supports against rails slipping beyond the bridge were assumed. Otherwise condition (14.49) would not be satisfied.

The coefficients of deflection and of the depth of the bridge are assumed according to Table 14.6, which are generally current values for half the standard load (approximately traffic load).

On the basis of these data, the expansion lengths L according to equation (14.49) were found. Since the displacement of weaker rails of S 49 and T types is smaller than the displacement of the type R 65, according to equation (14.46), the expansion lengths L are greater for types S 49 and T than for the type R 65 of rails.

TABLE 14.5. Coefficient of horizontal longitudinal fastening of rails (in extreme winter conditions)

Bridges	Railway bed	Fastening and sleepers		k_2	k_3
				N mm^{-2}	
Steel	With ballast	Wooden sleepers		5	10
	Without ballast	Free fastening	(1)	2	10
		Indirect fastening	(2)	7	10
		Direct fastening	(3)	8	10
Composite	With ballast	Wooden sleepers		5	10
		Prestressed sleepers		6	12
	Without ballast	Direct fastening	(3)	8	10
Concrete and prestressed	With ballast	Wooden sleepers		5	10
		Prestressed sleepers		6	12
	Without ballast	Direct fastening	(3)	8	10

Notes: (1) rail, sole-plate, wooden sleepers – can slide along stringers,
(2) rail, sole-plate, wooden sleepers, stringers,
(3) rail, sole-plate, bridge deck.

TABLE 14.6. Coefficients of depth and of deflection of bridges

Coefficient of		Bridges	
		Steel	Composite, concrete, prestressed
Depth	m_1	12	15
Deflection	m_2	2000	4000

14.3.4 Mutual displacement condition

The next condition for the long-welded rail on bridges is the condition that the relative displacement u between the rail and the bridge should be limited by the maximum admissible value b. The omission to comply with this condition results in plastic deformation and rupture of clamps, sole-plates, screws and other parts of the permanent way and the bridge deck.

This condition is expressed mathematically

$$|u| \leq b, \qquad (14.50)$$

where u the maximum mutual rail and bridge displacement. The condition (14.50) yields the expansion length for the individual cases.

Condition (14.50) is only applied to bridges without ballast for the absolutely maximum temperature ranges ΔT. The coefficient k is considered according to Table 14.4, but the coefficients of thermal extension α_0 according to Table 14.7. These values are nearer reality and are based on our experiments [78] and on reports [160] and characterize the fact that concrete bridges and particularly the bridges with ballast are considerably influenced by friction.

TABLE 14.7. Coefficient of thermal extension

Bridges	Railway bed	α_0 ($10^{-6}/°C$)
Steel	With ballast	6
	Without ballast	9
Composite	With ballast	5
	Without ballast	6
Concrete, prestressed	With ballast	5
	Without ballast	6
Rails	α	12

The maximum admissible mutual displacement b is found from the equation

$$b = \frac{2F}{kd}, \qquad (14.51)$$

where $F = 19$ kN, according to laboratory experiments of Sackmauer [182], is the ultimate strength of a bolt joining the sleeper to the stringer or of rail anchoring clamp; $k = 7$ or 8 N mm^{-2} is the coefficient for indirect or for direct fastening of rails, respectively, see Table 14.4, and $d = 600$ mm is the spacing of sleepers.

The condition (14.51) considerably reduces the expansion length L which, once again, is smaller for the type R 65 of rails than for types S 49 and T. This is due to the smaller mutual displacement of weaker rails with reference to the bridge. If new types of sole-plates and clamps are used, which allow the longitudinal displacement of the rails along the bridge, the condition (14.51) could be omitted.

14.3 Expansion length of bridges with long-welded rail

a) Fixed bearing on one bridge end

The greatest mutual displacement occurs above the movable bearing and, according to Fig. 14.3 and equations (14.6), attains the value of

$$u = u_4(L) - u_2(L) = \frac{\alpha_0 \Delta T}{2\lambda}(1 + \lambda L - e^{-\lambda L}). \qquad (14.52)$$

b) Fixed bearing in the middle of the bridge

For this case Fig. 14.4 and equations (14.21) yield the maximum relative displacement above the movable bearing for a greater expansion length

$$u = u_6(L) - u_3(L) = \frac{\alpha_0 \Delta T}{2\lambda}\left[1 + \lambda L - \left(1 + \frac{\lambda L}{1+r}\right)e^{-\lambda L(2+r)/(1+r)}\right]. \qquad (14.53)$$

Once again, the ratio of spans of a three-span bridge is denoted by $1 : r : 1$.

c) Alternate bearings on an intermediate pier

According to the notation used in Fig. 14.5 and equations (14.24) the greatest mutual displacement in the case of a two-span bridge of equal spans is above the second movable bearing

$$u = u_6(L) - u_3(L) = \frac{\alpha_0 \Delta T}{2\lambda}\left[1 + \lambda L - (\lambda L + e^{-\lambda L})e^{-\lambda L}\right]. \qquad (14.54)$$

d) Two movable bearings on one pier

The bridge of two equal spans with movable bearings on one pier, see Fig. 14.6 and equations (14.27), possesses the maximum relative displacement above the pier

$$u = u_6(0) - u_3(0) = -\alpha_0 \Delta T\, L \qquad (14.55)$$

so that the expansion length can be found directly from the condition (14.50), i.e.

$$L \leq \frac{b}{\alpha_0 \Delta T} \qquad (14.56)$$

e) Three-span bridge with alternate bearings

The bridge of three and more equal spans with alternate bearings has the maximum mutual displacement always above the last movable bearing. For a three span bridge, see Fig. 14.7 and equations (14.31), the displacement is

$$u = u_8(L) - u_4(L) = \frac{\alpha_0 \Delta T}{2\lambda}\left\{1 + \lambda L - \left[\lambda L + (\lambda L + e^{-\lambda L})e^{-\lambda L}\right]e^{-\lambda L}\right\} \qquad (14.57)$$

and for n fields the equation (14.35) applies.

14.3.5 Stability condition

The long-welded rail must also be secured against buckling in summer. According to our assumptions given in Sect. 14.3.2 the maximum axial force in the long-welded rail on the bridge can be

$$N = \sigma_{adm} A = \begin{cases} 1.74 \text{ MN} \\ 1.07 \text{ MN} \end{cases} \text{for types} \begin{cases} \text{R } 65 \\ \text{S } 49, \text{ T} \end{cases} \text{of rails.} \qquad (14.58)$$

According to [182] the ultimate force on the stability limit of the long-welded rail at the end of the bridge is

$$N_{cr} = \begin{cases} 5.65 \text{ MN} \\ 4.65 \text{ MN} \end{cases} \text{for types} \begin{cases} \text{R } 65 \\ \text{S } 49, \text{ T} \end{cases} \text{of rails} \qquad (14.59)$$

and beyond the safety angles in the track

$$N_{cr} = \begin{cases} 4.74 \text{ MN} \\ 4.24 \text{ MN} \end{cases} \text{for types} \begin{cases} \text{R } 65 \\ \text{S } 49, \text{ T} \end{cases} \text{of rails.} \qquad (14.60)$$

The analysis of an endless bar on an elastic foundation gives the minimum critical force

$$N_{cr} = 2(EI_y k_y)^{1/2}, \qquad (14.61)$$

where k_y is the coefficient of elastic foundation of rails in vertical or horizontal transverse directions. For a very low estimation of $k_y = 10$ N mm^{-2} and for horizontal buckling we obtain for the force (14.61)

$$N_{cr} = \begin{cases} 9.80 \text{ MN} \\ 7.43 \text{ MN} \end{cases} \text{for types} \begin{cases} \text{R } 65 \\ \text{S } 49, \text{ T} \end{cases} \text{of rails.} \qquad (14.62)$$

In any case the safety against buckling is

$$N_{cr}/N > 2, \qquad (14.63)$$

which is required for the long-welded rail.

Consequently, the condition of stability is complied with automatically by keeping to the admissible stress σ_{adm}.

14.3.6 Admissible expansion length of bridges

For every condition from Sects 14.3.2 to 14.3.5 and for every case of bridge bearing arrangement the maximum admissible expansion length of bridges of various material and with various arrangement of ballast, of superstructure and of floor system was calculated. The calculation was based on the data given in Sects 14.3.1 to 14.3.5 and carried out by an iterative method with the step of $L = 1$ m according to the block diagram shown in Fig. 14.16.

Table 14.8 contains those values of expansion length L which were obtained as a minimum for the each case. In this way it has been ensured that the lengths L satisfy all criteria described in Sects 14.3.2 to 14.3.5.

14.3 Expansion length of bridges with long-welded rail

For steel bridges with ballast and with free fastening the decisive criterion is the gap condition in case of R 65 rails. In case of types S 49 and T of rails it is the strength condition. The expansion length of steel bridges with indirect or direct fastening is controlled by the mutual displacement condition.

In composite steel, concrete and prestressed bridges with ballast the strength condition is decisive, similarly for the same bridges with direct fastenings of rails S 49 and T. However, for direct fastening of rails R 65 the mutual displacement condition is decisive.

If sliding fastening of rails, which allows longitudinal horizontal displacement, is used on the bridge, the condition of mutual displacement could be omitted. Then in the case of bridges without ballast, the strength condition may be used, which yields higher values of L. These data are given in brackets in Table 14.8.

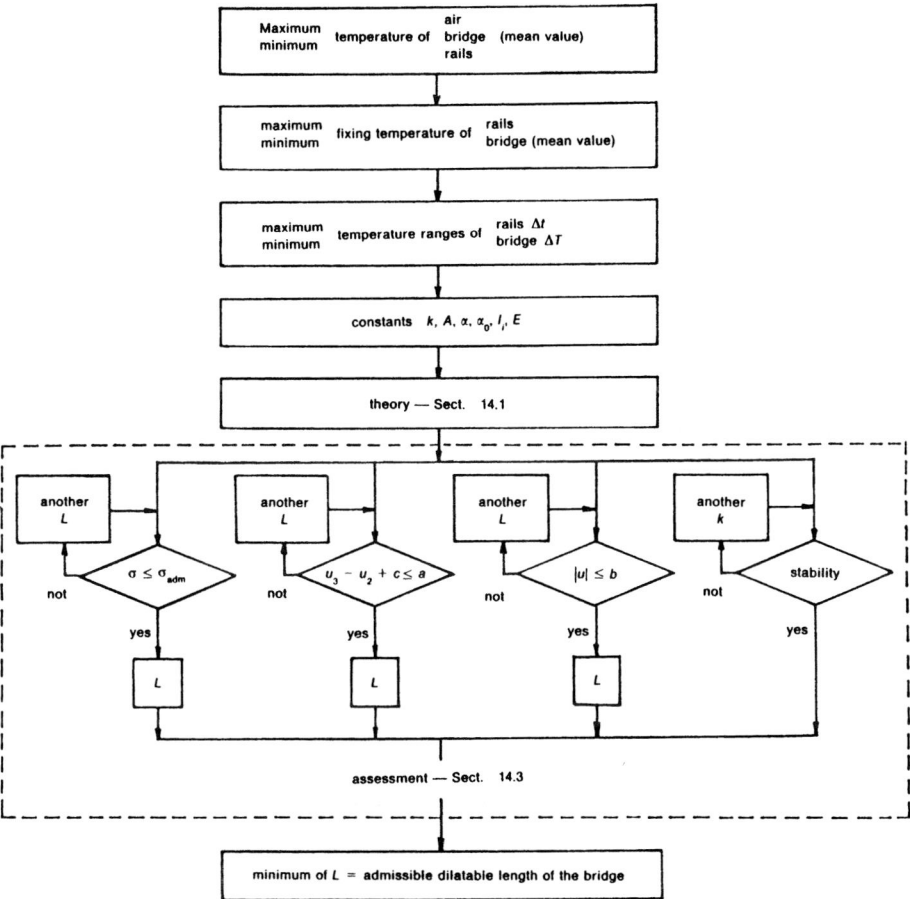

Fig. 14.16. Block diagram for the calculation of the admissible expansion length L of a bridge with long-welded rail.

TABLE 14.8. Admissible expansion length L of railway bridges with continuous welded rails

			Railway bridges										
			Steel				Composite				Concrete, prestressed		
			Ballast				Ballast				Ballast		
			With Wooden sleeper	Without Fastening			With Sleepers		Without Fastening	With Sleepers		Without Fastening
Case No.	Arrangement of bearings and expansion length L, see Figure	type of rails		Free (1)	Indirect (2)	Direct (3)	Wooden	Prestr.	Direct (3)	Wooden	Prestr.	Direct (3)
							L (m)						
1	14.3	R65	129	71	23	20	167	118	50	(59)	184	131	59	(66)
		S 49, T	86	58	24	21	100	71	36		109	77	39	
2	14.4 a, b, c $r = 0$	R 65	129	71	22	19	169	120	49	(60)	186	132	58	(66)
		S 49, T	91	61	23	20	104	73	37		112	80	40	
3	14.4d $r = 0, L = 2l_3$	R65	129	71	44	38	323	240	98	(120)	323	264	116	(132)
		S 49, T	153	102	46	40	208	146	74		224	160	80	
4	14.5	R 65	129	71	26	23	162	114	52	(57)	180	127	60	(64)
		S 49, T	77	51	26	23	92	65	33		102	72	36	
5	14.6	R 65	65	36	19	17	104	74	34	(37)	114	80	39	(40)
		S 49, T	56	38	19	17	64	46	23		70	50	25	
6	14.4b $r = 1$	R 65	129	71	24	20	168	120	50	(60)	186	132	58	(66)
		S 49, T	88	60	24	22	102	72	36		112	80	40	
7	14.4b $r = 1.3$	R 65	129	71	23	21	168	120	51	(60)	186	131	60	(67)
		S 49, T	90	60	25	21	101	74	37		110	78	39	
8	14.7 3 and more spans	R 65	129	71	26	23	161	114	52	(57)	180	127	60	(64)
		S 49, T	75	50	27	24	90	64	32		101	71	36	

Notes: (1), (2), (3), – see Table 14.4,
values of L in brackets are applicable if mutual displacement condition is omitted (special rail anchoring devices),
case No. 3 can be applied for bridges without fixed bearings (rubber or sliding bearings etc.),
△ fixed bearing, ○ movable bearing.

14.3 Expansion length of bridges with long-welded rail

The maximum admissible expansion lengths of bridges with long-welded rail in Table 14.8. are sufficiently safe, because their computation was based on fully worn rails with the coefficient of thermal extension $\alpha = 12 \times 10^{-6}/°C$ in the majority of cases. The non-linear relation between the resistance force and displacement, the use of safety angles, thermal inertia of bridges and hysteresis were not taken into account.

14.4 Horizontal forces in bridges due to temperature changes

When the rails are jointed or provided with expansion devices at the ends of railway bridges, no horizontal forces due to temperature changes are induced into bridges, according to theoretical assumptions. In the case of the long-welded rail, however, horizontal longitudinal forces are generated by temperature ranges which affect the bridge, the bearings, piers and abutments. Additionally horizontal transverse loads occur if the long-welded rail lies in curve.

Let us consider the rail on the bridge of a span l in a curve of a radius r (see Fig. 14.17). After the separation of the bridge and rails, the rails are affected by the forces N_F and the bridge by the force X_F on the side of fixed (F) bearings, and by the forces N_M and X_M, respectively, on the side of movable (M) bearings. Fig. 14.17 shows the positive forces affecting the bridge. The abutments are subjected to positive forces in opposite directions as shown in Figs. 14.18 to 14.25.

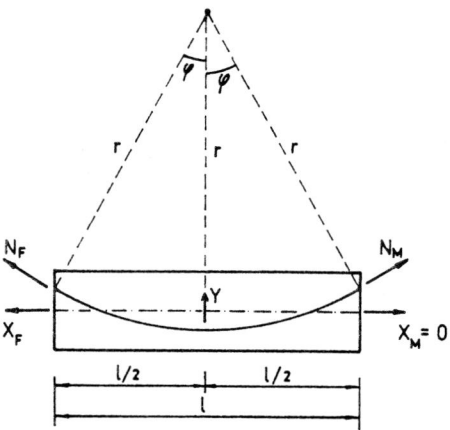

Fig. 14.17. External forces affecting the bridge and the rail in curve, plan.

According to assumption (a) in Sect 14.1.1 the forces in the movable bearing is
$$X_M = 0.\qquad(14.64)$$

The condition of equilibrium of forces in horizontal longitudinal and horizontal transverse directions according to Fig. 14.17 yields
$$X_F + N_F \cos\varphi = N_M \cos\varphi,\qquad(14.65)$$

$$Y = (N_F + N_M) \sin \varphi, \qquad (14.66)$$

where φ is half the central angle according to Fig. 14.17. This angle is determined from the relation

$$\sin \varphi = \frac{l}{2r}. \qquad (14.67)$$

As $r \gg l$, $\cos \varphi \approx 1$.

Therefore, the horizontal forces under the fixed bearing is, from equation (14.65),

$$X_F = N_M - N_F \qquad (14.68)$$

and the horizontal transverse force is, from equation (14.66),

$$Y = (N_F + N_M) \frac{l}{2r}. \qquad (14.69)$$

The force X_F can also be found from Fig. 14.1 as the sum of forces affecting every bridge element which has to be transmitted by the fixed bearing

$$X_F = -\int_0^{l_2} k\big[u_4(x) - u_2(x)\big] \, dx. \qquad (14.70)$$

Equations (14.68) and (14.70) give identical results.

In practice, the horizontal transverse force Y is replaced with horizontal transverse informly distributed load

$$p = Y/l = (N_F + N_M)/(2r), \qquad (14.71)$$

which affects the bridge in the positive direction shown in Figs. 14.19 to 14.25.

If the rail on the bridge is straight, $r \to \infty$, the horizontal longitudinal forces is determined by equation (14.68) or equation (14.70), while for the horizontal transverse load equation (14.71) gives

$$p = 0. \qquad (14.72)$$

According to equations (14.68) to (14.71) the load affecting the bridge depends on the normal forces N_F and N_M in the rails which must be calculated for the individual cases.

a) Bridge with expansion devices above both bearings (Fig. 14.18)

If the rails are provided with expansion devices at both bridge ends, all forces due to temperature equal zero. Hence

$$X_F = 0, \quad \text{and} \quad p = 0. \qquad (14.73)$$

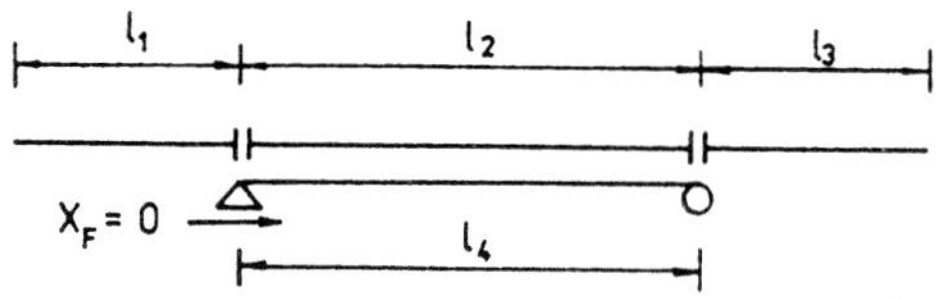

Fig. 14.18. Bridge with expansion devices above both bearings, elevation.

14.4 Horizontal forces in bridges due to temperature changes

b) Bridge with an expansion device above a movable bearing (Fig. 14.19)

If the rails are provided with an expansion device above a movable bearing, the temperature-produced forces are

$$X_F = -N_2(0) = X_{F0} - EA \, e^{-\lambda l_2} \left[\alpha \, \Delta t - \frac{1}{2} \alpha_0 \, \Delta T (2 - e^{-\lambda l_2}) \right], \quad (14.74)$$

$$pr = \frac{1}{2} N_2(0) = -\frac{1}{2} X_F. \quad (14.75)$$

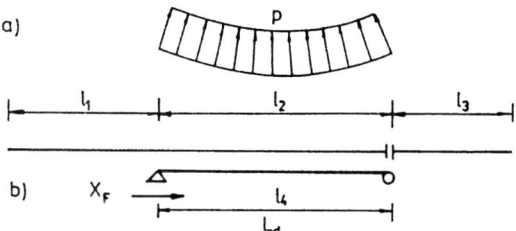

Fig. 14.19. Bridge with an expansion device above a movable bearing, a) plan, b) elevation.

where

$$X_{F0} = EA \left(\alpha \, \Delta t - \frac{1}{2} \alpha_0 \, \Delta T \right). \quad (14.76)$$

The forces X_{F0} and

$$P_0 = -\frac{1}{2} X_{F0} \quad (14.77)$$

for the constants from Tables 14.2 and 14.4 are given in Table 14.9.

The dimensionless relation

$$X = X_F / X_{F0} = pr / P_0 \quad (14.78)$$

is plotted as a function of the bridge span l_2 in Figs 14.26 and 14.27 for two basic cases: the bridge with ballast (Fig. 14.26) and the bridge without ballast (Fig. 14.27).

From the data of Table 14.9 and Figs 14.26 and 14.27 it is possible to find the forces X_F and p for the given span l_2 as follows:

$$X_F = X X_{F0}, \quad (14.79)$$

and

$$p = X P_0 / r. \quad (14.80)$$

Figures 14.26 and 14.27 show that in the case at b) there is a limit to forces X_F and p with reference to the bridge span l_2.

TABLE 14.9. Horizontal longitudinal forces X_{F0} and transverse forces P_0 (kN) in railway bridges due to temperature changes

Case No.	Position of bearings and notation according to the Figure	Type of rails	Temperature extreme	Steel — With ballast, Wooden sleepers		Steel — Without ballast, Fastening Free (1)		Steel — Without ballast, Fastening Indirect (2)		Steel — Without ballast, Fastening Direct (3)		Composite — With ballast, Wooden sleepers		Composite — With ballast, Prestressed sleepers		Concrete, prestressed — With ballast, Wooden sleepers		Concrete, prestressed — With ballast, Prestressed sleepers	
				X_{F0} (4)	P_0 (5)	X_{F0} (4)	P_0 (5)	X_{F0} (4)	P_0 (5)	X_{F0} (4)	P_0 (5)	X_{F0} (4)	P_0 (5)	X_{F0} (4)	P_0 (5)	X_{F0} (4)	P_0 (5)	X_{F0} (4)	P_0 (5)
1	14.18	R 65 S 49	Summer Winter	1669 -1899	-835 949	1486 -1560	-743 780	1486 -1560	-743 780	1486 -1560	-743 780	1704 -2072	-852 1036	1912 -2072	-956 1036	1704 -2116	-852 1058	1912 -2116	-956 1058
2	14.19	R 65	Summer Winter	1269 -1444	-635 722	1131 -1186	-565 593	1131 -1186	-565 593	1131 -1186	-565 593	1296 -1576	-648 788	1455 -1576	-727 788	1296 -1609	-648 805	1455 -1609	-727 805
		S 49	Summer Winter	-1669 1899	-835 949	-1486 1560	-743 780	-1486 1560	-743 780	-1486 1560	-743 780	-1704 2072	-852 1036	-1912 2072	-956 1036	-1704 2116	-852 1058	-1912 2116	-956 1058
3	14.20	R 65	Summer Winter	-1269 1444	-635 722	-1131 1186	-565 593	-1131 1186	-565 593	-1131 1186	-565 593	-1296 1576	-648 788	-1455 1576	-727 788	-1296 1609	-648 805	-1455 1609	-727 805
		S 49	Summer Winter	-389 973	-1942 2580	-534 1175	-1949 2577	-260 572	-1900 2468	-231 509	-1896 2460	-448 881	-1967 2598	-472 944	-2186 2620	-506 885	-1990 2617	-518 907	-2203 2625
4	14.21	R 65	Summer Winter	-212 530	-1456 1911	-397 873	-1480 1956	-244 537	-1452 1893	-218 480	-1448 1884	-220 441	-1462 1908	-217 435	-1620 1907	-250 437	-1470 1913	-259 453	-1631 1918

Notes: (1), (2), (3) — see Table 14.4.
(4) X_F; (5) p.

14.4 Horizontal forces in bridges due to temperature changes

c) Bridge with expansion device above fixed bearing (Fig. 14.20)

Should the expansion device be placed above a fixed bearing (which would be unusual), the horizontal forces generated by temperature would be

$$X_F = N_3(0) = \left\{ X_{F0} - \frac{1}{2} EA\alpha_0 \Delta T \left[\lambda l_2 + (1 + \lambda l_2) e^{-\lambda l_2} \right] \right\} (1 - e^{-\lambda l_2}) \quad (14.81)$$

and

$$pr = \frac{1}{2} N_3(0) = -\frac{1}{2} X_F, \quad (14.82)$$

where

$$X_{F0} = EA \left(-\alpha \Delta t + \frac{1}{2} \alpha_0 \Delta T \right) \quad (14.83)$$

and

$$P_0 = \frac{1}{2} X_{F0}. \quad (14.84)$$

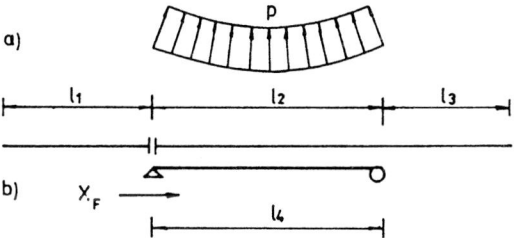

Fig. 14.20. Bridge with an expansion device above fixed bearing, a) plan, b) elevation.

The values of X_{F0} and P_0 are calculated in Table 14.9 from the values of Tables 14.2 and 14.4. The dependence (14.78) of the dimensionless quantity X on the expansion length l_2 for two basic cases is plotted in Figs 14.28 and 14.29. In the case at c) the forces X_F and p increase with increasing span l_2, which reveals the unsuitability of such location of the expansion device. The actual load for the given span l_2 could be found from equations (14.79) and (14.80).

d) Single-span bridge with long-welded rail (Fig. 14.21)

If the long-welded rail passes along the bridge with a fixed bearing at one end, the force

$$X_F = N_3(0) - N_2(0) = -\frac{1}{2} EA\alpha_0 \Delta T \lambda l_2 (1 - e^{-\lambda l_2}) \quad (14.85)$$

originates under the fixed bearing, according to equations (14.17) and (14.68).

The transverse load according to equations (14.17), (14.69) and (14.71) appears in the following form

$$pr = -\frac{1}{2} EA\alpha\Delta t \left\{ 2 + \frac{\alpha_0 \Delta T}{2\alpha \Delta t} \left[-2(1 - e^{-\lambda l_2}) + \lambda l_2 (1 + e^{-\lambda l_2}) \right] \right\}, \quad (14.86)$$

where $N_F = N_2(0)$ and $N_M = N_3(0)$.

The forces denoted as

$$X_{F0} = X_F \quad (14.87)$$

and

$$P_0 = pr \quad (14.88)$$

were found for the data of Tables 14.2 to 14.8 and are given in Table 14.9. The calculation used equations (14.85) and (14.86), where $l_2 = L$ was substituted.

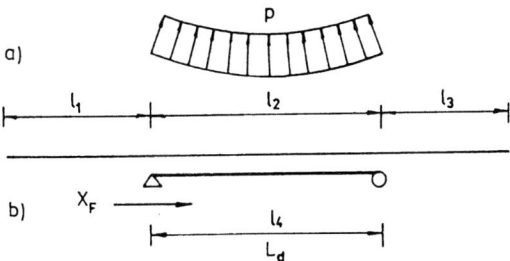

Fig. 14.21. Single-span bridge with long-welded rail, a) plan, b) elevation.

Figures 14.30 and 14.31 show

$$X = X_F / X_{F0} \quad (14.89)$$

and

$$P = pr / P_0 \quad (14.90)$$

plotted against the span l_2 for the bridge with ballast. The force X_F for the given span l_2 is then found from equation (14.79) and the load p from equation (14.90)

$$p = PP_0 / r. \quad (14.91)$$

It can be stated, very approximately, that the quantity X increases in proportion with increasing span, while the quantity P does not change much with increasing span. This conclusion is used in Table 14.10 which gives only the value of

$$X_F / L \quad (14.92)$$

and pr for which $l_2 = L$. The values from Tables 14.2 to 14.8 have been used for the calculation of the forces in Table 14.10. For other values of $l_2 \leq L$ the force X_F may be found approximately from the relation

$$l_2 X_F / L \quad (14.93)$$

while the load pr remains approximately independent of l_2.

14.4 Horizontal forces in bridges due to temperature changes

e) Continuous bridge with long-welded rail (Figs 14.22 and 14.23)

If the long-welded rail is mounted on a continuous bridge with a fixed bearing on an intermediate pier, temperature changes produce the forces

$$X_F = N_4(0) - N_2(0) = -\frac{1}{2} EA\alpha_0 \Delta T \lambda (l_3 - l_2)\left[1 - e^{-\lambda(l_2+l_3)}\right] \quad (14.94)$$

and

$$pr = -\frac{1}{2} EA\alpha \Delta t \left\{ 2 + \frac{\alpha_0 \Delta T}{2\alpha \Delta t}\left[-2(1 - e^{-\lambda(l_2+l_3)})\right.\right.$$
$$\left.\left. + \lambda(l_2 + l_3)(1 + e^{-\lambda(l_2+l_3)})\right]\right\} \quad (14.95)$$

derived from equations (14.21).

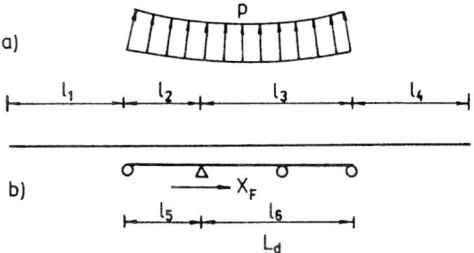

Fig. 14.22. Continuous bridge with long-welded rail, a) plan, b) elevation.

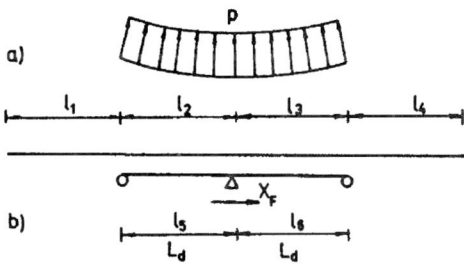

Fig. 14.23. Continuous bridge with long-welded rail or two simply supported beams with fixed bearings on central pier, a) plan, b) elevation.

The forces (14.90) and (14.91) were calculated for the two cases:

$$l_2 = l_3/2 \quad \text{- see Fig. 14.22}$$

and

$$l_2 = l_3 \quad \text{- see Fig. 14.23}$$

and the results are tabulated in Table 14.10. However, the relation (14.92) has been replaced with the dependence on the difference of $L = l_3$ and l_2 according to Fig. 14.22

$$X_F/(L - l_2). \quad (14.96)$$

TABLE 14.10. Horizontal longitudinal forces X_F and horizontal transverce load p of railway bridges due to temperature changes

Case No.	Position of bearings and notation according to Figure	Type of rails	Temperature extreme	Steel								Composite				Concrete, prestressed			
				With ballast		Witout ballast						With ballast				With ballast			
				Wooden sleepers		Fastening						Wooden sleepers		Prestressed sleepers		Wooden sleepers		Prestressed sleepers	
						Free(1)		Indirect (2)		Direct (3)									
				X_F/L	pr	X_F/L	pr	X_F/L	pr	X_F/L	pr	X_F/L	pr	X_F/L	pr	X_F/L	pr	X_F/L	pr
				kN m^{-1}	kN	kN m^{-1}	kN	kN m^{-1}	kN	kN m^{-1}	kN	kN m^{-1}	kN	kN m^{-1}	kN	kN m^{-1}	kN	kN m^{-1}	kN
4	14.21	R 65	Summer	−3.113	−1942	−7.631	−1949	−11.298	−1900	−11.574	−1896	−2.753	−1967	−3.935	−2186	−2.809	−1990	−3.985	−2203
			Winter	7.783	2580	16.788	2577	24.855	2468	25.463	2460	5.506	2598	7.870	2620	4.916	2617	6.974	2625
		S 49	Summer	−2.495	−1456	−6.611	−1480	−10.619	−1452	−10.917	−1448	−2.203	−1462	−3.106	−1620	−2.269	−1470	−3.233	−1631
			Winter	6.237	1911	14.543	1956	23.362	1893	24.017	1884	4.407	1908	6.212	1907	3.970	1913	5.659	1918
5	(4) 14.22	R 65	Summer	−3.391	−2023	−8.625	−2050	−13.819	−1937	−14.313	−1927	−2.898	−2067	−4.114	−2295	−2.918	−2108	−4.131	−2324
			Winter	8.477	2784	18.976	2800	30.401	2550	31.489	2530	5.796	2798	8.228	2838	5.107	2823	7.229	2837
		S 49	Summer	−2.827	−1496	−7.493	−1555	−12.713	−1490	−13.221	−1480	−2.432	−1506	−3.434	−1664	−2.467	−1522	−3.502	−1686
			Winter	7.068	2010	16.485	2119	27.968	1976	29.087	1956	4.864	1997	6.868	1994	4.318	2005	6.129	2014
5	14.23	R 65	Summer	0	−2121	0	−2180	0	−1991	0	−1974	0	−2178	0	−2415	0	−2236	0	−2086
			Winter		3028		3085		2670		2633		3022		3077		3047		3067
		S 49	Summer	0	−1547	0	−1651	0	−1544	0	−1527	0	−1561	0	−1717	0	−1585	0	−1751
			Winter		2138		2330		2095		2058		2106		2102		2114		2128

14.4 Horizontal forces in bridges due to temperature changes

6	14.24	R 65	summer	−3.294	−1499	−7.052	−1967	−13.362	−1969	−13.770	−1962	−3.656	−2001	−5.140	−2207	−3.854	−2020	−5.534	−2235
			winter	8.236	2017	15.514	2616	29.396	2621	30.294	2606	7.312	2666	10.280	2662	6.746	2670	9.682	2681
		S 49	summer	−3.170	−1482	−6.866	−1513	−13.002	−1515	−13.422	−1509	−2.496	−1480	−3.682	−1644	−2.780	−1494	−3.958	−1654
			winter	7.928	1975	15.106	2027	28.606	2032	29.528	2019	4.992	1944	7.364	1954	4.864	1956	6.926	1958
7	(5) 14.25	R 65	summer	−2.084	−1860	−4.933	−1811	−7.514	−1799	−7.964	−1799	−1.937	−1892	−2.708	−2097	−2.030	−1913	−2.851	−2118
			winter	5.211	2376	10.854	2274	16.530	2246	17.522	2247	3.875	2449	5.416	2441	3.552	2482	4.989	2476
		S 49	summer	−1.511	−1397	−3.888	−1369	−7.029	−1368	−7.451	−1367	−1.381	−1407	−1.896	−1564	−1.454	−1412	−2.046	−1570
			winter	3.778	1764	8.553	1710	15.464	1708	16.392	1708	2.761	1799	3.791	1795	2.544	1812	3.580	1811
7	(6) 14.25	R 65	summer	−0.374	−2014	−1.424	−2034	−3.855	−1941	−4.160	−1929	−0.183	−2062	−0.277	−2261	−0.133	−2107	−0.198	−2309
			winter	0.934	2760	3.133	2764	8.481	2559	9.152	2534	0.365	2789	0.554	2769	0.232	2822	0.346	2809
		S 49	summer	−0.551	−1475	−1.546	−1505	−3.061	−1494	−3.315	−1482	−0.369	−1486	−0.564	−1638	−0.315	−1500	−0.453	−1657
			winter	1.377	1957	3.401	2009	6.734	1985	7.293	1960	0.738	1956	1.129	1942	0.552	1967	0.793	1964
7	(7) 14.25	R 65	summer	−1.074	−2104	−2.963	−2187	−5.576	−2059	−5.953	−2055	−0.828	−2143	−1.183	−2341	−0.786	−2190	−1.121	−2391
			winter	2.684	2986	6.519	3102	12.268	2818	13.098	2809	1.655	2950	2.366	2930	1.375	2967	1.961	2953
		S 49	summer	−0.985	−1532	−2.616	−1609	−4.894	−1593	−5.230	−1590	−0.808	−1539	−1.151	−1689	−0.795	−1556	−1.126	−1713
			winter	2.462	2101	5.755	2238	10.767	2204	11.507	2196	1.616	2062	2.301	2045	1.391	2064	1.971	2061

Notes: (1), (2), (3) – see Table 14.4,
(4) $X_F/(L-l_2)$, (5) X_{F1}/L, $p_1 r$,
(6) X_{F2}/L, $p_2 r$, (7) X_{F3}/L, $p_3 r$.

In the case of $l_2 = l_3$ shown in Fig. 14.23, according to equation (14.94)
$$X_F = 0.$$

f) Two-span bridge with long-welded rail (Fig. 14.24)

If the long-welded rail is mounted on a bridge with movable bearings on the intermediate pier, the forces are:

$$X_{F1} = N_3(0) - N_2(0) = -\frac{1}{2} EA\alpha_0 \Delta T \lambda (l_2 + l_3)$$
$$\cdot \left[1 - \frac{1 - e^{-\lambda l_3}}{\lambda (l_2 + l_3)} \right] (1 - e^{-\lambda l_2}), \qquad (14.97)$$

$$X_{F2} = N_3(0) - N_4(0) = \frac{1}{2} EA\alpha_0 \Delta T \lambda (l_2 + l_3)$$
$$\cdot \left[1 - \frac{1 - e^{-\lambda l_2}}{\lambda (l_2 + l_3)} \right] (1 - e^{-\lambda l_3}), \qquad (14.98)$$

$$pr = -\frac{1}{2} EA\alpha \Delta t \left\{ 2 + \frac{\alpha_0 \Delta T}{2\alpha \Delta t} \left[-3 + e^{-\lambda l_2} \right. \right.$$
$$\left. \left. + \left[\lambda (l_2 + l_3) + e^{-\lambda l_3} \right] (1 + e^{-\lambda l_2}) \right] \right\}. \qquad (14.99)$$

These quantities, derived from equation (14.28) have been found from the data of Tables 14.2 to 14.8 for the case of $l_2 = l_3 = L$. The results are given in Table 14.10. In this case $X_{F2} = -X_{F1}$.

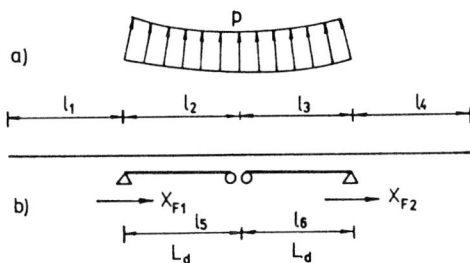

Fig. 14.24. Two-span bridge with long-welded rail and movable bearings on central pier, a) plan, b) evation.

g) Three-span bridge with a long-welded rail (Fig. 14.25)

If the long-welded rail passes along a three-span bridge with alternate bearings, it is possible to derive, according to equation (14.32), the following temperature generated forces

$$X_{F1} = -\frac{1}{2} EA\alpha_0 \Delta T (1 - e^{-\lambda l_2}) \left[-1 + \lambda_2 l_2 + \lambda l_3 e^{-\lambda l_3} \right.$$
$$\left. + (1 + \lambda l_4) e^{-\lambda (l_3 + l_4)} \right], \qquad (14.100)$$

$$X_{F2} = -\frac{1}{2} EA\alpha_0 \Delta T(1 - e^{-\lambda l_3}) \left[\lambda(l_3 - l_2) - e^{-\lambda l_2} + (1 + \lambda l_4) e^{-\lambda l_4}\right], \tag{14.101}$$

$$X_{F3} = -\frac{1}{2} EA\alpha_0 \Delta T(1 - e^{-\lambda l_4}) \left[1 + \lambda(l_4 - l_3) - (\lambda l_2 + e^{-\lambda l_2}) e^{-\lambda l_3}\right], \tag{14.102}$$

$$p_1 r = -\frac{1}{2} EA\alpha \Delta t \left\{ 2 + \frac{\alpha_0 \Delta T}{2\alpha \Delta t} \left[-3 + e^{-\lambda l_2} \right.\right.$$
$$\left.\left. + \left[\lambda l_2 + \lambda l_3 e^{-\lambda l_3} + (1 + \lambda l_4) e^{-\lambda(l_3+l_4)}\right](1 + e^{-\lambda l_2})\right]\right\}. \tag{14.103}$$

$$p_2 r = -\frac{1}{2} EA\alpha \Delta t \left\{ 2 + \frac{\alpha_0 \Delta T}{2\alpha \Delta t} \left[-4 + \lambda(l_2 + l_3) + e^{-\lambda l_2} \right.\right.$$
$$\left.\left. + \left[\lambda(l_2 + l_3) + e^{-\lambda l_2}\right] e^{-\lambda l_3} + (1 + \lambda l_4)(1 + e^{-\lambda l_3}) e^{-\lambda l_4}\right]\right\}. \tag{14.104}$$

$$p_3 r = -\frac{1}{2} EA\alpha \Delta t \left\{ 2 + \frac{\alpha_0 \Delta T}{2\alpha \Delta t} \left[-3 + \lambda(l_3 + l_4) + [1 + \lambda(l_3 + l_4)] \right.\right.$$
$$\left.\left. \cdot e^{-\lambda l_4} + (\lambda l_2 + e^{-\lambda l_2})(1 + e^{-\lambda l_4}) e^{-\lambda l_3}\right]\right\}. \tag{14.105}$$

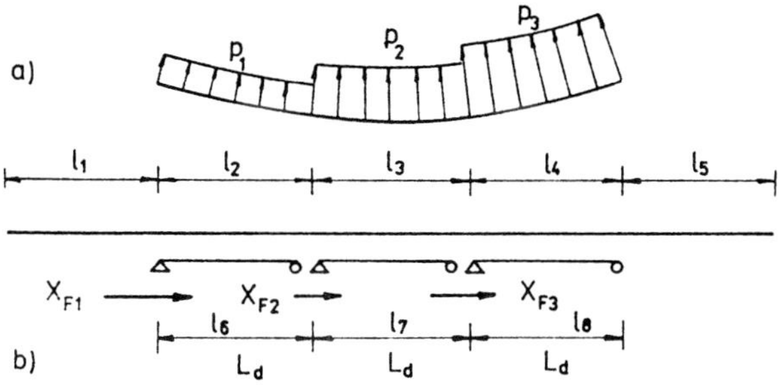

Fig. 14.25. Three-span bridge with long-welded rail and alternated movable and fixed bearings on intermediate piers, a) plan, b) elevation.

The forces (14.100) to (14.105), represented in Fig. 14.25, have been calculated for $l_2 = l_3 = l_4 = L$. The results are given in Table 14.10.

The data may also be applied approximately to bridges of two or four and more spans will alternate bearings with the understanding that the forces X_{F1} and p_1 will be used for the first span with a fixed bearing on the abutment, the forces X_{F3} and p_3 for the last span with a movable bearing on the abutment, and the forces X_{F2} and p_2 for the intermediate spans.

Otherwise, it holds generally for a bridge of n spans with alternate bearings, using the notation according to Fig. 14.25, that the horizontal longitudinal forces X_n under fixed bearings and the horizontal transverse force Y_n can be calculated using the following formulae:

$$X_n = N_{n+2}(0) - N_{n+1}(0),$$
$$Y_n = \left[N_{n+2}(0) + N_{n+1}(0)\right] \frac{l_{n+1}}{2r}. \tag{14.106}$$

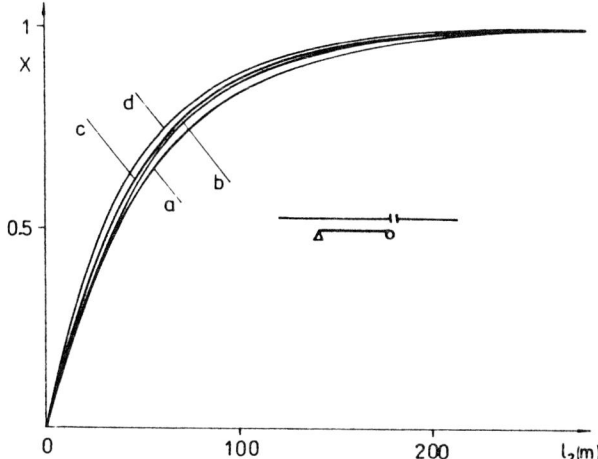

Fig. 14.26. Dimensionless quantity $X = X_F/X_{F0} = pr/P_0$ plotted against bridge span l_2; expansion device above movable bearing (Fig. 14.19); steel bridge with ballast, timber sleepers;
R 65 rails: a) summer, b) winter,
S 49 rails: c) summer, d) winter temperature extremes.

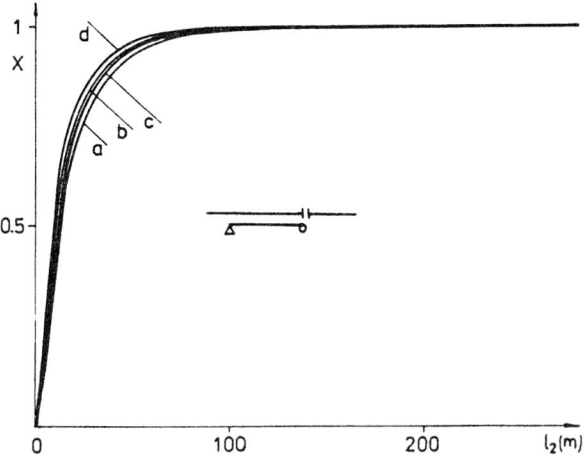

Fig. 14.27. Dimensionless quantity $X = X_F/X_{F0} = pr/P_0$ plotted against bridge span l_2; expansion device above movable bearing (Fig. 14.19); steel bridge without ballast, direct fastening of
R 65 rails: a) summer, b) winter,
S 49 rails: c) summer, d) winter temperature extremes.

14.4 Horizontal forces in bridges due to temperature changes

The temperature-generated forces are calculated for extreme summer and winter temperature ranges Δt and ΔT in Tables 14.9 and 14.10. The direction of the positive action of force X_F on the abutment and the piers, or of the uniformly distributed load p on the bridge is shown in Figs 14.18 to 14.25.

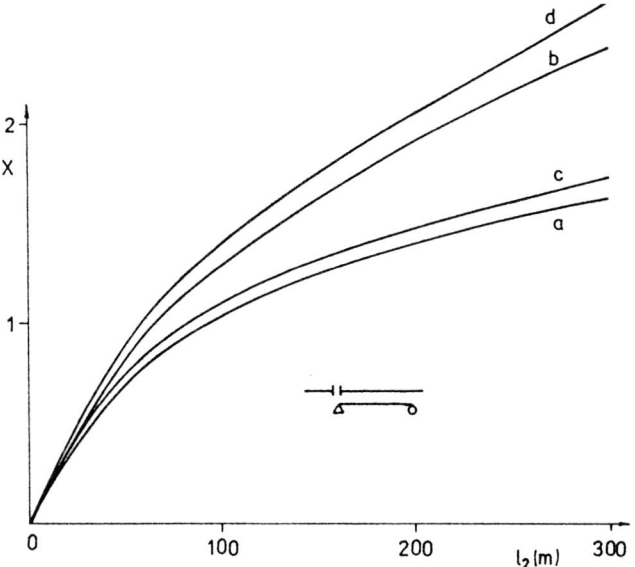

Fig. 14.28. Dimensionless quantity $X = X_F/X_{F0} = pr/P_0$ plotted against bridge span l_2; expansion device above fixed bearing (Fig. 14.20); steel bridge with ballast, timber sleepers;
R 65 rails: a) summer, b) winter,
S 49 rails: c) summer, d) winter temperature extremes.

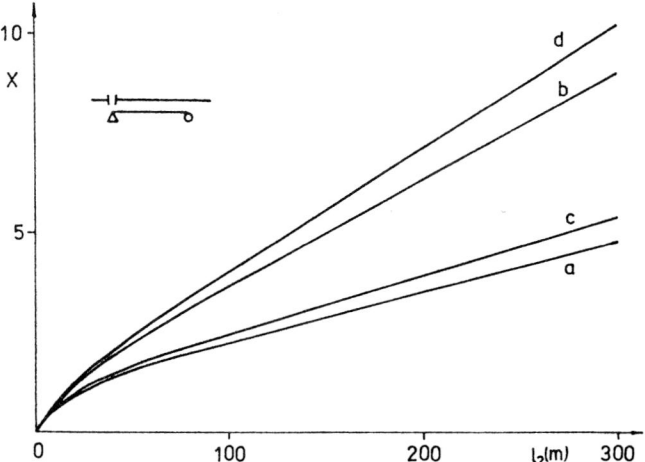

Fig. 14.29. Dimensionless quantity $X = X_F/X_{F0} = pr/P_0$ plotted against bridge span l_2; expansion device above fixed bearing (Fig. 14.20); steel bridge without ballast, direct fastening of
R 65 rails: a) summer, b) winter,
S 49 rails: c) summer, d) winter temperature extremes.

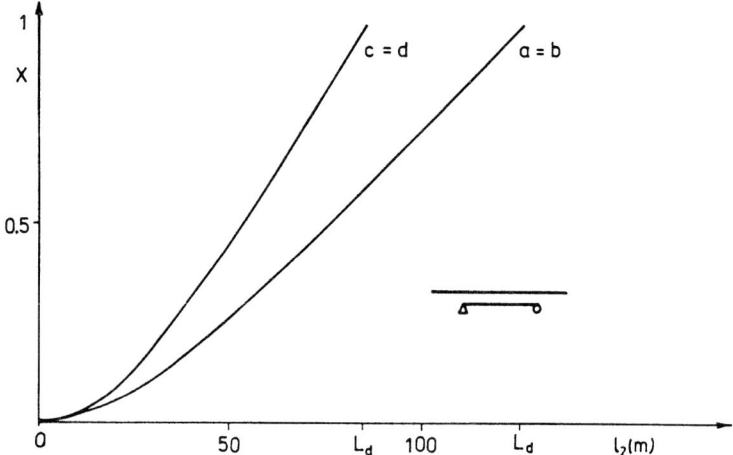

Fig. 14.30. Dimensionless quantity $X = X_F/X_{F0}$ plotted against bridge span l_2 with long-welded rail (Fig. 14.21); steel bridge with ballast, timber sleepers;
R 65 rails: a) summer, b) winter,
S 49 rails: c) summer, d) winter temperature extremes.

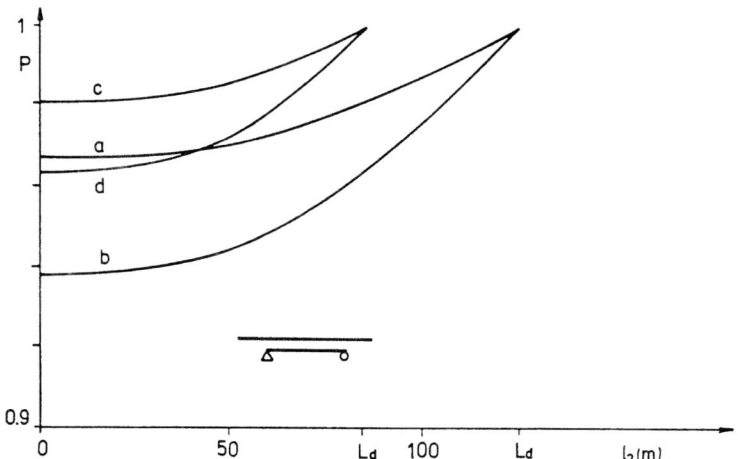

Fig. 14.31. Dimensionless quantity $P = pr/P_0$ plotted against bridge span l_2 with long-welded rail (Fig. 14.21); steel bridge with ballast, timber sleepers;
R 65 rails: a) summer, b) winter,
S 49 rails: c) summer, d) winter temperature extremes.

14.5 Effect of some parameters

The influence of some dimensionless parameters has been studied on steel bridges without ballast (i.e. cases (a) and (e) from Sect. 14.1.2). The following data were considered:

$$\Delta t = -58°C, \quad \alpha = 12 \times 10^{-6} / °C, \quad E = 2.1 \times 10^5 \text{ MPa},$$
$$\Delta T = -55°C, \quad \alpha_0 = 9 \times 10^{-6} / °C, \quad A = 6\,297 \text{ mm}^2 \; (S \; 49).$$

The force (14.18) and the stress in the long-welded rail for this case are as follows

$$N = +2.1 \times 10^2 \times 6\,297 \times 12 \times 10^{-6} \times 58 = +920.4 \text{ kN},$$
$$\sigma = N/A = +146.2 \text{ MPa}.$$

14.5.1 Rail displacement

The dimensionless rail displacement above a movable bearing (14.6), $l_4 = L$, Fig. 14.3,

$$u_3(0)/L = \frac{\alpha_0 \Delta T}{2\lambda L} (\lambda L - 1 + e^{-\lambda L}) \tag{14.107}$$

is shown in Fig. 14.32 as a function of λL.

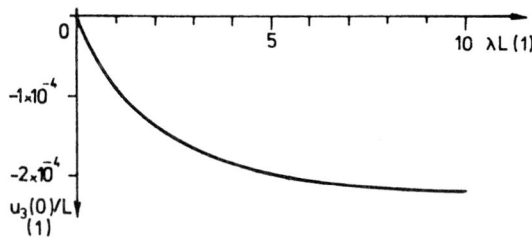

Fig. 14.32. Dimensionless rail displacement above movable bearing of a simply supported beam as a function of λL.

14.5.2 Rail force

The force in the long-welded rail above the last movable bearing attains, according to equations (14.17), (14.25), (14.32), (14.33) and Figs 14.3, 14.5 and 14.7, the following value

$$\frac{N_{n+1}(L)}{N} = 1 + \frac{\alpha_0 \Delta T}{2\alpha \Delta t} (\lambda L - 1 + C_n). \tag{14.108}$$

This function is shown in Fig. 14.33 for n equal spans.

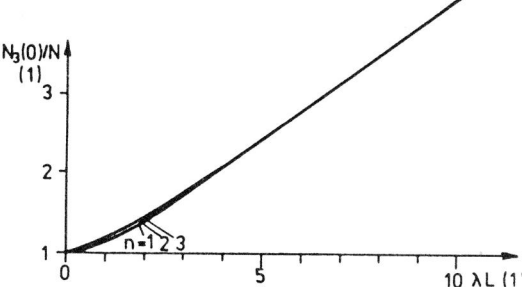

Fig. 14.33. Dimensionless force in long-welded rail above the last movable bridge bearing of a n-span bridge with alternate bearings as a function of λL.

14.5.3 Mutual displacement

Mutual displacement of the bridge and the rail for the case of a bridge with n equal spans with alternate bearings is, from equation (14.35):

$$\frac{u}{L} = \frac{u_{2(n+1)}(L) - u_{n+1}(L)}{L} = \frac{\alpha_0 \Delta T}{2\lambda L}(1 + \lambda L - C_n) \quad (14.109)$$

(see Fig. 14.34).

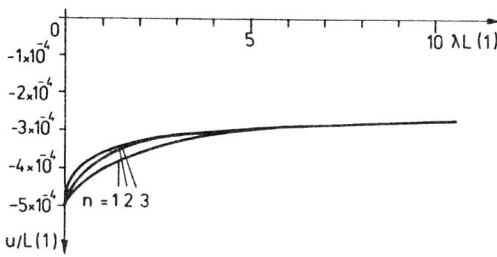

Fig. 14.34. Dimensionless mutual displacement above movable bearing of a n-span bridge with alternated bearings as a function of λL.

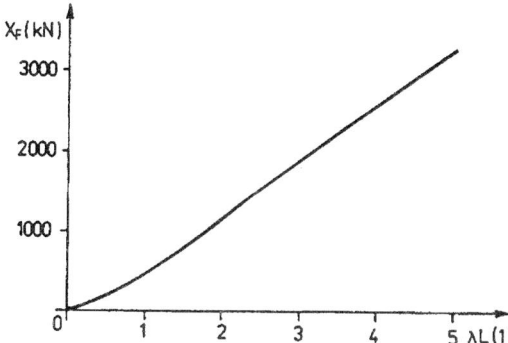

Fig. 14.35. Force X_F transmitted to fixed bearings from two long-welded rails as a function of λL.

14.5.4 Force in fixed bearings

The force transmitted from two rails into fixed bearings is found from equation (14.85), Fig. 14.3a. It is shown as a function of the dimensionless quantity λL in Fig. 14.35.

14.5.5 Uniform load subjecting the bridge with curved rail

The horizontal transverse load p (uniformly distributed) applied to the bridge (from two rails, Fig. 14.21) with long-welded rail in curve of radius r is found from equation (14.86):

$$p = \frac{Y}{l} = -\frac{EA\alpha\Delta t}{2r}\left\{2 + \frac{\alpha_0 \Delta T}{2\alpha\Delta t}\left[-2(1 - e^{-\lambda L}) + \lambda L(1 + e^{-\lambda L})\right]\right\}$$

(14.110)

This equation is shown in Fig. 14.36 as a function of λL and the radius of curvature of the track r.

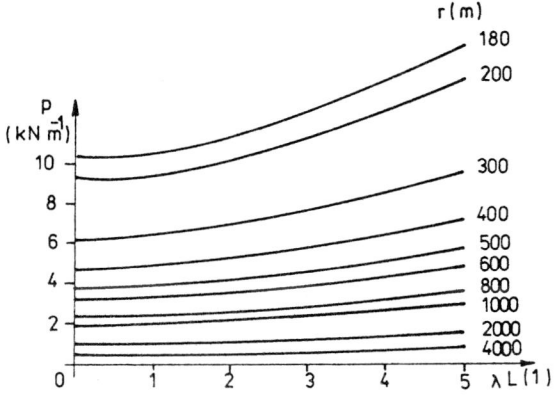

Fig. 14.36. Uniformly distributed horizontal transverse load p affecting two long-welded rails on a bridge with the track in curve as a function of λL for various radii of curvature r.

14.5.6 Strength condition

The stress (14.37) has been calculated for several practical cases and is shown in Fig. 14.37.

14.5.7 Gap condition

The gap condition in the case of rail rupture above a movable bearing (14.48) was adjusted to the form of

$$a = -2\alpha\Delta t\,\frac{L}{\lambda L} + \frac{4L}{m_1 m_2}. \qquad (14.111)$$

It is represented graphically in Fig. 14.38 for several cases of L, m_1, and m_2.

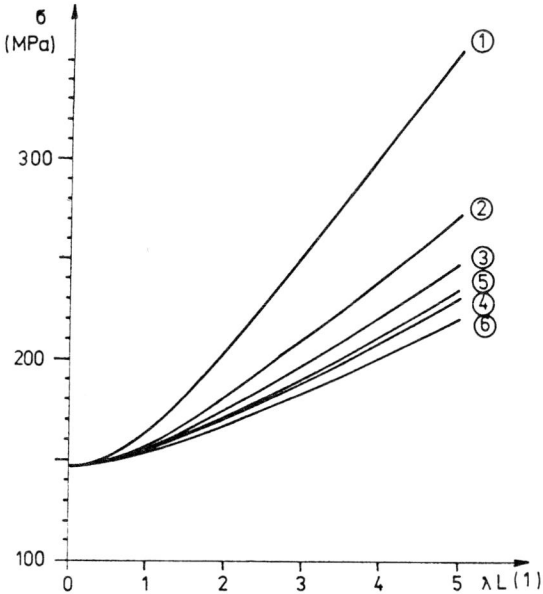

Fig. 14.37. Stress σ (14.37) as a function of λL:
1 – steel bridges without ballast, *2* – steel bridges with ballast, *3* – composite bridges without ballast, *4* – composite bridges with ballast, *5* – reinforced or prestressed concrete bridges without ballast, *6* – reinforced of prestressed concrete bridges with ballast.

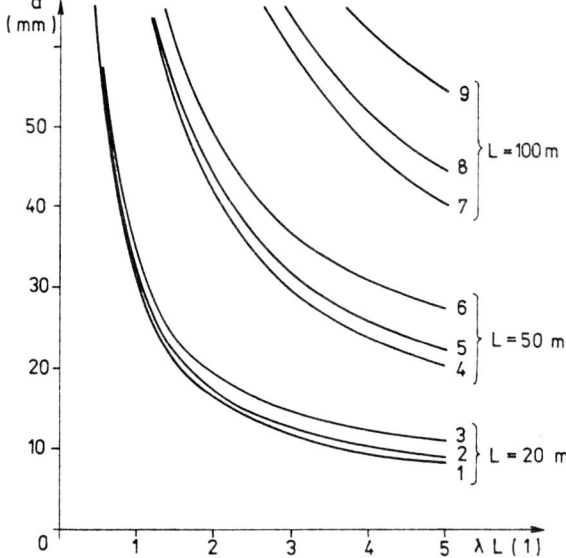

Fig. 14.38. Gap *a* (14.111) as a function of λL:
$1 - L = 20$ m, $m_1 = 8$, $m_2 = 4000$; $2 - L = 20$ m, $m_1 = 12$, $m_2 = 2000$; $3 - L = 20$ m, $m_1 = 15$, $m_2 = 1000$; $4 - L = 50$ m, $m_1 = 8$, $m_2 = 4000$; $5 - L = 50$ m, $m_1 = 12$, $m_2 = 2000$; $6 - L = 50$ m, $m_1 = 15$, $m_2 = 1000$; $7 - L = 100$ m, $m_1 = 8$, $m_2 = 4000$; $8 - L = 100$ m, $m_1 = 12$, $m_2 = 2000$; $9 - L = 100$ m, $m_1 = 15$, $m_2 = 1000$.

14.5.8 Mutual displacement condition

The condition of mutual displacement between the bridge and the long-welded rail (14.50), adjusted to the form of

$$b = \frac{\alpha_0 \Delta T L}{2\lambda L} (1 + \lambda L - e^{-\lambda L}), \qquad (14.112)$$

is represented in Fig. 14.39 for several practical cases.

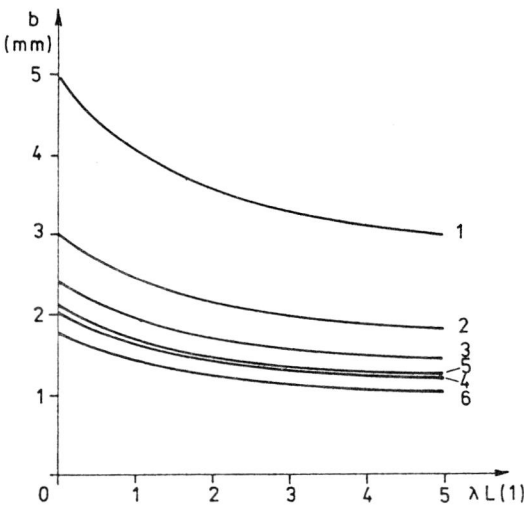

Fig. 14.39. Mutual displacement b (14.112) of a bridge with span $L = 10$ m as a function of λL (for other spans, b should be multiplied by $L/10$):
1 – steel bridges without ballast, *2* – steel bridges with ballast, *3* – composite bridges without ballast, *4* – composite bridges with ballast, *5* – reinforced or prestressed concrete bridges without ballast, *6* – reinforced or prestressed concrete bridges with ballast.

14.6 Conclusions for the application of long-welded rail on bridges

Railway bridges influence the stresses and displacements of long-welded rail, especially above movable bridge bearings. For these reasons, it is necessary to critically assess the use of long-welded rail, the limiting factor being the expansion length of the bridge.

The maximum admissible expansion bridge length is determined on the basis of four conditions: strength, gap in the case of rail rupture, mutual rail and bridge displacement, and stability. The strength and gap conditions decide the admissible expansion length of bridges with ballast, while the condition of mutual displacement determines it in the case of bridges without ballast. The stability condition is complied with, if a certain stress in the rail has not been exceeded.

The admissible expansion length of bridges depends on the fastening of rails and on its maintenance, on rail cross section area, on the presence of ballast, on the material of the bridge and on the maximum difference of temperature of the rails and of the bridge from the fixing temperature.

During temperature changes the long-welded rail also introduces horizontal longitudinal forces into the fixed bearings of bridges, in piers and abutments. If the track on the bridge is curved, a horizontal transverse load also originates which is inversely proportional to the radius of curvature of the track with long-welded rail.

Bibliography

[1] ABDEL-ROHMAN, M., NAYFEH, A. H.: Active control of nonlinear oscillations in bridges. Journal of Engineering Mechanics, 113 (1987), No. 3, 335–348.
[2] ANDREEV, V. G., GLYBINA, G. K.: Response of railway bridges to longitudinal forces. Transportnoe stroiteľstvo, 1973, No. 5, 42–43. (Андреев, В. Г., Глыбина, Г. К.: Работа железнодорожных мостов на продольные силы. Транспортное строительство, 1973, № 5, 42–43.).
[3] ARBABI, F., LI, F.: Effect of nonlinear parameters on stresses in railroad tracks. Journal of Structural Engineering, 114 (1988), No. 1, 165–183.
[4] AYRE, R. S., FORD, G., JACOBSEN, L. S.: Transverse vibration of a two-span beam under action of a moving constant force. Journal of Applied Mechanics, 17 (1950), No. 1, 1–12.
[5] AYRE, R. S., JACOBSEN, L. S.: Transverse vibration of a two-span beam under the action of a moving alternating force. Journal of Applied Mechanics, 17 (1950), No. 3, 283–290.
[6] BARCHENKOV, A. G.: Dynamic analysis of highway bridges. Transport, Moscow 1976. (Барченков, А. Г.: Динамический расчет автодорожных мостов. Транспорт, Москва 1976).
[7] BARSOM, J. M.: Fatigue behaviour of weathered steel components. Transport Research Records, 1984, No. 950/2, 1–10.
[8] BAŤA, M., PLACHÝ, V.: Analysis of dynamic effects on engineering structures. Elsevier, Amsterdam 1987.
[9] BAUMANN, M., BRIANZA, M., ENSNER, K., THÜRLIMANN, B.: Versuche an der ersten vorgespannten Eisenbahnbrücke der Schweiz. Edigenossiche Technische Hochschule Zürich, 1987, Bericht Nr. 8301-1.
[10] BETZHOLD, CH.: Erhöhung der Beanspruchungen des Eisenbahnoberbaues durch Wechselwirkungen zwischen Fahrzeug und Oberbau. Glasers Annalen, 81 (1957), No. 3, 76–82, No. 4, 108–115, No. 5, 137–145.
[11] BHATTI, M. H., GARG, V. K., CHU, K. H.: Dynamic interaction between freight train and steel bridge. Journal of Dynamic Systems, Measurement and Control, Transactions of ASME, 107 (1958), No. 1, 60–66.
[12] BIGGERS, S. B., WILSON, J. F.: Dynamic interactions of high speed tracked air cushion vehicles with their guideways – a parametric study. AIAA Paper, 1971, No. 386, 12 p.
[13] BILLING, J. R.: Dynamic loading and testing of bridges in Ontario. Can. J. Civ. Eng., 11 (1984), No. 11, 833–843.
[14] BIRGER, I. A., PANOVKO, Y. G. (eds): Strength, stability, vibrations. Mashinostroenie, Moscow 1968, Vol. 1, 2, 3. (Биргер, И. А., Пановко, Я. Г. (ред.): Прочность, устойчивость, колебание. Машиностроение, Москва 1968, том 1, 2, 3).
[15] BIRMANN, F.: Federung und Dämpfung im Oberbau. Neuere Erkenntnisse aus dem In- und Ausland. Leichtbau der Verkehrsfahrzeuge, 23 (1979), No. 5, 113–121.
[16] BLEICH, J., ROSENKRANS, R., VINCENT, G. S., COLLOUGH, C. B.: The mathematical theory of vibration in suspension bridges. U. S. Gov. Print. Office, Washington 1950.
[17] BOGACZ, R., POPP, K.: Dynamics and stability of train-track-system. 2nd International Conference on Recent Advances in Structural Dynamics, ISVR Southampton, 1984.
[18] BOGDANOFF, J. L.: A new cumulative damage model. Journal of Applied Mechanics, 45 (1978), No. 2, 246–250, 251–257, No. 4, 733–739.

[19] BOLOTIN, V. V.: Statistical methods in structural mechanics. Strojizdat, Moscow 1965, 2nd edition. (Болотин, В. В.: Статистические методы в строителной механике. Стройиздат, Москва 1965, 2. изд.).
[20] BOLOTIN, V. V.: Random vibrations of elastic systems. Nauka, Moscow 1979. (Болотин, В. В.: Случайные колебания упругих систем. Наука, Москва 1979).
[21] BOLOTIN, V. V.: A unified approach to damage accumulation and fatigue crack growth. Eng. Fract. Mech., 22 (1985), No. 3, 387–398.
[22] BONDAR, N. G., KAZEI, I. I., LESOKHIN, B. F., KOZMIN, Y. G.: Dynamics of railway bridges. Transport, Moscow 1965. (Бондарь, Н. Г., Казей, И. И., Лесохин, Б. Ф., Козьмин, Ю. Г.: Динамика железнодорожных мостов. Транспорт, Москва 1965).
[23] BONDAR, N. G., KOZMIN, YU. G., ROĬTBURD, Z. G., TARASENKO, V. P., YAKOVLEV, G. N.: Interaction of railway bridges with vehicles. Transport, Moscow 1984. (Бондарь, Н. Г., Козьмин, Ю. Ф., Ройтбурд, З. Г., Тарасенко, В. П., Яковлев, Г. Н.: Взаимодействе железнодорожных мостов с подвыжным составом. Транспорт, Москва 1984).
[24] BRAMALL, B.: Longitudinal forces on bridges. Railway Engr. Int., 4 (1979), No. 1, 45–46.
[25] BRAUNE, W., WEBER, W.: Zur Dämpfung einfeldriger, verkehrslastfreier Eisenbahnbrücken. Bericht, Bundesbahn-Zentralamt, München 1976, 1977.
[26] Bridge Subcommittee Reports. Government of India Central Publication Branch, Calcutta. Techn. Paper, 1926, No. 247.
[27] BRÜCKMANN, B.: Brückenmesswesen, Brückenschwingungen und Brückenbelastbarkeit. Der Eisenbahnbau, 3 (1950), No. 10, 230–235, No. 11, 250–254.
[28] BYERS, W. G.: Impact from railway loading on steel girder spans. J. Struct. Div., Proc. ASCE, 96 (1970), No. 6, 1083–1103, errata 97 (1971), No. 4, p. 1365.
[29] BYERS, W. G.: Frequency of railway bridge damage. Journal of Structural Engineering, 111 (1985), No. 8, 1635–1646.
[30] CANTIENI, R.: Dynamische Belastungsversuche an der Bergspurbrücke Deibüel. EMPA, Dübendorf, 1988, Bericht Nr. 116/4 A, B.
[31] CASSÉ, M.: La détermination des effects dynamiques dans les ponts. Rév. gén. Chem. de fer, 77 (1958), No. 7/8, 406–408.
[32] CHAMBRON, E.: Les ouvrages d'art de la ligne nouvelle. Révue générale de Chemins de fer, 95 (1976), No. 11, 717–732.
[33] CHELOMEI, V. N. (ed.): Vibrations in technics. Mashinostroenie, Moscow 1978–1981, Vol. 1–6. (Челомей, В. Н. (ред.): Вибрации в технике. Машиностроение, Москва 1978–1981, том 1–6).
[34] CHILVER, A. H.: A note on the Mise-Kunii theory of bridge vibration. Quarterly Journ. Mech. and Applied Math., 9 (1956), Nr. 2. 207–211.
[35] CHRISTIANO, P. P., Culver, C. G.: Horizontally curved bridges subjected to moving load. J. Struct. Div., Proc. ASCE, 95 (1969), No. 8, 1615–1643, discussion 96 (1970), No. 4, 861–863, No. 11, 2524–2527.
[36] CHU, K. H., GARG, V. K., BHATTI, M. H.: Impact in truss bridge due to freight trains. Journal of Engineering Mechanics, 111 (1985), No. 2, 159–174.
[37] CHU, K. H., GARG, V. K., WANG, T. L: Impact in railway prestressed concrete bridges. Journal of Structural Engineering, 112 (1986), No. 5, 1036–1051.
[38] CLARK, R. A., DEAN, P. A., ELKINS, J. A., NEWTON, S. G.: An investigation into the dynamic effects of railway vehicles running on corrugated rails. Journal of Mechanical Engineering Science, 24 (1982), No. 2, 65–76.
[39] CLORMANN, U. H., SEEGER, T.: Rain-flow-HCM, ein Zählverfahren für Betriebsfestigkeitsnachweis auf werkstoffmechanischer Grundlage. Stahlbau, 55 (1986), No. 3, 65–71.
[40] COUSSY, O., MAHMOUD, S.: Dynamique des ouvrages d'art sous charges mobiles. Bull. liais. Lab. ponts et chaussées, 1984, No. 131, 117–121, 141, 143, 145, 148.

[41] CRANDALL, S. H. (ed.): Random vibration. Vol. I, II. The M.I.T. Press, Cambridge, Mass. 1958, 1963.
[42] CRANDALL, S. H., MARK, W. D.: Random vibration in mechanical systems. Academic Press, New York, London 1963.
[43] DAHLBERG, T.: Vehicle – bridge interaction. Vehicle System Dynamics, 13 (1984), No. 5, 187–206.
[44] DAHLBERG, T.: Dynamic interaction between train and track – a literature survey. Chalmers Tekniska Högskola, Skrift F 120, Göteborg 1989.
[45] DÄHN, J.: Schwingungen der Trägermitte des beiderseitig frei aufliegenden Trägers unter rollender schwingungsfähiger Last. Wiss. Zeitsch. Humboldt Univ. Berlin, Math.-naturwiss. Reihe, 13 (1964), No. 5, 869–880.
[46] DANIELSKI, L., RABIEGA, J.: Betriebsbelastungen von Eisenbahnbrücken. Proceedings of the IABSE Colloquium 'Fatigue of Steel and Concrete Structures', Lausanne 1982, 833–839.
[47] DANILENKO, E.I., FROLOV, L. N., ROMANOV, V. M., GRACHEV, A. V., MORAS, E.: Effect of vertical load on horizontal forces in the track. Vestnik VNIIZhT, 1979, No. 1, 41–44. (ДАНИЛЕНКО, Е. И., ФРОЛОВ, Л. Н., РОМАНОВ, В. М., ГРАЧЕВ, А. В., МОРАС, Е.: Влияние вертикальной нагрузки при измерении горизонтальных сил в пути. Вестник ВНИИЖТ, 1979, № 1, 41–44).
[48] DESAI, C. S., SIRIWARDANE, A. M.: Numerical models for track support structures. Journal of the Geotechnical Engineering Division, ASCE, 108 (1982), No. 3, 461–480.
[49] DIN 45 667: Klassierverfahren für das Erfassen regelloser Schwingungen. 1969.
[50] DOUGLAS, B. M., REID, W. H.: Dynamic tests and system identification on bridges. Journal of Structural Engineering, 108 (1982), No. 10, 2295–2312.
[51] DOWLING, N. E.: Fatigue failure predictions for complicated stress – strain histories. J. Mater. 7 (1972), No. 1, 71–87.
[52] Dynamics of steel elevated guideways – an overview. Journal of Structural Engineering, 111 (1985), No. 9, 1873–1898.
[53] The Dynamic behaviour of concrete structures. RILEM, Recommendations of good practice for methods of testing and design. Matériaux et constructions, 17 (1984), No. 101, 395–400.
[54] ENDO, T., SATOH, N., YAMAUCHI, T., KONDO, Y., TAKAYAMA, H.: Fatigue damage evaluation for steel bridges. Mitsubishi Heavy Ind. Techn. Rev., 25 (1988), No. 1, 30–36.
[55] Ergebnisse der experimentellen Brückenuntersuchungen in der USSR. Forchungsarbeiten des Wissenschaftlich-technischen Komitees des Volkskomissariats für Verkehrswesen, Band 89, Transpetschat, Moskau 1928.
[56] Eurocode 3, Chapter 9 Fatigue, 1988.
[57] Fatigue and fracture reliability: a state-of-the-art review. Journal of the Structural Division, Proceedings ASCE, 108 (1982), No. 1, 1–88.
[58] FILIPPOV, A. P., KOKHMANYUK, S. S.: Dynamic effects of moving loads on bars. Naukova dumka, Kiev 1967, (ФИЛИППОВ, А. П., КОХМАНЮК, С. С.: Динамическое воздействие подвижных нагрузок на стержин. Наукова думка. Киев 1967).
[59] FISHER, J. W., ALBRECHT, P. A., YEN, B. T., KLINGERMANN, D. J., MCNAMEE, B. M.: Fatigue strength of steel beams with welded stiffeners and attachments. National cooperative highway research program report 147. Transportation Research Board, Washington 1974.
[60] FISHER, J. W., MERTZ, D. R., ZHONG, A.: Steel bridge members under variable amplitude long life fatigue loading. Nat. Coop. Highway Res. Program Rep., 1983, No. 267, 26 p.
[61] FISHER, J. W.: Fatigue and fracture in steel bridges. John Wiley and Sons, New York 1984.
[62] FRANKE, L.: Schadensakkumulationsregel für dynamisch beanspruchte Werkstoffe und Bauteile. Bauingenieur, 60 (1985), No. 7, 271–279.
[63] FRÝBA, L.: Schwingungen des unendlichen, federnd gebetteten Balkens unter der Wirkung eines unrunden Rades. Zeitschrift für angewandte Mathematik und Mechanik, 40 (1960), No. 4, 170–184.

[64] FRÝBA, L.: Les efforts dynamiques des ponts-rails métalliques. Bulletin mensuel de l'Association Internationale du Congrès des Chemins de fer, 40 (1963), No. 5, 367–403.
[65] FRÝBA, L.: Příčníkový efekt ocelových železničních mostů (Cross-beam effect on steel railway bridges.) Inženýrské stavby, 14 (1966), No. 3, 115–118.
[66] FRÝBA, L.: Dynamischer Einfluss der unrunden Räder auf die Eisenbahnbrücken. Monatschrift der Internationalen Eisenbahn-Kongress-Vereinigung, 44 (1967), No. 5, 353-389.
[67] FRÝBA, L.: Impacts of two-axle system traversing a beam. International Journal of Solids and Structures, 4 (1968), No. 11, 1107–1123.
[68] FRÝBA, L.: Vibration of solids and structures under moving loads. Academia, Prague, Noordhoff International Publishing, Groningen 1972.
[69] FRÝBA, L.: Effect of a force moving at variable speed along a beam. Acta technica ČSAV, 18 (1973), No. 1, 54–68.
[70] FRÝBA, L.: Ohybové a podélné namáhání prutů (Bending and longitudinal stresses in bars). Stavebnícky časopis SAV, 22 (1974), No. 10, 593–602.
[71] FRÝBA, L.: Response of a beam to a rolling mass in the presence of adhesion. Acta technica ČSAV, 19 (1974), No. 6, 673–687.
[72] FRÝBA, L.: Quasi-static distribution of braking and starting forces in rails and bridge. Rail International, 5 (1974), No. 11. 698–716.
[73] FRÝBA, L.: Non-stationary response of a beam to a moving random force. Journal of Sound and Vibration, 46 (1976), No 3, 323–338.
[74] FRÝBA, L: Stationary response of a beam to a moving continuous random load. Acta technica ČSAV, 22 (1977), No. 4, 444–479.
[75] FRÝBA, L.: Rozjezdové a brzdné síly na železničních mostech (Starting and braking forces in railway bridges). Stavebnícky časopis SAV, 27 (1979), No. 12, 865–894.
[76] FRÝBA, L.: Estimation of fatigue life of railway bridges under traffic loads. Journal of Sound and Vibration, 70 (1980), No. 4, 527–541.
[77] FRÝBA, L.: Railway bridges subjected to traffic loads and their design for fatigue. Rail International, 11 (1980), No. 10, 573–598.
[78] FRÝBA, L.: Experimentální výzkum termického spolupůsobení bezstykové koleje a železničních mostů (Experimental research of thermal interaction of long welded rail with railway bridges). Sborník prací VÚŽ, 1983, No. 27, 25–74.
[79] FRÝBA, L.: Thermal interaction of long welded rails with railway bridges. Rail International, 16 (1985), No. 3, 5–24.
[80] FRÝBA, L.: Dynamic interaction of vehicles with tracks and roads. Vehicle System Dynamics, 16 (1987), No. 3, 129–138.
[81] Garg, V. K., Dukkipati, R. V.: Dynamics of railway vehicle systems. Academic Press, Toronto 1984.
[82] GARG, V. K., CHU, K. H., WANG, T. L.: A study of railway bridge/vehicle interaction and evaluation of fatigue life. Earthquake Engineering and Structural Dynamics, 13 (1985), No. 6, 689–709.
[83] GENIN, J., CHUNG, Y. I.: Response of a continuous guideway on equally spaced supports traversed by moving vehicle. Journal of Sound and Vibration, 67 (1979), No. 2, 245–251.
[84] GENIN, J., TING, E. C.: Vehicle-guideway interaction problems. Shock and Vibration Digest, 11 (1979), No. 12, 3–9.
[85] GENIN, J., TING, E. C.: Curved bridge response to a moving vehicle. Journal of Sound and Vibration, 81 (1982), No. 4, 469–475.
[86] GERASCH, W. J.: Schwingungsreduziering durch den Einbau eines Schwingungstigers in eine Spannbetonbrücke, Bauingenieur, 60 (1985), No. 2, 59–64.
[87] GESUND, H., YOUNG, D.: Dynamic response of beams to moving loads. Mém. Ass. intern. ponts et charpentes, 21 (1961), 95-110.

[88] GHOSN, M., MOSES, F: Markov renewal model for maximum bridge loading. Journal of Engineering Mechanics, 111 (1985), No. 9, 1093–1104.
[89] GIORGIO, D., FEDERICO, C.: A numerical method to define the dynamics behaviour of a train running on a deformable structure. Meccanica, 23 (1988), special issue, 27–42.
[90] GŁOMB, J.: O pracy dynamicznej mostów drogowych (Dynamic response of railway bridges). Archiwum inżynierii ladowej, 10 (1964), No. 1, 19–33.
[91] GŁOMB, J., WESELI, J.: Wpływ efektów dynamicznych w mostach metalowych na stawy graniczne nosnosci (Dynamic effects in steel bridges on their limit states). Archiwum inżynierii ladowej, 32 (1986), No. 1, 63–76.
[92] GOGELIYA, T. I.: Dynamic analysis of structures under moving loads using the finite element method. Soobshcheniya AN GSSR, 115 (1984), No. 1, 121–124. (Гогелия, Т. И.: Динамический расчет конструкций на подвиые нагрузки с применением метода конечных элементов. Сообщения АН ГССР, 115 (1984), № 1, 121–124).
[93] GÜNTHER, W. und Koll.: Schwingfestigkeit. Deutscher Verlag für Grundstoffindustrie, Leipzig 1973.
[94] HÄNEL, B.: Der Betriebsfestigkeitsnachweis in- und ausländischer sowie internationaler Berechnungsvorschriften des Stahlbaues. IfL-Mitt., 23 (1984), No. 6, 180–194.
[95] HARRIS, C. M., CREDE, C. E. (eds): Shock and vibration handbook. McGraw-Hill, New York, Toronto, London 1961, Vol. 1, 2, 3.
[96] HELLAN, K.: Introduction to fracture mechanics. McGraw-Hill, New York 1984.
[97] HERZOG, M.: Abschätzung der Restlebensdauer älterer genieteter Eisenbahnbrücken. Stahlbau, 54 (1985), No. 10, 309–312.
[98] HILLERBORG, A.: Dynamic influences of smoothly running loads on simply supported girders. Kungl. Tekniska Högskolan, Stockholm 1951.
[99] HINO, J., YOSHIMURA, T., ANANTHANARAYANA, N.: Vibration analysis of non-linear beams subjected to a moving load using the finite element method. Journal of Sound and Vibration, 100 (1985), No. 4, 477–491.
[100] HIRT, M. A., HAUSAMMANN, H.: Betriebsfestigkeit von Eisenbahnbrücken in Verbundweise am Beispiel der Morobbia-Brücke. École polytechnique fédérale de Lausanne, report ICOM 027, 1976
[101] HIRT, M. A.: Neue Erkenntnisse auf dem Gebiet der Ermüdung und deren Berücksichtigung bei der Bemessung von Eisenbahnbrücken. Bauingenieur, 52 (1977), 255–262.
[102] HIRT, M. A., KUMMER, E.: Die Ermüdungswirkung der Betribslasten von Eisenbahnbrücken aus Stahl anhand der Messungen an der Brücke Oberrüti. École polytechnique fédérale de Lausanne, report ICOM 063, 1979.
[103] HOBBACHER, A.: Cumulative fatigue by fracture mechanics. Journal of Applied Mechanics, 44 (1977), No. 4, 769–771.
[104] HOSHIYA, M., MARUYAMA, O.: Identification of running load and beam system. Journal of Engineering Mechanics, 113 (1987), No. 6, 813–824.
[105] INBANTHAN, M. J., WIELAND, M.: Bridge vibration due to vehicle moving over rough surface. Journal of Structural Engineering, 113 (1987), No. 9, 1994–2008.
[106] INGLIS, C. E.: A mathematical treatise on vibration in railway bridges. The University Press, Cambridge 1934.
[107] ITO, M., KITAGAWA, M.: Response of multi-span suspension bridges to a moving concentrated load. Annu. Rept. Eng. Res. Inst. Fac. Eng. Univ. Tokyo, 29 (1970), 11–18.
[108] ITOH, F.: Fatigue damage estimation of railway bridge members. Quarterly Reports, 11 (1970), No. 1, 13–18.
[109] IWANKIEWICZ, R., SNIADY, P.: Vibration of a beam under a random stream of moving forces. Journal of Structural Mechanics, 12 (1984), No. 1, 13–26.
[110] IZUKA HIROSHI: A statistical study on life time of bridges. Proc. Jap. Soc. Civ. Eng., 1988, No. 392, 73–82.

[111] JEFFCOTT, H. H.: On the vibration of beams under the action of moving loads. Philosophical Magazine, ser. 7, 8 (1929), No. 48, 66–97.
[112] JEZEQUEL, L.: Response of periodic systems to a moving load. Journal of Applied Mechanics, 48 (1981), 613–618.
[113] KAMKE, E.: Differentialgleichungen, Lösungsmethoden und Lösungen, I. Gewöhnliche Differentialgleichungen. Akademische Verlagsgesellschaft Geest & Portig K.–G., Leipzig 1956, 5th ed.
[114] KAZAKEVICH, M.I.: Aerodynamics of bridges. Transport, Moscow 1987. (КАЗАКЕВИЧ, М. И.: Аэродинамика мостов. Транспорт, Москва 1987).
[115] KAZEI, I. I.: Dynamic analysis of railway bridges. Transzheldorizdat, Moscow 1960. (КАЗЕЙ, И. И.: Динамический расчет пролетных строений железнодорожных мостов. Трансжелдориздат, Москва 1960).
[116] KEIJI, Y., KATSUHIDE, T.: On analysis of lateral forced vibration of rail vehicle truck running on rails with sinusoidal irregularities. Bull. JSME, 28 (1985), No. 235, 139–147.
[117] KERR, A. D., SHENTON III, H. W.: Railroad track analysis and determination of parameters. Journal of Engineering Mechanics, 112 (1986), No. 11, 1117–1134.
[118] KERR, A. D., EL-SIBAIE, M. A.: On the new equations for the lateral dynamics of rail-tie structure. J. Dyn. Syst., Meas. and Contr., 109 (1987), No. 2. 180–185.
[119] KLESNIL, M., LUKÁŠ. P.: Fatigue of metalic materials. Academia, 2nd revised edition, Prague 1992.
[120] KOLOUŠEK, V.: Dynamic in Engineering Structures. Academia, Prague, Butterworth, London 1973.
[121] KOMÍN, S.: Snižování účinků železničního provozu na kolej (Reducing of effects of railway traffic on track). NADAS, Prague 1964.
[122] KORENEV, B. G., RABINOVICH, I. M. (edited): Handbook of dynamics of structures. Strojizdat, Moscow 1972. (КОРЕНЕВ, Б. Г., РАБИНОВИЧ, И. М. (ред.): Справочник по динамике сооружений. Стройиздат. Москва 1972).
[123] KORN, G. A., KORN, T. M.: Mathematical handbook for scientists and engineers. McGraw-Hill, New York 1968, 2nd ed.
[124] KORTÜM, W., WORMLEY, D. N.: Dynamic interactions between travelling vehicles and guideway systems. Vehicle System Dynamics, 10 (1981), 285–317.
[125] KORTÜM, W.: Vehicle response on flexible track. IMechE, C 405/84, 47–58, 1984.
[126] KORTÜM, W., SCHIEHLEN, W.: General purpose vehicle system dynamics software based on multibody formalisms. Vehicle System Dynamics, 14, (1985), No. 4–6, 229–263.
[127] KOVAŘÍK, R., KOMÍN, S., LÍNEK, O., MAREK, J., SLADKÝ, R.: Bezstyková kolej (Long welded rail). NADAS, Prague 1970.
[128] I. Krajowa konferencja naukowa w sprawie dynamiki mostow. (1st national scientific conference on dynamics of bridges). Gliwice 1965. Zeszyty naukowe Politechniki Slaskiej, No. 201, seria Budownictwo, No. 20, 1967, 1–224.
[129] KROPÁČ, O.: Náhodné jevy v mechanických soustavách (Random events in mechanical systems). SNTL, Prague 1987.
[130] KRYLOV, A. N.: Über die erzwungenen Schwingungen von gleichförmigen elastischen Stäben. Mathematische Annalen, 61 (1905), 211.
[131] LANGER, J., KLASZTORNY, M.: Dynamiczne wytezenie pomostów w belkowych mostach kolejowych (Dynamic effects on railway girder bridges). Arch. inž. lad., 29 (1983), No. 3, 243–262.
[132] LEVY, S., WILKINSON, J. P. D.: The component element method in dynamics. McGraw-Hill, New York 1976.
[133] LIN, Y. K.: Probablilistic theory of structural dynamics. McGraw-Hill, New York 1967.
[134] LIN, Y. K., YANG, J. N.: A stochastic theory of fatigue crack propagation. AIAA Journal, 23 (1985), No. 1, 117–124.

[135] LINDGREN, G., RYCHLIK, I.: Rain flow cycle distributions for fatigue life prediction under Gaussian load processes. Fatigue and Fracture of Engineering Materials and Structures, 10 (1987), No. 3, 251-260.

[136] LUTES, L. D., CORAZAO, M., HU, S. J., ZIMMERMANN, J.: Stochastic fatigue damage accumulation. J. Struct. Eng., 110 (1984), No. 11, 2585–2601.

[137] MAARSCHALKERWAART, H. M. C. M. VAN: Fatigue-crack-control of riveted steel bridges and brittle fracture risk of Thomas-steel components. Proceedings of the conference Traffic Effects on Structures and Environment, High Tatras, 1987, Vol. 1, 53–62.

[138] MARQUARD, E.: Zur Berechnung von Brückenschwingungen unter rollenden Lasten. Ingenieur-Archiv, 23 (1955), No. 1, 19–35.

[139] MATSUISHI, M., ENDO, T.: Fatigue of metals subjected to varying stress. Kyushu District meeting of Japan Society of Mechanical Engineers, Japan, March 1968.

[140] MATSUURA, A.: A theoretical analysis of dynamic response of railway bridge to the passage of rolling stock. Quarterly Report of Railway Technical Research Institute, Tokyo, 11 (1970), No. 1, 18–21.

[141] MATSUURA, A.: Dynamic behaviour of bridge girder for high speed railway bridge. Transactions of JSCE, 9 (1977), 294–298.

[142] MELCER, J.: Závislosť dynamického súčinitela mosta od rýchlosti pohybu vozidla (Dependance of the dynamic impact factor of a bridge on the speed of the vehicle). Stavebnícky časopis SAV, 36 (1988), No. 1, 3–18.

[143] MERCER, C. A., LIVESAY, J.: Statistical counting methods as a means of analysis the load histories of light bridges. Journal of Sound and Vibration, 27 (1983), No. 3, 399–410.

[144] MILDNER, M., QUOOS, V.: Betriebsfestigkeitsnachweiss von Eisenbahnbrücken aus Stahlbeton und Spannbeton nach TGL 42702/01. Signal und Schiene, 29 (1985), No. 2, 68–70, 75.

[145] MILDNER, K., QUOOS, V.: Zum Betriebsfestigkeitsnachweiss von Eisenbahnbrücken aus Stahlbeton und Spannbeton. Die Strasse, 25 (1985), No. 5, 152–157.

[146] MINER, M. A.: Cumulative damage in fatigue. Journal of Applied Mechanics, 12 (1945), No. 3, 159–164.

[147] MISE, K., KUNII, S.: A theory for the forced vibrations of railway bridge under the action of moving loads. Quart. Journ. Mech. Applied Math., 9 (1956), No. 2, 195–206.

[148] MOHAMAD, A. R., MAYFEH, A. H.: Active control of nonlinear oscillations in bridges. J. Eng. Mech., 113 (1987), No. 3, 335–348.

[149] MOHAMMADI, J., ADELI, H.: Structural fatigue failure under random vibration. Civ. Eng. Pract. and Des. Eng., 4 (1985), No. 7, 551–562.

[150] Napetvaridze, Sh. G. (editor): Seismicity of transport and net structures. Nauka, Moscow 1986. (НАПЕТВАРИДЗЕ, Ш. Г. (ред.): Сейсмостойкость транспортных и сетевых сооружений. Наука, Москва 1986.)

[151] NOWACKI, W.: Dynamika budowli (Dynamics of structures). Arkady Budownictwo-Sztuka-Architektura, Warszawa 1961.

[152] NOWAK, A. S., THARMABALA, T.: Bridge reliability evaluation using load tests. Journal of Structural Engineering, 114 (1988), No. 10, 2268–2279.

[153] NYMAN, W. E., MOSES, F.: Calibration of bridge fatigue design model. J. Struct. Eng., 111 (1985), No. 6, 1251–1266.

[154] O'CONNOR, C., CHAN, T. H. T.: Dynamic wheel loads from bridge strains. Wheel loads from bridge strains: laboratory studies. Journal of Structural Engineering. 114 (1988), No. 8, 1703–1740.

[155] OLSON, P. E., JOHNSSON, S.: Seitenkräfte zwischen Rad und Schiene. Glasers Annalen, 1959, No. 5, 153–161.

[156] OLSSON, M.: Finite element, modal co-ordinate analysis of structures subjected to moving loads. Journal of Sound and Vibration, 99 (1985), No. 1, 1–12.

[157] OLSSON, M.: Analysis of structures subjected to moving loads. Lund Institute of Technology, Division of Structural Mechanis, Report TVSM-1003, Lund, Sweden 1986.
[158] OLSSON, M.: Dynamisk brodimensionering Väg- och vattenbyggaren, 1988, no. 2, 23–25.
[159] ORE D 23: Determination of dynamic forces in bridges. Reports 1–17, Utrecht 1955-1970.
[160] ORE D 101: Braking and acceleration forces on bridges and interaction between track and structure. Reports 1–28, Utrecht 1969–1985.
[161] ORE D 105: Noise abatement on bridges. Reports 1–3, Utrecht 1966–1971.
[162] ORE D 128: Statistical distribution of axle-loads and stresses in railway bridges. Reports 1–10, Utrecht 1973–1979.
[163] ORE D 154: Stresses and strength of longitudinal girders and cross girders of bridges. Reports 1–6, Utrecht 1981–1985.
[164] ORE D 154.1: Stresses and strength of orthotropic decks. Reports 1–6, Utrecht 1984–1989.
[165] ORE D 160: Permissible deflection of bridges. Reports 1–6, Utrecht 1983–1988.
[166] PALAMAS, J.: Comportement dynamiques des ponts sous l'effet de charges roulantes. Influence des imperfections de profil. Constr. mét., 21 (1984), No. 3, 35–43.
[167] PALAMAS, J., COUSSY, O., BAMBERGER, Y.: Effects of surface irregularities upon the dynamic response of bridges under suspended moving loads. Journal of Sound and Vibration, 99 (1985), No. 2, 235–245.
[168] PALMGREN, A.: Die Lebensdauer von Kugellagern. Zeitschrift des Vereines der deutschen Ingenieure, 68 (1924), 339–341.
[169] PARIS, P. C., ERDOGAN, F.: A critical analysis of crack propagation laws. Journal of Basic Engineering. ASME, 85 (1963), No. 3, 528–534.
[170] POPP, K., HABECK, R., BREINL, W.: Untersuchungen zur Dynamik von Magnetschwebefahrzeugen auf elastischen Fahrwegen. Ingenieur-Archiv, 46 (1977), 1–19.
[171] POPP, K.: Näherungslösung für die Durchsenkungen eines Balkens unter einer Folge von wandernden Lasten. Ingenieur-Archiv, 46 (1977), 85–95.
[172] POPP, K., BOGACZ, R.: Dynamik des Systems Fahrzeug-Gleis. Zeitschrift für angewandte Mathematik und Mechanik, 65 (1985), No. 4, T 89–T 91.
[173] REBIEGA, J.: Podstawy nowego sposobu wymiarowanija stalowych mostów kolejowych na zmeczenie (New method for the design of steel railway bridges to fatigue). Archivum inżynierii ladowej, 30 (1984), No. 1, 93–117.
[174] REKTORYS, K., ET AL.: Survey of applicable mathematics. Iliffe Books, London 1969.
[175] Report of the Bridge Stress Commitee. Department of Scientific and Industrial Research. H. M. Stationery Office, London 1928.
[176] ROBSON, J. D.: An introduction to random vibration. Edinburgh Univ. Press, Edinburgh, Elsevier, Amsterdam 1964.
[177] ROHMAN, M. A., NAYFEH, A. H.: Passive control of nonlinear oscillations in bridges. Journal of Engineering Mechanics, 113 (1987), No. 11, 1694–1708.
[178] ROLFE, S. T., BARSOM, J. M.: Fracture and fatigue control in structures. Prentice-Hall, Englewood Cliffs 1977.
[179] ROOKE, D. P., CARTWRIGHT, D. J.: Compendium of stress intensity factors. H. M. Stationery Office, London.
[180] RUBLE, E. J.: Impact in railroad bridges. J. Struct. Div., Proc. ASCE, 81 (1955), No. 736,1–36.
[181] SACHS, L.: Statistische Auswertungsmethoden. Springer-Verlag, Berlin 1969, 2nd ed.
[182] SACKMAUER, L.: Stanovenie základných parametrov a navrhnutie bezstykovej kolaje na mostoch a vypracovanie podkladov pre predpis o bezstykovej kolaji na mostoch (Estimation of basic parameters, design of long welded rail on bridges and basis for a standard of long welded rail on bridges). Report R-O-Ž/VÚD-19/2, Prague 1968.
[183] SADIKU, S., LEIPHOLZ, H. H. E.: On the dynamics of elastic systems with moving concentrated masses. Ingenieur-Archiv, 57 (1987), 223–242.

[184] SAIGAL, S.: Dynamic behaviour of beam structures carrying moving masses. Journal of Applied Mechanics, 53 (1986), No. 1, 222–224.
[185] SALLER, H.: Einfluss bewegter Last auf Eisenbahnoberbau und Brücken. Kreidels Verlag, Berlin 1921.
[186] SCHIEHLEN, W. O.: Probabilistic analysis of vehicle vibrations. Probabilistic Engineering Mechanics, 1 (1986), No. 2, 99–104.
[187] SCHILLING, C. G.: Stress cycles for fatigue design of steel bridges. Journal of Structural Engineering, 110 (1984), No. 6, 1222–1234.
[188] SCHULZE, H.: Das dynamische Zusammenwirken von Lokomotive und Brücke. Deutsche Eisenbahntechnik, 7 (1959), No. 5, 229–236.
[189] SEDLACEK, G., JACQUEMOUD, J.: Herleitung eines Lastmodells für den Betriebsfestigkeitsnachweis von Strassenbrücken. Forschung Strassenbau und Strassenverkehrstechnik, Heft 430, 1984.
[190] SIH, G. C.: Handbook of stress intensity factors. Lehigh University 1973.
[191] SLÁMA, J.: Pohyb v čase náhodně proměnných sil po soustavách nosníků (Movement of forces randomly varying in time along a system of beams). Sborník prací VÚŽ, Prague 1983, No. 24, 49–81.
[192] SLAVIK, M.: Spezielle dynamische Einflusslinien. Die Strasse, 28 (1988), No. 8. 233–236.
[193] SLOSS, J. M., ADALI, S., SADEK, I. S., BRUCH Jr., J. C.: Displacement feedback control of beams under moving loads. Journal of Sound and Vibration, 122 (1988), No. 3, 457–464.
[194] SMITH, I. F. C., HIRT, M. A.: Fatigue design concepts. IABSE Periodica 4/1984, IABSE Surveys S-29/84, 57–72.
[195] SNIADY, P.: Vibration of a beam due to a random stream of moving forces with random velocity. Journal of Sound and Vibration, 97 (1984), No. 1, 23–33.
[196] SONNTAG, P. E., STREESE, D.: Zur Steifigkeit von Eisenbahnbrücken für Schnellbahnen. Eisenbahntechnische Rundschau, 25 (1976), No. 10, 623–629.
[197] SOROKIN, E. S.: To the theory of internal damping in the vibration of elastic systems. Gosstrojizdat. Moscow 1960. (Сорокин, Е. С.: К теории внутреннего трения при колебаниях упругих систем. Гостройиздат, Москва 1960).
[198] STANIŠIČ, M. M., EULER, J. A., MONTGOMERY, S. T.: On a theory concerning the dynamical behaviour of structures carrying moving masses. Ingenieur-Archiv, 43 (1974), 295–305.
[199] STANIŠIČ, M. M.: On a new theory of the dynamic behaviour of the structures carrying moving masses. Ingenieur-Archiv, 55 (1985), 176–185.
[200] STASSEN, H. G.: Random lateral motions of railway vehicles. Doctor thesis, Technische Hogeschool, Delft 1967.
[201] STOKES, G. G.: Discussion of a differential equation relating to the breaking of railway bridges. Transactions of the Cambridge Philosophical Society, 8 (1849), Part 5, 707–735. Also in: Mathematical and Physical Papers, 2 (1883), 178–220.
[202] Summary of tests on steel girder spans. American Railway Engineering Association – Bulletin, 61 (1959), No. 551, 51–78.
[203] SZELISKI, Z. L., ELKHOLY, I. A.: Fatigue investigation of a railway truss bridge. Can. J. Civ. Eng., 11 (1984), No. 3, 625–631.
[204] SHAPOSHNIKOV, N. N., KASHAEV, S. K., BABAEV, V. B., DOLGANOV, A. A.: Analysis of structure under moving loads using the finite element method. Stroitel'naya mekhanika i raschet sooruzhenij, 1986, N. 1, 50–54. (Шапошников, Н. Н., Кашаев, С. К., Бабаев, В. Б., Долганов, А. А.: Расчет конструкций на действие подвижной нагрузки с использованием метода конечных элементов. Строительная механика и расчет сооружений, 1986, № 1, 50–54)
[205] THOMAS, P. O.: Use of long welded rails over bridges. Rail International, 11 (1980), No. 6, 397–400.
[206] TING, E. C., YENER, M.: Vehicle-structure interactions in bridge dynamics. Shock and Vibration Digest, 15 (1983), No. 12, 3–9.
[207] TIMOSHENKO, S. P., YOUNG, D. H.: Vibration problems in engineering. D. van Nostrand, New York, 1955, 3-rd ed.

[208] Trudy Dnepropetrovskogo instituta inzhenerov zheleznodorozhnogo transporta. Volumes 25 (1956), 28 (1958), 31 (1961), 32 (1961), 38 (1962), 45 (1963), 49 (1965), 73 (1968), Transport, Moscow. (Труды Днепропетровского института инженеров железнодорожного транспорта. Том 25 (1956), 28 (1958), 31 (1961), 32 (1961), 38 (1962), 45 (1963), 49 (1965), 73 (1968), Транспорт, Москва).
[209] TUNNA, J. M.: Fatigue life prediction for Gaussian random loads at the design stage. Fatigue and Fracture of Engineering Materials and Structures, 9 (1986), No. 3, 169–184.
[210] TURNER, J. D., PRETLOVE, A. J.: A study of the spectrum of traffic-induced bridge vibration. Journal of Sound and Vibration, 122 (1988), No. 1, 31–42.
[211] UIC Merkblatt 702 V: Lastibild für die berechnung der Tragwerke der internationalen Strecken. Paris 1974, 2nd ed.
[212] UIC Fiche 776-1 R: Charges à prendre en consideration dans le calcul des ponts-rails. Paris 1979, 3rd ed.
[213] UIC Merkblatt 778-1 E: Empfehlungen zur Berücksichtigung der Ermüdung bei der Bemessung stählerner Eisenbahnbrücken. Paris 1981, 1st ed.
[214] USHKALOV, V. F., REZNIKOV, L. M., REDKO, S. F.: Stochastic dynamics of railway vehicles. Naukova dumka, Kiev 1982. (УШКАЛОВ, В. Ф., РЕЗНИКОВ, Л. М., РЕДКО, С. Ф.: Статистическая динамика рельсовых экипажей. Наукова думка. Киев 1982).
[215] VERIGO, M. F., KOGAN, A. Y.: Interaction of track and railway vehicles. Transport, Moscow 1986. (ВЕРИГО, М. Ф., КОГАН, А. Я.: Взаимодействие пути и подвижного состава. Транспорт Москва 1986).
[216] VLASOV, V. Z.: Thin-walled elastic bars. Fizmatgiz, Moscow 1959, 2nd ed. (ВЛАСОВ, В. З.: Тонкостенные упругие стержни. Физматгиз, Москва 1959, 2. изд.).
[217] VU-QUOC, L., OLSSON, M.: Interaction between high-speed moving vehicles and flexible structures: An analysis without assumption of known vehicle nominal motion. Structural Engineering Mechanics and Materials, Department of Civil Engineering. University of California, Berkeley, Report No. UCB/SEMM-87/10, 1987.
[218] VU-QUOC, L., OLSSON, M.: Formulation of a basic building-block model for interaction of high-speed vehicles on flexible structures. University of California, Berkeley, Report No. UCB/SEMM-88/03, 1988.
[219] WALLRAPP, O.: Elastic vehicle guideway structures. Proceedings of the third ICTS Seminar on Advanced vehicle system dynamics held at Amalfi, Italy, May 5–10, 1986. Supplement to Vehicle System Dynamics, 16 (1987), 215–232.
[220] WATSON, P., REBBECK, R. G.: Modern methods of fatigue assessment. Railway Engineering Journal, 4 (1975), No. 6, 10–23.
[221] WEBER, H. H.: Zur direkten Messung der Kräfte zwischen Rad und Schiene. Glasers Annalen, 1961, No. 6, 236–244.
[222] WILLIS, R., ET AL.: Preliminary essay to the Appendix B: Experiments for determining the effects produced by causing weights to travel over bars with different velocities. In GREY, G., ET AL.: Report of the commissioners appointed to inquire into the application of iron to railway structures. W. Clowes and Sons, London 1849. Also in BARLOW, P.: Treatise on the strength of timber, cast iron and malleable iron. London 1851.
[223] WU, J. S., DAI, C. W.: Dynamic response of multispan nonuniform beam due to moving loads. Journal of Structural Engineering, 113 (1987), No. 3, 458–478.
[224] WU, J. S., LEE, M. L., LAI, T. S.: The dynamic analysis of a flat plate under a moving load by the finite element method. International Journal for Numerical Methods in Engineering, 24 (1987), No. 4, 743–762.
[225] YAO, J. T. P., KOZIN, F., WEN, Y. K., YANG, J. N., SCHUELLER, G. I., DITLEVSEN, O.: Stochastic fatigue, fracture and damage analysis. Struct. Safety, 3 (1986), No. 3–4, 231–267.
[226] YOKOSE, K., THUCHIYA, K.: Analysis lateral forced vibration of rail vehicle truck running on

rails with sinusoidal irregularities. Bulletin of the Japan Society Mechanical Engineering, 28 (1985), No. 235, 139–147.

[227] YOSHIMURA, T., HINO, J., KAMATA, T., ANANTHANARAYANA, N.: Random vibration of a non-linear beam subjected to a moving load: a finite element method analysis. Journal of Sound and Vibration, 122 (1988), No. 2, 317–329.

[228] ZAKOR, A. L., KAZAKEVICH, M. I.: Damping of vibration of bridge structures. Transport, Moscow 1983. (ЗАКОР, А. Л., КАЗАКЕВИЧ, М. И.: Гашение колебаний мостовых конструкций. Транспорт, Москва 1983).

[229] ZIENCKIEWICZ, O.: The finite element method. McGraw-Hill, New York 1977.

[230] ZIMMERMANN, H.: Die Schwingungen eines Trägers mit bewegter Last. Centralblatt der Bauverwaltung, 16 (1896), No. 23, 249–251, No. 23 A, 257–260, No. 24, 264–266, No. 26, 288.

[231] ZWERNEMAN, F. J., FRANK, K. H.: Fatigue damage under variable amplitude loads. Journal of Structural Engineering, 114 (1988), No. 1, 67–83.

Author index

A

Abdel-Rohman, M. 311
Adali, S. 319
Adeli, H. 317
Albrecht, P. A. 217, 244, 313
Ananthanarayana, N. 21, 315, 321
Andreev, V. G. 140, 311
Arbabi, F. 311
Ayre, R. S. 311

B

Babaev, V. B. 319
Bamberger, Y. 318
Barchenkov, A. G. 311
Barlow, P. 320
Barsom, J. M. 255, 311, 318
Baťa, M. 21, 22, 67, 77, 311
Baumann, M. 21, 311
Betzhold, Ch. 311
Bhatti, M. H. 311, 312
Biggers, S.B. 11, 311
Billing, J. R. 311
Birger, I. A. 69, 311
Birmann, F. 311
Bleich, J. 311
Bogacz, R. 311, 318
Bogdanoff, J.L. 311
Bolotin, V. V. 26, 312
Bondar, N. G. 21, 59, 63, 92, 93, 111, 312
Bramall, B. 312
Braune, W. 21, 80, 94, 104, 106, 107, 110, 312
Breinl, W. 11, 21, 318
Brianza, M. 21, 311
Bruch Jr., J. C. 319
Brückmann, B. 21, 312
Byers, W. G. 21, 312

C

Cantieni, R. 21, 312
Cartwright, D. J. 254, 318
Cassé, M. 21, 312
Chambron, E. 21, 312
Chan, T. H. T. 317
Chelomei, V. N. 72, 312

Chilver, A. H. 312
Christiano, P. P. 312
Chu, K. H. 312, 314
Chung, Y. I. 314
Clark, R. A. 312
Clormann, U. H. 312
Collough, C. B. 311
Corazao, M. 317
Coussy, O. 312, 318
Crandall, S. H. 22, 26, 313
Crede, C. E. 78, 315
Culver, C. G. 312

D

Dahlberg, T. 21, 114, 313
Dähn, J. 313
Dai, C. W. 320
Danielski, L. 21, 313
Danilenko, E.I. 189, 313
Dean, P. A. 312
Desai, C. S. 41, 42, 313
Ditlevsen, O. 320
Dolganov, A. A. 319
Douglas, B. M. 313
Dowling, N. E. 200, 202, 203, 204, 313
Dukkipati, R. V. 37, 59, 125, 126, 129, 314

E

Elkoholy, I. A. 319
Elkins, J. A. 312
El-Sibaie, M. A. 316
Endo, T. 204, 313, 317
Ensner, K. 21, 311
Erdogan, F. 253, 255, 256, 318
Euler, J. A. 319

F

Federico, C. 315
Filippov, A. P. 313
Fisher, J. W. 217, 244, 313
Ford, G. 311
Frank, K. H. 321
Franke, L. 313
Frolov, L. N. 189, 313

Frýba, L. 11, 21, 31, 39, 41, 43, 44, 48, 49, 50, 51, 64, 79, 92, 96, 101, 111, 114, 116, 118, 119, 122, 123, 124, 129, 131, 132, 133, 139, 142, 156, 164, 170, 173, 187, 233, 275, 313, 314

G

Garg, V. K. 37, 59, 125, 126, 129, 311, 312, 314
Genin, J. 314
Gerasch, W. J. 314
Gesund, H. 314
Ghosn, M. 315
Giorgio, D. 315
Glomb, J. 21, 315
Glybina, G. K. 140, 311
Gogeliya, T.I. 315
Grachev, A.V. 189, 313
Grey, G. 320
Günther, W. 200, 315

H

Habeck, R. 11, 21, 318
Hänel, B. 315
Harris, C. M. 78, 315
Hausammann, H. 21, 315
Hellan, K. 315
Herzog, M. 21, 315
Hillerborg, A. 21, 315
Hino, J. 21, 315, 321
Hirt, M. A. 21, 315, 319
Hobbacher, A. 315
Hoshiya, M. 315
Hu, S. J. 317

I

Inbanthan, M. J. 315
Inglis, C. E. 21, 48, 315
Ito, M. 315
Itoh, F. 21, 315
Iwankiewicz, R. 315
Izuka Hiroshi 315

J

Jacobsen, L. S. 311
Jacquemoud, J. 233, 319
Jeffcott, H. H. 316
Jezequel, L. 316
Johnsson, S. 317

K

Kamata, T. 21, 321
Kamke, E. 103, 316

Kashaev, S. K. 319
Katsuhide, T. 316
Kazakevich, M. I. 11, 316, 321
Kazei, I. I. 21, 92, 93, 312, 316
Keiji, Y. 316
Kerr, A. D. 316
Kitagawa, M. 315
Klasztorny, M. 316
Klesnil, M. 205, 253, 316
Klingermann, D.J. 217, 244, 313
Kogan, A. Y. 320
Kokhmanyuk, S. S. 313
Koloušek, V. 21, 22, 35, 36, 42, 48, 67, 77, 105, 316
Komín, S. 183, 280, 316
Kondo, Y. 313
Korenev, B.G. 94, 104, 316
Korn, G. A. 42, 316
Korn, T. M. 42, 316
Kortüm, W. 11, 21, 316
Kovařík, R. 280, 316
Kozin, F. 320
Kozmin, Y. G. 21, 59, 63, 92, 93, 111, 312
Kropáč, O. 204, 316
Krylov, A. N. 21, 316
Kummer, E. 21, 315
Kunni, S. 21, 317

L

Lai, T. S. 320
Langer, J. 316
Lee, M. L. 320
Leipholz, H. H. E. 318
Lesokhin, B.F. 21, 92, 93, 312
Levy, S. 316
Li, F. 311
Lin, Y. K. 316
Lindgren, G. 317
Línek, O. 280, 316
Livesay, J. 200, 203, 317
Lukáš, P. 205, 253, 316
Lutes, L.D. 317

M

Maarschalkerwaart, H. M. C. M. van 21, 255, 317
Mahmoud, S. 312
Marek, J. 280, 316
Mark, W. D. 22, 26, 313
Marquard, E. 21, 317
Maruyama, O. 315
Matsuishi, M. 204, 317
Matsuura, A. 21, 112, 115, 317
Mayfeh, A. H. 317

McNamee, B. M. 217, 244, 313
Melcer, J. 21, 114, 317
Mercer, C. A. 200, 203, 317
Mertz, D. R. 217, 244, 313
Mildner, M. 317
Miner, M. A. 246, 258, 317
Mise, K. 21, 317
Mohamad, A. R. 317
Mohammadi, J. 317
Montgomery, S. T. 319
Moras, E. 189, 313
Moses, F. 315, 317

N

Napetvaridze, S. G. 11, 317
Nayfeh, A. H. 311, 318
Newton, S. G. 312
Nowacki, W. 317
Nowak, A. S. 317
Nyman, W. E. 317

O

O'Connor, C. 317
Olson, P. E. 317
Olsson, M. 21, 317, 318, 320

P

Palamas, J. 318
Palmgren, A. 246, 258, 318
Panovko, Y. G. 69, 311
Paris, P. C. 253, 255, 256, 318
Plachý, V. 21, 22, 67, 77, 311
Popp, K. 11, 21, 311, 318
Pretlove, A. J. 320

Q

Quoos, V. 317

R

Rabiega, J. 21, 313, 318
Rabinovich, I. M. 94, 104, 316
Rebbeck, R. G. 204, 320
Redko, S. F. 320
Reid, W. H. 313
Rektorys, K. 30, 67, 318
Reznikov, L. M. 320
Robson, J. D. 22, 26, 318
Rohman, M. A. 318
Roitburd, Z. G. 21, 59, 63, 111, 312
Rolfe, S. T. 255, 318

Romanov, V. M. 189, 313
Rooke, D. P. 254, 318
Rosenkrans, R. 311
Ruble, E. J. 21, 318
Rychlik, I. 317

S

Sachs, L. 83, 318
Sackmauer, L. 286, 288, 318
Sadek, I. S. 319
Sadiku, S. 318
Saigal, S. 319
Saller, H. 21, 319
Satoh, N. 313
Schiehlen, W. 11, 316, 319
Schilling, C. G. 319
Schueller, G. I. 320
Schulze, H. 21, 319
Sedlacek, G. 233, 319
Seeger, T. 312
Shaposhnikov, N. N. 319
Shenton III, H. W. 316
Sih, G. C. 254, 319
Siriwardane, A. M. 41, 42, 313
Sladký, R. 280, 316
Sláma, J. 21, 63, 162, 175, 319
Slavik, M. 21, 319
Sloss, J. M. 319
Smith, I. F. C. 319
Sniady, P. 162, 315, 319
Sonntag, P. E. 21, 319
Sorokin, E. S. 94, 104, 105, 106, 319
Stanišič, M. M. 319
Stassen, H. G. 21, 319
Stokes, G. G. 21, 319
Streese, D. 21, 319
Szeliski, Z. L. 319

T

Takayama, H. 313
Tarasenko, V .P. 21, 59, 63, 111, 312
Tharmabala, T. 317
Thomas, P. O. 282, 319
Thuchiya, K. 21, 320
Thürlimann, B. 21, 311
Timoshenko, S. P. 21, 319
Ting, E. C. 21, 314, 319
Tunna, J. M. 320
Turner, J. D. 320

U

Ushkalov, V.F. 320

Author index

V

Verigo, M. F. 320
Vincent, G. S. 311
Vlasov, V. Z. 170, 171, 320
Vu-Quoc, L. 320

W

Wallrapp, O. 320
Wang, T. L. 312, 314
Watson, P. 204, 320
Weber, H. H. 185, 320
Weber, W. 21, 80, 94, 104, 106, 107, 110, 312
Wen, Y. K. 320
Weseli, J. 315
Wieland, M. 315
Wilkinson, J. P. D. 316
Willis, R. 21, 320
Wilson, J. F. 11, 311
Wormley, D. N. 11, 21, 316
Wu, J.S. 320

Y

Yakovlev, G. N. 21, 59, 63, 111, 312
Yamauchi, T. 313
Yang, J. N. 316, 320
Yao, J. T. P. 320
Yen, B. T. 217, 244, 313
Yener, M. 21, 319
Yokose, K. 21, 320
Yoshimura, T. 21, 315, 321
Young, D. 314
Young, D. H. 319

Z

Zakor, A. L. 321
Zhong, A. 217, 244, 313
Zienkiewicz, O. 321
Zimmermann, H. 21, 321
Zimmermann, J. 317
Zwerneman, F. J. 321

Subject index

A

acceleration 117
adhesion 130
algorithm for a computer 208
alternate bearings on one pier 268, 282, 287
arch 36
assessment for fatigue 251
axle forces 185, 193
    spacing 195, 197

B

ballast 41, 107, 285, 286
bar, curved 35
    thin-walled 170, 175
beam 29, 67, 162
    continuous 31, 72–76
    clamped 69, 70
    mass 29
    massless 30
    simple 68, 70
bearing
    alternate 268
    fixed 158, 264, 307
    hinged 32
    moving 103, 159, 270
bending moment 42, 217, 222, 242
    ranges 224–227
Bernoulli–Euler beam 29, 60, 62, 130, 162, 163
Bessel function 103
block diagram 210, 289
body rigid 53
bouncing 53
boundary conditions 265, 267, 268, 270, 271
BR 278
braking force 138
bridge
    composite 279, 285, 286, 289
    concrete 82, 83, 87, 107, 109, 110, 279, 285, 289
    deck 39, 157
    deflection 283, 285
    depth 283, 285
    inspections 260, 261
    plate girder 82, 86
    prestressed concrete 43
    steel 80, 85, 108, 244, 279, 285, 286
    steel truss 81, 85
    suspension 31
    with n-spans and with alternate bearings 272

C

cantilever 69, 71
CD 113
characteristic train 217
coefficient of adhesion 133, 137, 158, 159, 160
    thermal extension 263, 264, 286
    variation 166, 167, 168, 169, 170, 172, 177
computer program 211
condition
    gap 278, 282, 307
    mutual displacement 278, 285, 306, 309
    stability 278, 288
    strength 278, 280, 307
conditions
    boundary 32, 142
    dynamic 32
    geometric 32
    initial 32, 33
continuous beam 242
    bridge 297
    welded rails 262, 288, 309
Coriolis acceleration 65
correlation table 204
corrugation 123
Coulomb friction 57, 99, 132, 141
counting
    methods 199, 212
    range 203
    range-mean 203
    range-pair 203
    rules 207
covariance 27, 164, 165, 166, 168, 169, 172, 174, 178
crack length 258, 261
cross section, constant 29, 67
    variable 42
cross-beam 37
cross-beam effect 38, 39, 124
CSD 140, 156, 183, 184, 218, 221, 222, 229, 232, 252, 276, 279
ČSN 246, 278
curve 181, 291

D

D'Alembert principle 22, 47, 64
damage accumulation 244
    fatigue 248, 249, 250
damper 57
damping 22, 179
    complex 104
    elements 54
    internal 104
    Kelvin–Voigt 22
    non-linear 106
    of railway bridges 94
    proportional to the rate of stress variations 98
    ratio 23
    subcritical 23, 95, 107, 112
    viscous 29, 57, 95, 130
DB 80, 114, 115, 116, 156
deceleration 117
deflection
    dynamic 97, 113, 119
    of the deck grillage 39
    static 35, 96
    vertical 29, 33
Dirac function 63
displacement
    horizontal 131
    horizontal longitudinal 263
    vertical 131
distribution exponential 233
    Gauss 109, 213
    Student 83, 109
dynamic
    amplification 23
    coefficient 23, 114, 216
        statistical 215
    impact factor 23
    magnification 23

E

effect of span 146
    rail length 148
    velocity 136, 195, 197
effects dynamic 19
    inertial 43
    spatial 47
    vertical 19
elastic foundation 43
elastic layer 139, 262
element
    damping 57
    elastic-plastic 58
    mass 51
    spring 54

empirical formulae 90, 110, 231
end clamped 32
    free 32
equation
    differential 29
    frequency 67
    integral-differential 30
    of motion 61
ERRI 11, 22
Eurocode 247
expansion admissible 288, 290
    device 292, 293, 295
    length 278, 284
experiments 80, 107, 156, 175, 185, 228, 230, 275–277

F

fastening of rails 281, 285
fatigue assessment 251
    crack 253, 257
    damage 248, 249, 250
        accumulation 258
    failure 246
    life 258
fixed bearing in the middle of the bridge 266, 281, 287
    on one bridge end 264, 281, 287
flat wheel 123, 280
flow diagram 210, 289
force
    axle 185, 193
    braking 158
    centrifugal 181
    constant 48
    contact 131
    critic 103, 288
    distribution 144
    harmonic variable 49
    horizontal 170
        due to temperature changes 291, 294, 298
        longitudinal 139, 187, 294, 298
            friction 103
        transverse 294, 298
    in fixed bearing 307
    longitudinal normal 263
    moving 48, 94, 112
        constant 48, 112
        harmonic variable 49
        horizontal 162
        vertical 48, 162
    starting 158, 161
    vertical axle 185, 189, 194
formulae empirical 90, 110, 231

foundation elastic 43, 262
    Winkler 139
Fourier series 121
FRA 125, 129
fracture mechanics 244, 260
frame 35
frequency natural 23, 66, 168, 176
    of bridges 85, 88, 89, 91
    of loaded bridges 77
Fresnel integrals 118
friction 57, 99
function
    autocorrelation 27
    correlation 27
    Gamma 233

G

Galerkin method 103
Gamma function 233
gap condition 278, 282, 307
Gauss distribution 213
geometric position of rails 125
gravity acceleration 35
Green's function 30
growing traffic loads 240

H

Heaviside unit function 142
history 19
Hooke's law 30, 263
horizontal forces due to temperature changes 291, 294, 298
    longitudinal effects 130
    transverse effect 162
hunting 53
hysteresis 205

I

impact 123
    lateral 162
    of flat wheels 123
influence function 30, 229, 242
inspection of bridges 258, 260, 261
integral transformations 142, 173
interaction thermal 262
interval of inspections 258, 260
irregularities
    isolated 126
    periodic 121
    random 126
    track 19, 120, 162, 175

K

Kelvin–Voigt damping 95

L

lateral impact 162
life of bridges 258
limit states theory 250, 251
load 29, 33
    continuous 49
    factor 159, 252
    random 50
    traffic 183, 191, 192, 221, 223, 232
    uniform in curved rail 307
loading length 159
locomotive
    diesel-electric 39
    electric 39
    steam 49
logarithmic decrement of damping 23, 107, 110
long-welded rails 262, 288, 309
longitudinal girder 37, 51
lumped mass 22, 52

M

mass elements 51
    lumped 37, 52
    sprung 53
    unsprung 53
matrix
    damping 38
    mass 38, 66
    stiffness 38, 66
mean value 26, 83, 208
method
    counting 199
    Euler 133
    Galerkin 42
    Newton–Raphson 42
    of equivalent damage 247
    peak counting 201
    Ritz 42
    Runge–Kutta 133
    sampling 200, 215
    threshold 201
model
    Bernoulli–Euler 44
    of a bridge 29, 45, 59, 262
    of a train 59
    of long-welded rail 262
    of the bridge deck 38
    of the substructure 42
    of vehicles 47

Subject index

quasistatic 139
Rayleigh 44
shear 44
Timoshenko 44
theoretical 113
modulus of elasticity 29, 33
    in shear 168
moment of inertia 29, 33, 35
motion
    constant velocity 112
    disc 130
    longitudinal 53
    variable velocity 116
    sway 54
movement of the vehicle 63
multi-track bridges 252
mutual displacement 306
    condition 278, 285, 306, 309

N

Navier's hypothesis 30
Newton's law 22, 47
NS 156, 242
number of stress cycles 235–239

O

ÖBB 156
ORE 11, 12, 21, 22, 80, 112, 129, 204, 217, 243, 245
OSZhD 21
overloading 241

P

Palmgren–Miner theory 246, 258
parameter
    acceleration 118
    damping 96, 113
    deceleration 118
    velocity 96, 113
    study 129, 136, 146, 305
Paris–Erdogan equation 253, 255, 256
passenger 61
period of natural vibrations 23
perturbation method 103
pitching 54
plate 77
    orthotropic 33
    rigid 53
Poisson's ratio 33, 34
power spectral density 28, 127–129
probability 278
    density 231

distribution
    beta 233
    chi 233
    exponential 233
process
    ergodic 28
    non-stationary 28
    stationary 28
propagation of fatigue cracks 253, 257

Q

quasistatic solution 163, 166, 168, 171

R

rail 279, 280, 281, 286
    bed 41, 158
    displacement 285, 305, 306
    force 305
    geometric position 125
rain-flow counting method 204, 218, 219
RATP 156
regression 83, 85, 89
    linear 81, 85
    power 85, 231, 235
reliability 83, 85, 235
residual service life 258, 259, 260
resonance curve 25
rolling 54
    disc 130
rotation 131
rotatory inertia 43

S

Saint-Venant principle 30
SBB 156
service life 253
shear 43
    force 242
simple beam 242
single-span bridge 295
sinusoidal motion 162, 175
sleeper 39, 279, 281, 285
    effect 39, 41, 124
slipping 131
SNCB 156
SNCF 114, 115
solution, approximate 163, 166, 168, 171
    dynamic 164, 167, 169, 173
span 30, 146
spring, linear 55
    non-linear 55
stability condition 278, 288

standard deviation 26
statistical evaluation 213, 215
steel bridges 217, 228, 234, 244
stiffness, dynamic 80, 84
   of the plate 33
stop 56
strength condition 278, 280, 307
stress
   admissible 280
   cycles 232
   extremes 213
   intensity factor 253
   number of cycles 235
   range 200, 208, 217, 219, 229, 231, 244
   spectrum 209, 217, 219, 222, 229
   strain diagram 205
string 31
system complex 34
system with n degrees of freedom 37
   with one degree of freedom 22, 25
SZD 140

T

temperature change 263, 264
   fixing 279
   maximum 278, 279
   minimum 278, 279
theory of limit states 93, 250, 251
     permissible stresses 93
   Palmgren–Miner 246
thermal interaction 262
thin-walled bar 170
three-span bridge 300
   with alternate bearings 271, 282, 287
track ascending 140
   descending 140
   irregularities 120
traffic loads 183, 221, 223, 232, 261
     growing 240
train, characteristic 217, 220, 221
   freight 195, 197

   passenger 196, 197, 217, 223
truss 34
two movable bearings on one pier 270, 282, 287
two-span bridge 300

U

UIC 11, 20, 22, 80, 90, 91, 92, 112, 181, 185, 278

V

value, centred random 27
variance 26, 164, 165, 166, 167, 168, 172, 174
variation coefficient 27
vector of displacement 36, 38
   of forces 38
velocity 195, 197
   constant 112
   of trains 197
   of vehicles 112
   variable 116
vibration
   deterministic 22
   forced 23
   horizontal 166, 177
   stochastic 26
   torsional 168, 177
   vertical 163

W

weight of the bridge 145, 152
   of the vehicle 145, 152
Wheatstone bridge 186, 187, 188
white noise 165, 174
width of the gap 283
Wiener–Khinchin relations 28
Winkler foundation 141
Wöhler curve 245, 248, 249, 253, 259

Y

yawing 54